CLIMATE GOVERNANCE AND FEDERALISM

The majority of the world's largest carbon emitters are either federations or have adopted systems of decentralised governance. The realisation of the world's climate mitigation objectives therefore depends in large part on whether and how governments within federal systems can co-operate to reduce carbon emissions and catalyse the emergence of low-carbon societies. This volume brings together leading experts to explore whether federal or decentralised systems help or hinder efforts to mitigate and adapt to climate change. It reviews the opportunities and challenges federalism offers for the development and implementation of climate mitigation and adaption policies and identifies the conditions that influence the outcomes of climate governance. Including in-depth case studies of fourteen different jurisdictions, this is an essential resource for academics, policymakers, and practitioners interested in climate governance and the best practices for enhancing climate action. This title is also available as Open Access on Cambridge Core.

ALAN FENNA is Professor of Politics at the John Curtin Institute of Public Policy, Curtin University, Western Australia. He specialises in Australian and comparative federalism and public policy and has authored or co-authored a range of books, book chapters and journal articles on those subjects. He worked in the Federal Affairs Division of the Department of Premier and Cabinet, Western Australia; was an elected member of local government; and served as president of the Australian Political Studies Association (APSA) 2009–10.

SÉBASTIEN JODOIN is Associate Professor in the Faculty of Law of McGill University, where he holds the Canada Research Chair in Human Rights, Health, and the Environment. He is the founding director of the Disability-Inclusive Climate Action Research Programme, a pioneering initiative to generate, co-produce, and translate knowledge at the intersections of disability and climate justice. His work in the field of climate change has been cited by the Intergovernmental Panel on Climate Change and the Office of the High Commissioner for Human Rights.

JOANA SETZER is Assistant Professor at the Grantham Research Institute, at the London School of Economics and Political Sciences. She leads the Institute's research on climate litigation as well as the Climate Change Laws of the World project – the most comprehensive global resource on climate policy, legislation, and litigation. She regularly advises a range of international organisations, governments, and non-governmental organisations in the field of climate law and policy.

CLIMATE GOVERNANCE AND FEDERALISM

A Forum of Federations Comparative Policy Analysis

Edited by

ALAN FENNA
Curtin University

SÉBASTIEN JODOIN
McGill University

JOANA SETZER
London School of Economics and Political Science

Secrétariat aux relations canadiennes Québec

Forum of Federations
The Global Network on Federalism and Devolved Governance

CAMBRIDGE
UNIVERSITY PRESS

CAMBRIDGE
UNIVERSITY PRESS

Shaftesbury Road, Cambridge CB2 8EA, United Kingdom

One Liberty Plaza, 20th Floor, New York, NY 10006, USA

477 Williamstown Road, Port Melbourne, VIC 3207, Australia

314–321, 3rd Floor, Plot 3, Splendor Forum, Jasola District Centre, New Delhi – 110025, India

103 Penang Road, #05-06/07, Visioncrest Commercial, Singapore 238467

Cambridge University Press is part of Cambridge University Press & Assessment, a department of the University of Cambridge.

We share the University's mission to contribute to society through the pursuit of education, learning and research at the highest international levels of excellence.

www.cambridge.org
Information on this title: www.cambridge.org/9781009249652

DOI: 10.1017/9781009249676

When citing this work, please include a reference to the DOI 10.1017/9781009249676

First published 2023

A catalogue record for this publication is available from the British Library.

Library of Congress Cataloging-in-Publication Data
Names: Fenna, Alan, 1958- editor. | Jodoin, Sébastien, 1982- editor. | Setzer, Joana, editor.
Title: Climate governance and federalism : a forum of federations comparative policy analysis / edited by Alan Fenna, Sébastien Jodoin, Joana Setzer.
Description: Cambridge ; New York, NY : Cambridge University Press, 2023. | Includes bibliographical references and index.
Identifiers: LCCN 2022045295 (print) | LCCN 2022045296 (ebook) | ISBN 9781009249652 (hardback) | ISBN 9781009249676 (epub)
Subjects: LCSH: Climate change mitigation–Government policy–Case studies. | Federal government–Case studies. | Decentralization in government–Case studies.
Classification: LCC TD171.75 .C635 2023 (print) | LCC TD171.75 (ebook) | DDC 363.738/740723073–dc23/eng/20221110
LC record available at https://lccn.loc.gov/2022045295
LC ebook record available at https://lccn.loc.gov/2022045296

ISBN 978-1-009-24965-2 Hardback

Contents

List of Figures *page* vii
List of Tables viii
Notes on Editors and Contributors ix
Foreword xvii
PHILLIP GONZALEZ, DIANA CHEBENOVA, AND RUPAK CHATTOPADHYAY
Acknowledgements xx

1 Climate Governance and Federalism: An Introduction 1
ALAN FENNA, SÉBASTIEN JODOIN, AND JOANA SETZER

2 Climate Governance and Federalism in Australia 14
ALAN FENNA

3 Climate Governance and Federalism in Brazil 41
FABIANA BARBI SELEGUIM AND FERNANDO REI

4 Climate Governance and Federalism in Canada 64
KATHRYN HARRISON

5 Climate Governance and Quasi-Federalism in China 86
HONGTAO YI AND SHUAI CAO

6 Climate Governance and Federalism in Ethiopia 109
YALEMSEW ADELA, ADEFIRES WORKU, AND TILAYE NIGUSSIE

7 Climate Governance and Federalism in the European Union 128
MARIACHIARA ALBERTON

8 Climate Governance and Federalism in Germany 150
PETER ECKERSLEY, KRISTINE KERN, WOLFGANG HAUPT, AND HANNAH MÜLLER

9 Climate Governance and Federalism in India 177
ADITYA VALIATHAN PILLAI AND NAVROZ K. DUBASH

10 Climate Governance and Decentralization in Indonesia 198
MONICA DI GREGORIO AND MOIRA MOELIONO

11 Climate Governance and Federalism in Mexico 218
MARCO HEREDIA AND BEATRIZ CORRAL

12 Climate Governance and Federalism in South Africa 241
JAAP DE VISSER AND ANÉL DU PLESSIS

13 Climate Governance and Federalism in Spain 263
SUSANA GALERA RODRIGO, MERCEDES ALDA-FERNANDEZ, AND MARIO KÖLLING

14 Climate Governance and Federalism in Switzerland 285
MARLENE KAMMERER, SEAN MUELLER, KARIN INGOLD, AND MARIA GALLMANN

15 Climate Governance and Federalism in the United States 306
BARRY RABE AND HANNAH SMITH

16 Reflections on Climate Governance and Federalism 328
ALAN FENNA

Index 337

Figures

4.1	Total GHG emissions (kt CO_2eq) by province or territory	*page* 66
4.2	GHG Emissions per person (tonnes CO_2eq/yr) by province or territory	66
5.1	Annual CO_2 emissions of China and the world	87
5.2	China's total CO_2 and energy-related CO_2 emissions	88
5.3	Annual fossil CO_2 emissions in China	89
5.4	Structures of environmental federalism in China	91
8.1	Energy-based typology of the German *Länder*	162
9.1	State dependence on central transfers in 2016–17	183
14.1	Switzerland's overall GHG footprint, 1996–2015	289
14.2	Swiss GHG emissions per sector, 1990–2018	290

Tables

3.1 Climate change policies in Brazilian States *page* 47
3.2 Climate change policies in Brazilian municipalities 51
3.3 Institutions responsible for the development and implementation of climate change policies in Brazil 53
5.1 Analytical framework for case studies of Chinese environmental quasi-federalism 96
5.2 Three rounds of pilot low-carbon initiatives 97
5.3 Pioneering low-carbon provinces and cities and their innovative measures 101
8.1 GHG emissions enshrined in *Land* and federal climate protection acts 153
8.2 Climate responsibilities across tiers of government in Germany 156
8.3 Population, GHG emissions and renewable energy in the German *Länder* 160
8.4 Energy-based typology of the German *Länder* 161
13.1 Spanish national evolution of aggregate emissions 266
13.2 Main climate change plans and objectives by AC 275
13.3 Spanish non-state governments in transnational networks 279
14.1 Vertical division of climate change powers in Switzerland 287
14.2 Factors potentially explaining cantonal climate adaptation and municipal climate mitigation 296
14.3 Solution pathways to the adoption of a cantonal climate adaptation strategy 297
14.4 Solution pathways to the adoption of a municipal energy label 299

Notes on Editors and Contributors

Editors

Alan Fenna is Professor of Politics at the John Curtin Institute of Public Policy, Curtin University, Western Australia. He specialises in Australian and comparative federalism and public policy. He is co-author, most recently, of *Interrogating Public Policy Theory: A Political Values Perspective* (2019) and *Comparative Federalism: A Systematic Inquiry* (2015), and co-editor of *Australian Government and Politics* (2021). He worked in the Federal Affairs division of the Department of Premier and Cabinet of the Government of Western Australia; was an elected member of local government; and served as President of the Australian Political Studies Association (APSA) 2009–10.

Sébastien Jodoin is Associate Professor in the Faculty of Law of McGill University, where he holds the Canada Research Chair in Human Rights, Health, and the Environment. He is the founding director of the Disability-Inclusive Climate Action Research Programme, a pioneering initiative to generate, co-produce, and translate knowledge at the intersections of disability and climate justice. His work in the field of climate change has been cited by the Intergovernmental Panel on Climate Change and the Office of the High Commissioner for Human Rights.

Joana Setzer is Assistant Professor at the Grantham Research Institute, at the London School of Economics and Political Sciences. She leads the Institute's research on climate litigation as well as the Climate Change Laws of the World project – the most comprehensive global resource on climate policy, legislation, and litigation. She regularly advises a range of international organisations, governments, and non-governmental organisations in the field of climate law and policy.

Contributors

Yalemsew Adela is a senior professional specialized in development studies and environmental technology, who is working as an innovation analyst at the Forest Sector Transformation Unit, at Ethiopian Forestry Development. Previously he has held research-post lectureships in environmental engineering, and leadership roles at the Ethiopian Environment and Forest Research Institute, Jimma University as well as at the University of Rostock. His interests in research and development are in the areas of sustainability, climate and environmental governance, as well as rural livelihoods and development.

Mariachiara Alberton is a senior researcher in environmental law at the Institute for Comparative Federalism, Eurac Research in Bolzano/Bozen, Italy. Her research focuses on the multi-level governance of the environment and climate change. She investigates in particular how intergovernmental arrangements and dynamics at multiple levels enable or hinder environmental protection, participatory processes, and the emergent transition and transformation processes toward sustainability and climate resilience.

Mercedes Alda-Fernandez is Associate Professor of Political Science and Public Administration at the Rey Juan Carlos University in Madrid (Spain). Her main research interests are intergovernmental relations, public policies, particularly environmental, climate, and energy policies, electoral behaviour, governments, and legislative studies. She has taught, published, and participated in different projects in these areas.

Fabiana Barbi Seleguim is a postdoctoral researcher at the University of São Paulo–USP, Brazil. She obtained her PhD in Environment and Society from the University of Campinas–Unicamp. She is a research fellow of the Earth System Governance Research Network and Urban Climate Change Research Network. She has consulted for national and international organisations such as ICLEI and GIZ.

Shuai Cao is a PhD student at the John Glenn College of Public Affairs, Ohio State University. His research interests focus on network analysis, collective action, public decision-making, and policy process in the context of energy and the environment. He has expertise in research on water governance and policy.

Beatriz Corral is Director of Evaluations Planning and Management at the National Institute of Ecology and Climate Change in Mexico. She has a doctorate in Government and Public Administration, and experience in the analysis, monitoring, and evaluation of federal public programmes and policies. Between 2006 and 2009 she worked at COLEF, a Mexico–United States cross-border study centre, and since then she has worked as an official of the federal public administration in Mexico where she has collaborated in the social, health, finance, and environmental sectors.

Monica Di Gregorio is Associate Professor of Environmental Politics and Governance and the Co-director of the Sustainability Research Institute at the School of Earth and Environment, University of Leeds. Her research examines contentious environmental politics, governance, and development themes including multi-actor and multi-level governance of climate change and natural resources in the Global South. She is a Research Associate of the Priestley International Centre for Climate and the Centre for Climate Change Economics and Policy, and has been a Senior Associate of the Center for International Forestry Research. The research that informed her contribution to this book was funded by a grant of the German Federal Ministry for Environment.

Navroz K. Dubash is a professor at the Centre for Policy Research, a New Delhi-based think tank, and an adjunct senior research fellow at the Lee Kuan Yew School of Public Policy, National University of Singapore (NUS). He has been engaged in global and national debates on climate change, air quality, energy, and water as a researcher, policy advisor, and activist for over twenty-five years. Navroz is a co-ordinating lead author for the chapter on policies and institutions for the Intergovernmental Panel on Climate Change 6th Assessment Report and has worked to inform and advise Indian government policymaking over the last decade.

Peter Eckersley is a senior research fellow in public policy and management at Nottingham Trent University, UK, a Research Associate at the Leibniz Institute for Research on Society and Space, Erkner, Germany, and Managing Editor of *Local Government Studies*. As an interdisciplinary researcher who draws on political science, public administration, and geographical perspectives, he focuses on multi-level governance, climate change, public policy, austerity, and public accountability, with a particular focus on Germany and the UK.

Susana Galera Rodrigo is Associate Professor of Administrative Law at the Rey Juan Carlos University in Madrid and collaborates with other academic and research institutions and publication boards. From her earliest works she has focused on the European integration approach and particularly on the implementation of EU law at the national level. From this perspective she has explored many specific fields such as customs law or judicial protection, also environmental and energy transition policies. In addition to her teaching and research activities, she also occasionally acts as a public law expert both for national and international bodies.

Maria Gallmann completed her masters in Climate Sciences at the Oeschger Center for Climate Change Research (OCCR) of the University of Bern in Switzerland, specialising in research on climate change policy networks. She previously worked with Skat Consulting Ltd. in international development cooperation and currently works with South Pole, developing and monitoring carbon compensation projects in Switzerland. She is passionate about understanding the world's complex socio-ecological systems through interdisciplinary perspectives and driving positive action leading to a sustainable and inclusive future.

Kathryn Harrison is Professor of Political Science at the University of British Columbia. Before entering academia, Harrison worked as an engineer in the oil industry and a policy analyst for Environment Canada and the United States Congress. She has published widely on Canadian and US climate policy. Harrison is currently chair of the Mitigation Expert Advisory Panel of the Canadian Climate Institute, and a member of the British Columbia Climate Solutions Council.

Wolfgang Haupt works as research associate and postdoctoral fellow at the Leibniz Institute for Research on Society and Space in Erkner, Germany. He has a background in geography and urban regional development and holds a PhD in urban studies from the Gran Sasso Science Institute in L'Aquila and Sant'Anna School of Advanced Studies in Pisa. His research focuses on local and regional climate governance (mitigation and adaptation) as well as the governance of climate change within multi-level systems. Further research interests and foci include transnational municipal climate networks, urban resilience, and policy learning.

Marco Heredia is an environmental and climate law and policy expert. He has expertise in policy design, implementation, and evaluation at local, federal, and international settings. He served several years as programme manager for environmental law at the Commission for Environmental Co-operation in North America. He holds bachelor, masters, and doctorate degrees in Law, and a bachelor degree in English Education from the National University in Mexico.

Karin Ingold is Professor at the Institute of Political Science at the University of Bern. She heads the research group of Policy Analysis and Environmental Governance that is affiliated with the Oeschger Center of Climate Change Research at the University of Bern and with the Environmental Social Science Department of Eawag, the Water Research Institute of the ETH Domain, in Zurich, Switzerland. In her research she is interested in public policies and how they originate, with a special focus on political network analysis, advocacy coalitions, and policy instruments and mixes.

Marlene Kammerer is a postdoctoral researcher at the Institute of Political Studies, University of Bern. She is also affiliated with the Oeschger Center of Climate Change Research at the University of Bern. She obtained her PhD from the Centre for Comparative and International Studies at the University of Zurich and ETH Zurich in 2018. Her main research interest is the intersection between sub-national, national, and international environmental policy, particularly focusing on contemporary problems such as climate change and plastics pollution.

Kristine Kern is a professor and head of the research group Urban Sustainability Transformations at the Leibniz Institute for Research on Society and Space in Erkner (Germany) and is also affiliated with Åbo Akademi University (Finland). Her research interests concentrate on environmental governance and sustainability transformations in multi-level systems, in particular on local and regional climate and energy governance, the sustainable development of cities and regions, and transnational city networks.

Mario Kölling is Assistant Professor at the Department of Political Science at the Spanish National Distance University (UNED), and senior researcher at the Manuel Giménez Abad Foundation, Zaragoza. From 2011 to 2014 he was Garcia Pelayo Researcher at the Centre for Political and Constitutional Studies (CEPC) in Madrid. He works and publishes on issues related to comparative federalism, the EU budget, and national and sub-national parliaments in EU affairs.

Moira Moeliono is a Senior Associate at the Center for International Forestry Research. She works on community management of forest resources, the social aspects of resource uses, and adaptive collaborative management of forest resources, decentralisation, tenure, and climate change policy issues.

Sean Mueller is Assistant Professor at the Institute of Political Studies, University of Lausanne, funded by the Swiss National Science Foundation ('Eccellenza' grant). He obtained his PhD in Politics and Government from the University of Kent/UK (2013), and his MA from the University of Fribourg/CH (2006). His main research areas are Swiss and comparative federalism as well as sub-national politics and multi-level governance more broadly.

Hannah Müller is an urban researcher and activist, currently completing her masters in Historical Urbanism at the Technical University of Berlin. Her main areas of interest encompass the critical analysis of urban, ecological transformations, and social movements with a focus on housing rights and the new municipalism, as well as feminist and activist research methods.

Tilaye Nigussie is currently Director of the Forest Sector Transformation Unit, providing programmatic, policy, and strategic support to Ethiopian Forestry Development through the United Nations Development Programme. Previously he worked with several national and international organisations in senior leadership and management positions inside and outside Ethiopia, including Burundi, DR Congo, Rwanda, and Tanzania. He has trained in Ethiopia, the USA, and the UK in forestry and environmental management. He was a lecturer and academic dean at the Wondo College of Forestry and Natural Resources.

Aditya Valiathan Pillai is Associate Fellow at the Centre for Policy Research, a think tank in New Delhi. He studies climate change policies and politics in India and South Asia. His recent work focuses on institutional themes, covering areas such as the history of Indian climate governance, the state and adaptation politics, the emergence of climate federalism, and challenges to building a South Asian power pool. In his previous role as a programme officer at The Asia Foundation, he managed support to South Asian civil society working in Himalayan basins, and wrote about the political economy of water and electricity in the region.

Anél du Plessis is Professor of Law and the South African Research Chair in Cities, Law, and Environmental Sustainability, Faculty of Law, North-West University (South Africa). She has published extensively on local government, climate change law and governance, sustainability, and cities. She is editor and co-editor of *Environmental Law and Local Government in South Africa* (2021) and *The Globalisation of Urban Governance* (2019). She is also editor of the Special Series on Cities, Law and Environmental Sustainability of the *Potchefstroom Electronic Law Journal.*

Barry Rabe is the J. Ira and Nicki Harris Family Professor of Public Policy and the Arthur Thurnau Professor of Environmental Policy at the Gerald Ford School of Public Policy at the University of Michigan. He is also a non-resident senior fellow at the Brookings Institution and an elected fellow of the National Academy of Public Administration. Rabe is the author or co-author of six books, including *Can We Price Carbon?* (MIT 2018) and *Trump, the Administrative Presidency and Federalism* (Brookings 2020).

Fernando Rei is Associate Professor in the PhD Program in International Environmental Law at the Catholic University of Santos, and Professor of Environmental Law in Armando Alvares Penteado Foundation. His research examines environmental politics, governance, and the paradiplomacy of global climate change. He is also Scientific Director of the Brazilian Society of International Environmental Law, and he was twice President of the Environmental Agency of the State of São Paulo.

Hannah Smith is an urban planner who received a bachelor's degree in Environmental Studies and a master's degree in Urban and Regional Planning from the University of Michigan, focused on sustainability and land use planning. During her time at Michigan, she was a policy analyst and research assistant focused on climate policy research as well as the role of renewable energy in communities. She is currently an urban planning consultant in southeast Michigan.

Jaap de Visser is Professor in Public Law at the Dullah Omar Institute at the University of the Western Cape. He has published widely on multi-level government, local government, and federalism in Africa. He has consulted for many national and international organisations, including the World Bank, the United Nations Human Settlements Programme, United States Agency for International Development, and the Forum of Federations. He is vice-president of the International Association of Centres for Federal Studies, and an alternate board member of the Commonwealth Local Government Forum.

Adefires Worku is Director for Climate Science Research at the Ethiopian Environment and Forest Research Institute. His main focus is on generating and adopting technologies, knowledge, and information on climate change monitoring, adaptation, and mitigation, and examining the potentials of strategic integration of environmental resources such as forests and products and traditional ecological knowledge systems for enhancing the social-ecological resilience of smallholder farmers, pastoral communities, private sectors, and beyond.

Hongtao Yi is Associate Professor and Director of Graduate Professional Studies at the John Glenn College of Public Affairs, Ohio State University. His research interests focus on energy and environmental policy, collaborative governance, and policy process. His works appear in journals such as *Public Administration Review, Journal of Public Administration Research and Theory*, *Policy Studies Journal, Governance*, *Global Environmental Change*, *Environmental Politics*, *and Energy Policy*.

Foreword

A publication on climate governance and federal systems is an important, welcome, and timely contribution to understanding in this policy area of federative governance. This book's relevance is enormous when one considers the gravity of the challenge that lies ahead and the role that federated systems play in addressing it. After all, a number of the world's largest emitters are either federations (e.g., the United States, India) or have adopted forms of decentralised governance. A number of other federations, such as Australia and Canada, have very high per capita emissions. Comparative global research helps us understand some of the ways in which federal or decentralised systems help or hinder efforts to mitigate and adapt to climate change. Federal and federal-type systems are distinctive in that their various and varied constituent units provide policymaking opportunities that may not be present in unitary systems. This book explores that dimension of decision-making and policy choices that its fourteen case studies have undertaken in their efforts to address the challenge of climate change. We hope that the experiences and findings presented in this publication will support progress in advancing policy options in this vital area.

Federal and federal-type systems of governance are integral to global efforts to address the climate crisis around the world. These efforts depend on policy leadership, innovation, and implementation and involve major energy and infrastructure changes that are often difficult and politically contentious.

The relevance of federalism and decentralisation to climate governance has gained increased recognition with the adoption of the Paris Agreement, which expressly acknowledges the role the different orders of government have to play in contributing to the realisation of climate mitigation and adaptation objectives.

The constituent units – states, provinces, cantons, *Länder* – not only often have responsibility for many of the policy and governance domains where climate change mitigation and adaptation can be best achieved, but they also provide an opportunity for a range of approaches and measures to be trialled. Their initiatives can build on what is done by central government, but also fill in gaps in national policymaking, as central governments may not be as well equipped to address and implement climate change policy within these systems.

In federal and decentralised countries, the formulation of national policies on climate change is a complex process where intergovernmental co-operation is required. At times, constituent units drive policy action in climate governance within those powers they are allocated. In addressing climate policy, constituent units have powers and policy levers that they can and do employ. Here, as often occurs also in other policy areas in multi-level systems, tensions and co-ordination challenges between the levels of government exist.

It was at the intersection of these considerations and dynamics that the Forum of Federations undertook this research initiative, an initiative that provides a combined global perspective and country case studies. The research aimed to provide insights into policy options, approaches, and mechanisms used by governments, collating the story of climate governance in federal and federal-type systems. It is an opportunity to learn from one another, which is the core of the Forum's mandate.

More specifically, this research initiative was formed to address some fundamental questions: What is the potential of federal and decentralised systems in addressing climate change? Are federated systems well suited to grappling with the complexity and uncertainty of climate change by providing policymakers with opportunities to develop solutions tailored to different scales and circumstances, experiment with innovative policy measures, and engender policy learning and convergence over time? Or conversely, are there regulatory overlaps and co-ordination deficits inherent in many federated states which may hinder the pursuit of effective climate action? What policy options do the different governments of federal systems pursue within their respective powers and institutions? What policy actions can these actors take to address the complex political and policy issues of climate change? What co-ordination institutions are important to the success of climate change governance? Are there policy insights and successes that can be adopted and applied?

At the programme's inception in 2018, the Forum discussed this initiative with the Government of Quebec, who responded enthusiastically. It was very much their encouragement and support that has enabled this project to proceed and to succeed. Since the programme's launch in 2019 with our partner the Université du Québec à Montreal, the Government of Quebec has provided generous organisational and financial support.

The Forum and partners assembled leading academic experts and practitioners from thirteen federal and decentralised countries – Australia, Brazil, Canada, China, Ethiopia, Germany, India, Indonesia, Mexico, South Africa, Spain,

Switzerland, and the United States – along with the European Union (EU) to produce this major comparative analysis on the dynamics within climate change governance and federal systems.

A significant part of the project was the international authors' conference and the Canadian roundtable discussion, which the Forum and its partners, the Université du Québec à Montreal and the Secrétariat du Québec aux relations canadiennes, hosted in Montreal. At this event, over thirty-five renowned experts and practitioners, including provincial and federal policymakers, participated in the two days of discussions.

At the international conference, guided by the project editorial team and the Forum staff, experts presented their country case studies and the preliminary findings of their research, addressing core questions to help shape the project template guiding all authors with a set of common questions to address in detailing the dynamics involved.

The Canadian roundtable examined the state of play in Canadian climate policy governance, highlighting some of the ongoing policy puzzles facing actors across the country and its constituent units. Participants addressed the interplay of climate change policy issues, ranging from provincial government prioritisation and co-operation to challenges and conflicts within the Canadian federal system in addressing climate change. The closing session addressed those mechanisms that the federal government and the provinces employed in dealing with those inherent co-ordination challenges and conflicts.

Following the conference in Montreal, the project editorial team and the experts worked together over two-and-a-half years in producing the individual country chapters as well as the Introduction and comparative Conclusion chapters – a process made more challenging by the Covid-19 pandemic.

In sum, it is our hope that this volume will provide practitioners and researchers with comparative policy insights into the relationship between climate governance and federalism. Supporting a deeper appreciation of policy- and decision-making that has made an impact on the quest to address climate change, the volume's research is a guide to people who are interested in learning about the federal dynamic of climate governance.

Finally, it was our aim and sincere hope that the perspectives gathered through this volume will provide an impetus for continued theoretical and applied work on climate change policy within these and other federal and decentralised systems.

Phillip Gonzalez, Diana Chebenova, and Rupak Chattopadhyay

Acknowledgements

This book is the product of many people, most importantly, the contributing authors. We heartily thank the authors for their co-operative participation in seeing this book through several stages and some unexpected turns in development, including the many challenges associated with the Covid-19 pandemic. We enjoyed working with all our contributors and appreciated their eagerness to produce as comprehensive and up-to-date chapters as possible and to respond to many different rounds of editing.

We wish to thank the anonymous reviewers that carefully read through each of the chapters and provided helpful comments. These individuals, of course, are not responsible for any author's errors of commission or omission, but their observations provided great help in improving the book.

We are grateful to the Québec government for providing funding that made this book possible, including its release in an open-access format.

We are also very grateful to Sarah Lambert, Emma Kiddle, and the entire editing and production team at Cambridge University Press for supporting this book from the proposal stage to the delivery of the manuscript.

We thank, as well, Juliette Bourdeau de Fontenay, Elise Mallette, and Amanda Bowie-Edwards who helped to proofread and format the chapters.

Finally, we owe a debt of gratitude to our loved ones for supporting us throughout the production of this book, as well as the ups and downs of the last two years.

Alan Fenna, Sébastien Jodoin, and Joana Setzer

1

Climate Governance and Federalism

An Introduction

ALAN FENNA, SÉBASTIEN JODOIN, AND JOANA SETZER

Climate change is one of the great challenges of our time, but is it more so or less so for federal systems? Federalism could be an asset thanks to its more numerous sites for action and different governance scales. On the other hand, its division of powers and number of governments could well complicate and undermine climate governance through conflict, shirking, or poor co-ordination. This book brings together a diverse range of country experiences to provide some insights into that question.

That diversity reminds us, though, that 'federalism' is a generic term for a broad type of government and, as discussed later in this chapter, each federation embodies the federal principle in its own way. Each federation also has its own underlying social and economic character that determines the way it operates. And the institutional structures of federalism are likely to have varying and often contradictory effects with complex interactions depending on the issue at hand and the political dynamics at the time.

This book examines how climate governance has played out in an array of federations and decentralised systems, focusing on the role that constituent units – the states, provinces, cantons, *Länder*, and so on – play in fostering the emergence of low-carbon and climate-resilient societies. To set the scene for that, this chapter lays out some ways of thinking about federalism's implications for policymaking and explores its diversities.

1.1 The Challenges of Climate Change Governance

Climate governance has two 'fundamentally different' components: *mitigation* and *adaptation* (Biesbrook and Lesnikowski 2018, 306). Mitigation efforts address the causes of anthropogenic climate change through measures that reduce carbon and other greenhouse gas (GHG) emissions or remove carbon dioxide currently in the atmosphere. Adaptation measures address the effects of climate change through adjustments to infrastructure and practices. Both have been the subject of international negotiations and agreements and both are affected by the way a system of federal or devolved governance operates.

The Paris Agreement of 2015 set a goal of limiting the increase in global average temperatures to well below 2 degrees Celsius by 2100 and committed to efforts at limiting this increase to 1.5 degrees Celsius. For the world to achieve the latter objective, the Intergovernmental Panel on Climate Change (IPCC) estimated that global anthropogenic carbon emissions need to decline by about 45 per cent from 2010 levels by 2030 and reach net zero by 2050. To limit global warming to below 2 degrees Celsius requires reductions of 20 per cent by 2030 and the achievement of carbon neutrality around 2075. To that end, the Paris Agreement requires participating governments to prepare and communicate emissions reductions commitments known as Nationally Determined Contributions (NDCs) every five years, with a view to a progressive increase in their mitigation efforts.

Around 76 per cent of the world's GHG emissions result from electricity and heat production, industrial processes, and transportation. The remaining 24 per cent is produced by deforestation, the conversion of land for agriculture and human settlement, the degradation of soils, and agricultural activities (IPCC 2014). While the measures needed to limit global warming can yield significant co-benefits, such as improved air quality and human health from reductions in carbon emissions, the costs are substantial. Industrialised economies must remake themselves; developing countries must find a different path to the future they seek. In the industrialised countries, GHG emissions are embedded in the basic structures of production, consumption, and everyday life and disruptive change of these 'locked-in' patterns is required. This is being tackled on a variety of fronts (e.g., Fekete et al. 2021; IPCC 2022).

The costs of this energy transition are upfront and the intended benefits down the track – although any co-benefits would have more immediate effect. Costs, furthermore, may well be unevenly distributed, underpinning a clash of interests alongside ideological divisions. In addition, the uncertainties that characterise complex social and ecological systems and our imperfect ability to predict their future dynamics and effects adds further challenges to policymaking in this domain. For low-income countries, instead of developing energy infrastructure based on fossil fuels, there is some opportunity to 'leapfrog' straight to cleaner, low-carbon technologies, the potential for which is rapidly increasing in the electricity sector with the steep fall in the cost of renewables (Arndt et al. 2019).

There are a range of policy instruments governments may use to 'de-carbonise' their economies, many of them complementary. The simplest, and in some ways 'first-best', option is to change the economic incentives by altering the pricing structure: imposing a cost on carbon emissions through the creation of a tax or cap-

and-trade system (Nordhaus 2013; OECD 2019; Rabe 2018). However, taxes that will potentially terminate valuable industries and drive up costs for consumers are not going to be popular, especially in jurisdictions with carbon-intensive industries. A carbon tax sufficiently punitive to be effective is unlikely to be politically feasible (Dolphin, Pollitt, and Newbery 2020; Jenkins 2014; Mildenberger 2020, 24; Rabe 2018). The more politically viable alternative has often been de facto or surrogate carbon taxes of one form or another (Rabe and Borick 2012). Beyond such market-based instruments, governments have introduced a variety of other mitigation policies, such as subsidising the development and adoption of low-carbon forms of energy and transportation; banning certain carbon-intensive products; and investing in efforts to enhance carbon sequestration in forests and soils (Fekete et al. 2021).

Moreover, acting on the assumption that anthropogenic climate change is unlikely to be arrested, and to cope with changes that may already be occurring, there must be a second string to the bow: adaptation.[1] In general, adaptation planning must contend with both slow-onset alterations in ecological systems (such as gradual increases in temperatures or decreases in the availability of water) and rapid-onset events (such as floods or hurricanes). This requires the development of adaptive processes and pathways to predict and assess the risks and consequences that come with these different types of change, and to increase resilience to their direct and indirect effects. These risks and consequences will naturally vary in their nature and extent from region to region, and thus adaptation has a more inherently local character. Nonetheless, it will frequently require or benefit from learning and collaboration between governments. Under the Paris Agreement, states have committed to developing and implementing adaptation plans and actions and to provide the United Nations Framework Convention on Climate Change (UNFCCC) with periodic updates on their efforts in doing so.

1.2 The Intersection of Federalism and Climate Governance

Much has been written about climate governance as a multi-level enterprise spanning everything from the local village to the global community (e.g., Jänicke 2017). Here we are interested more specifically in climate governance in federal systems. These systems are characterised by the coexistence of two (or occasionally three) constitutionally entrenched orders of government, each accountable to, and acting upon, its population. Federalism thus has two lines of division: vertically between the central government and the constituent units, and horizontally between each of the latter. There are always local governments as well, but these generally have a distinctly subordinate or ambiguous status (Steytler 2009).

Depending on the division of powers, constituent units may well be an important component in the implementation of national climate change commitments. The fact alone that climate change is 'intrinsically multilevel' suggests that federalism provides for an appropriately dispersed range of actions and may be well suited to the task (Brown 2012, 324). This underpins the recognition given by the 2015 Paris Agreement to 'the importance of the engagement of all levels of government and various actors, in accordance with respective national legislations of Parties, in addressing climate change'. It is also consistent with the subsidiarity principle that underlies federalism – tasks should be carried out as locally as practicable – and the associated notion that broad scope for territorial diversity in policies should thereby exist. However, there are also reasons to think that federalism may also create impediments to effective climate governance.

1.2.1 Cutting Both Ways

Federalism often seems like a 'double-edged sword' in this regard: capable of exerting quite contrary effects (Karapin et al. 2020). Which one dominates in any given situation will be the consequence of specific circumstances. In addition, the effects of federalism often cross-cut and neutralise each other (Weaver 2020). Drawing on work by a number of scholars, we suggest three possible ways in which federalism might prove favourable for climate change governance, and, conversely, three ways in which it might prove less so (Balthasar, Schreurs, and Varone 2020; Hueglin and Fenna 2015, 41–6; Jordaan et al. 2019; Weaver 2020).

Federalism's vertical and horizontal divisions can *facilitate governance* by providing scope for:

1. Locally appropriate and responsive measures.
2. A 'fail-safe' degree of redundancy whereby the constituent units can step in and compensate for failure or inaction by the central government.
3. Policy experimentation and inter-jurisdictional learning.

On the other hand, the divisions and fragmentation of a federal system may *impede governance* in any or all of the following three ways:

1. The existence of multiple 'veto points' may obstruct policymaking or lead to blame-shifting.
2. Federalism may result in a patchwork of policies with varying effectiveness, poorly co-ordinated either vertically or horizontally.
3. Constituent units may be constrained by collective actions problems and a 'race-to-the-bottom' competitive dynamic.

1.2.2 Federalism as Facilitator

One of the very earliest advantages claimed for federalism was the way it allows policymaking to reflect the differences in conditions and preferences from region to region (de Tocqueville 1848, 262). With regard to climate change, there may be significant differences between the regions of a country in economic structure and associated carbon intensity, natural resource endowment, exposure to different climate risks, public institutions and capabilities, expertise on climate issues, and political dynamics and cultures. In federations, relevant powers such as those over land-use planning, natural resources, transportation, electricity supply, water management, and emergency preparedness are typically assigned to the constituent units. They are thus able to craft policies most aligned with their context, interests, and disposition.

A second possible virtue of federalism lies in the redundancy of two orders of government (Landau 1973, 188). Federalism is thus an 'opportunity structure' with the potential for constituent units to take up the slack by launching their own mitigation or adaptation programmes in what has been labelled 'compensatory federalism' (Derthick 2010). One way this occurs is through 'venue shopping' by activists targeting the most receptive and relevant government (Pralle 2003). This can be expected to reflect partisan differences when the two orders of government are in ideologically different hands (Bulman-Pozen 2014) – differences reflective of the alignment between environmental attitudes and the traditional Left–Right ideological spectrum. For such compensatory action to be possible, the constituent units must, of course, have the requisite powers and financial resources.

Conceivably, the collective effort of individual jurisdictions could even amount to much the same as that which would have been achieved by an effective national programme. Even if it does not, that collective effort could be a reasonable surrogate, with a diversity of initiatives having an incrementally additive nature (Ostrom 2012). Furthermore, constituent unit action may play an instigating role whereby a dynamic process of 'multilevel reinforcement of policy action' generates momentum that drives action at the national level (Balthasar, Schreurs, and Varone 2020, 6). In turn, constituent units may continue to advance overall policy goals with programmes that complement and augment those implemented by the central government (Buzbee 2015).

Finally, the existence of multiple jurisdictions creates the potential for governments to learn from each other. It was long ago identified as an advantage of federal as distinct from unitary government that it provides multiple sites for 'a people to try experiments in legislation and administration' (Bryce 1893, 353). In what thus came to be called the 'laboratory federalism' thesis, if these 'experiments' prove worthy of emulation, they will diffuse in either the horizontal or the vertical plane. In the latter event, the experiments can be seen as playing a 'catalysing' role for central government action (Bernstein and Hoffman 2018). We can distinguish here between lessons about which solutions are effective or efficient, and lessons about which are politically feasible or practicable – 'instrumental policy learning' and 'political learning' (May 1992).

1.2.3 Federalism as Hindrance

On the other hand, the inhibiting effects of federalism are potentially substantial. First, it is possible that federalism's division of powers may itself act as an obstacle to policymaking. Historically, this often seemed to be the case, as captured by Dicey's (1915, 167) declaration that 'Federal government means weak government'. One way to express this is in terms of the multiple 'veto points' through which federalism may stymie policymaking. These could lie in the way the division of powers denies either order of government authority to act or creates a misalignment between responsibilities and capabilities. One set of studies, for instance, concluded that mitigation action by the central governments of both Austria and Switzerland in one particular policy domain was hamstrung by constituent unit non-compliance (Casado-Asensio and Steurer 2016). And vice versa, it is possible that constituent units are constrained by jurisdictional limits.

A quite separate veto point exists if the constituent units enjoy input into central government decision-making through their representation in second chambers. In such circumstances, it is quite possible that a number of them could impose a lowest-common-denominator policy.

Second, mitigation or adaptation efforts may take varying forms and be implemented to varying degrees between constituent units and between orders of government, even working at cross-purposes. This can reduce both the efficacy and the efficiency of measures substantially. Intergovernmental co-ordination is an important component of policymaking and implementation in modern federalism where the division of powers and responsibilities has become less and less clear. Furthermore, if responsibility is carried in large part by the constituent units, advances made in some jurisdictions where conditions and attitudes are more conducive may well do little to compensate for inaction in others more closely tied to high-emissions industries. 'Attempts to reduce greenhouse gas (GHG) emissions by one jurisdiction are meaningless if others allow emissions to increase by an equal (or greater) amount', argues Gordon (2015, 122–3). If that is the case, it is quite possible that an 'effective response ... requires vertical co-ordination' – which in this case seems to be a euphemism for central government control.

Third, constituent units in general may tend to under-invest in environmental protection. Simply put, the incentives are not there for them to take action in a broader public interest that they perceive as being contrary to their material self-interest (Engel and Rose-Ackerman 2001; Weibust 2009). They carry the full cost of any measure they implement while others share the benefits. And, if there is any environmental issue where the gulf between local costs and collective benefits is a yawning one and the incentive to free ride or shirk responsibility compelling, surely climate change is it (Adler 2008). At the extreme, this might induce an environmental 'race to the bottom', where jurisdictions minimise actions perceived as deterring inward investment and growth (Woods 2021). At the very least, there are reasons why we might expect foot-dragging by constituent units, and, again, top-down policies would seem to be required. These dynamics represent 'a substantial risk for success of a national policy if the focus is mitigation, but that risk is significantly reduced if the focus is adaptation', since benefits of the latter are much more likely to be retained locally (Fowler 2020, 153).

1.3 Varieties of Federalism and Other Variables

To this point we have treated *federalism* as if it were a single or undifferentiated phenomenon. In reality, it is only an abstraction; the real world is made up a range of *federations* or federal systems each with its own characteristics, style, and underlying realities (Fenna 2019). Federations differ in several important ways, all of which have an impact on the way they are likely to function in climate change governance. This book showcases a good part of that diversity. Not all are unambiguously federations. Spain does not formally describe itself as federal, though it is regarded by many observers as being so. The European Union, meanwhile, must be seen as only proto-federal – still more confederal in the degree to which its constituent units retain sovereignty and exercise the bulk of policymaking responsibility. It is also a meta-federation of sorts, since some of its members are themselves federations. In addition, we include two major cases, China and Indonesia, that while not federations at all, have systems of devolved governance through which their climate change policies are implemented. With its extraordinary degree of societal and geographical diversity, Indonesia would seem a natural home for federalism, but a deep 'aversion' to such a divided form of government has prevailed since independence (Kingsbury 2013; Reid 2007). Those diversities were given recognition, though, in the devolutionary programme of 2001 whereby provinces and municipalities now jostle for roles and resources with the central government. Finally, as an authoritarian – indeed, in several ways totalitarian – regime, China is a particularly distinctive case here.

1.3.1 Varieties of Federalism

Dividing Powers. Included in this book are federations where powers are divided in a 'dualist' fashion and those characterised by 'administrative federalism'. In dualist federations such as Australia, Canada, and the United States, the two orders of government exercise full powers of policymaking, implementation, and administration within their assigned areas of jurisdiction. In federations where an administrative approach prevails, the central government exercises a broad policymaking power but responsibility for implementation and administration rests with the constituent units. Administrative federalism originated in Germany, but is characteristic now of Austria, Switzerland, and the EU as well. It also influenced the design of the 1996 South Africa constitution (Choudhry 2020).

The distinction between the two models is not a pure one, with administrative federations dividing some powers in the dualist fashion and the dualist federations increasingly taking on characteristics of administrative federalism; however, the distinction remains an important one (Mueller and Fenna 2022).

Decentralised/Centralised. The division of powers is also one of 'degree' as well as 'type'. The federations canvassed in this collection vary considerably in their degree of centralisation or decentralisation. This can be evident in either or both the kinds of powers available to the constituent units and the fiscal resources they have at their disposal (Dardanelli et al. 2019a). While the established federations have generally undergone considerable centralisation over the years, Canada and Switzerland remain relatively decentralised (Dardanelli et al. 2019b). India and South Africa have notably centralised features – accentuated in the South African case by one-party rule (Tapscott 2015).

In some federations, the division of powers regarding climate change policy is largely a settled issue, but in others it remains a live one. There is also always the possibility that jurisdictional conflict will arise and have to be resolved by the courts as they do in almost all federations for the constitutional division of powers more generally (Aroney and Kincaid 2017).

Bicameralism. In both the German and EU cases, the administrative division of powers is complemented by arrangements whereby the constituent units enjoy direct representation in the central government and thus a degree of 'co-determination' over that process of central government policymaking. In Germany's system of integrated federalism, the *Länder* exercise their co-decision power through the second chamber of the federal parliament, the *Bundesrat* (Federal Council). In the EU, Member States exercise their direct influence through the Council of Ministers. While Switzerland does not have an equivalent chamber, the country's powerful system of direct democracy plays an important role in regulating the respective roles of the federation and the cantons. The interlocking structures of German and EU federalism have been accused of creating conditions for a 'joint-decision trap' (*Politikverflechtungsfalle*), imposing a high threshold for policy change and thus entrenching the status quo or biasing the system toward lowest-common-denominator policymaking. Switzerland's system of direct democracy has been accused of having the same effect (e.g., Mueller 2020).

Constituent units do not enjoy council-style representation in other federations and so federal bicameralism is not a factor outside Germany and the EU (Hueglin and Fenna 2015, 205–37). However, there are other federations with strong second chambers – Australia and the United States, for instance – so even if not meaningfully 'federal', bicameralism can still represent a significant veto point.

Conflict and Co-operation. All federations have had to develop mechanisms and processes of intergovernmental relations (IGR) through which governments can work together (Hueglin and Fenna 2015, 238–74; Poirier and Saunders 2015; Schnabel 2020). These practices of co-operative federalism have become a crucial feature of systems where overlap between orders of government and degree of policy interdependence means an increasing need for co-ordination, even if they vary in how and how well they work from one policy area to another as well as from one country to another.

In parliamentary federations, IGR takes the form of 'executive federalism', typically structured in two tiers: a layer of portfolio-defined councils comprising the relevant cabinet ministers; and, at the peak level, a heads-of-government meeting. In a presidential system with its separation of powers such as the United States, the head of government cannot speak for the whole government, but only 'the administration', and thus peak intergovernmentalism tends to be absent.

Presidential or Parliamentary? As the above suggests, the way a federation functions is influenced by the mode of representative democracy in operation. While Australia, Canada, and the United States are all similar in being dualist federations, the United States is distinctive in its presidential, separation-of-powers system of government. By contrast with parliamentarism, which concentrates authority in the executive, presidentialism disperses it between the executive and legislative branches. In addition to affecting the style of intergovernmental relations, this makes policymaking subject to more veto points. As a number of studies (e.g., Greer 2010, 181) have noted, what can look at first blush like a policy-retarding effect of federalism in the United States, 'is mostly due to a federal government riddled with internal veto points'.

1.3.2 Societal and Economic Factors

Federations differ in manifold other ways, including their level of political and economic development, their degree of federal diversity, and their economic base and resource structure. The cases here encompass countries across a wide range of economic and political development, as well as ones with deep diversity and those with a single national identity. Canada's bicommunal nature is intrinsic to the operation of its federal system, for instance, while Ethiopia is an extreme form of ethno-federalism (Fiseha and Habib 2010; Gagnon and Simeon 2010). Australia and Germany are far more homogeneous.

Countries such as Australia and Canada have a strong basis in the resource sector, including in hydrocarbons. At the other extreme are countries such as Switzerland without any significant petroleum or coal resources. Germany's high level of industrialisation was made possible historically by its coal reserves and it still relies on coal for one-quarter of its electricity generation. Not only does a substantial endowment of high-emissions resources affect the approach a country might take to climate change, but it is rare for such resources to be equally distributed *within* any federation, and thus the political geography of their resource economy will have significant consequences for the operation of federalism when it comes to climate governance (Brown 2012, 324). Similarly, not only do countries vary significantly in the range of renewable energy sources they can tap, but such potential often varies substantially within those federations.

1.4 Looking Ahead

The chapters that follow provide a picture of the way systems of federal or decentralised governance function in managing responses to this one particular – and particularly significant – policy issue. Has federalism enhanced climate governance by allowing policies to be tailored to regional conditions and preferences, by providing a fail-safe redundancy, and/or by multiplying opportunities for policy experimentation and learning? To what extent, by contrast, has it hindered climate governance by multiplying veto points, inviting a dysfunctional patchwork of policies, or imposing collective action constraints? The wide range of cases that follow provide an opportunity to reflect on the way these dynamics might work in very different institutional, political, economic, and societal circumstances.

Notes

1 Adaptation policy is defined by the IPCC as 'adjusting to the effects of both anthropogenic and natural climate change through initiatives that prevent or minimize harms as well as exploit opportunities generated by changes to the climatic system'.

References

Adler, Jonathan H. 2008. 'Hothouse Flowers: the vices and virtues of climate federalism'. *Temple Political & Civil Rights Law Review* 17 (2): 443–64.

Arndt, Channing, Doug Arent, Faaiqa Hartley, Bruno Merven, and Alam Hossain Mondal. 2019. 'Faster Than You Think: renewable energy and developing countries'. *Annual Review of Resource Economics* 11 (1): 149–68.

Aroney, Nicholas, and John Kincaid, eds. 2017. *Courts in Federal Systems: federalists or unitarists?* Toronto: University of Toronto Press.

Balthasar, Andreas, Miranda A. Schreurs, and Frédéric Varone. 2020. 'Energy Transition in Europe and the United States: policy entrepreneurs and veto players in federalist systems'. *Journal of Environment & Development* 29 (1): 3–25.

Bernstein, Steven, and Matthew Hoffman. 2018. 'The Politics of Decarbonisation and the Catalytic Impact of Subnational Climate Experiments'. *Policy Sciences* 51 (2): 189–211.

Biesbrook, Robbert, and Alexandra Lesnikowski. 2018. 'Adaptation: the neglected dimension of polycentric climate governance?' In *Governing Climate Change: polycentricity in action?* eds. Andrew Jordan, Dave Huitema, Harro van Asselt and Johanna Forster, 303–19. Cambridge: Cambridge University Press.

Brown, Douglas M. 2012. 'Comparative Climate Change Policy and Federalism: an overview'. *Review of Policy Research* 29 (3): 322–33.

Bryce, James. 1893. *The American Commonwealth*. 3rd ed. 2 vols. Vol. 1. London: Macmillan.

Bulman-Pozen, Jessica. 2014. 'Partisan Federalism'. *Harvard Law Review* 127 (4): 1077–1146.

Buzbee, William W. 2015. 'Climate Federalism, Regulatory Failure and Reversal Risks, and Entrenching Innovation Incentives'. In *The Law and Policy of Environmental Federalism: a comparative analysis*, eds. Kalyani Robbins, 145–68. Cheltenham: Edward Elgar.

Casado-Asensio, Juan, and Reinhard Steurer. 2016. 'Mitigating Climate Change in a Federal Country Committed to the Kyoto Protocol: how Swiss federalism further complicated an already complex challenge'. *Policy Sciences* 49 (3): 257–79.

Choudhry, Sujit. 2020. 'German Federalism in Translation: the German Grundgesetz's impact on South Africa's 1996 constitution'. In *Reformbaustelle Bundestaat*, eds. Felix Knüpling, Mario Kölling, Sabine Kropp and Henrik Scheller, 229–46. Wiesbaden: Springer.

Dardanelli, Paolo, John Kincaid, Alan Fenna et al. 2019a. 'Conceptualizing, Measuring, and Theorizing Dynamic De/Centralization in Federations'. *Publius* 49 (1): 1–29. https://doi.org/10.1093/publius/pjy036.

Dardanelli, Paolo, John Kincaid, Alan Fenna et al. 2019b. 'Dynamic De/Centralization in Federations: comparative conclusions'. *Publius* 49 (1): 194–219. https://doi.org/10.1093/publius/pjy037.

De Tocqueville, Alexis. 1848. *De la démocratie en Amérique*. 2nd ed. Paris: Pagnerre.

Derthick, Martha. 2010. 'Compensatory Federalism'. In *Greenhouse Governance: addressing climate change in America*, eds. Barry G. Rabe, 58–72. Washington DC: Brookings Institution.

Dicey, A. V. 1915. *Introduction to the Study of the Law of the Constitution*. 8th ed. London: Macmillan.

Dolphin, Geoffrey, Michael G. Pollitt, and David M. Newbery. 2020. 'The Political Economy of Carbon Pricing: a panel analysis'. *Oxford Economic Papers* 72 (2): 472–500.

Engel, Kirsten, and Susan Rose-Ackerman. 2001. 'Environmental Federalism in the United States: the risks of devolution'. In *Regulatory Competition and Economic Integration: comparative perspectives*, eds. Daniel C. Esty and Damien Geradin, 135–53. New York: Oxford University Press.

Fekete, Hanna, Takeshi Kuramochi, Mark Roelfsema et al. 2021. 'A Review of Successful Climate Change Mitigation Policies in Major Emitting Economies and the Potential of Global Replication'. *Renewable and Sustainable Energy Reviews* 137 (110602): 1–18.

Fenna, Alan. 2019. 'What Hope for Comparative Federalism?' In *A Research Agenda for Federalism Studies*, ed. John Kincaid, 76–92. Cheltenham: Edward Elgar.

Fiseha, Assefa, and Mohammed Habib. 2010. 'Ethiopia'. In *Diversity and Unity in Federal Countries*, eds. Luis Moreno and César Colino, Series: *A Global Dialogue on Federalism*, 139–66. Montreal & Kingston: McGill-Queen's University Press.

Fowler, Luke. 2020. *Environmental Federalism: old legacies and new challenges*. Abingdon: Routledge.

Gagnon, Alain-G., and Richard Simeon. 2010. 'Canada'. In *Diversity and Unity in Federal Countries*, eds. Luis Moreno and César Colino. Montreal & Kingston: McGill-Queen's University Press.

Gordon, David J. 2015. 'An Uneasy Equilibrium: the coordination of climate governance in federated systems'. *Global Environmental Politics* 15 (2): 121–41.

Greer, Scott L. 2010. 'How Does Decentralisation Affect the Welfare State? Territorial politics and the welfare state in the UK and US'. *Journal of Social Policy* 39: 181–201.

Hueglin, Thomas O., and Alan Fenna. 2015. *Comparative Federalism: a systematic inquiry*. 2nd ed. Toronto: University of Toronto Press.

IPCC. 2014. *Climate Change 2014: Synthesis Report. Contribution of Working Groups I, II and III to the Fifth Assessment Report of the Intergovernmental Panel on Climate Change.* Geneva: Intergovernmental Panel on Climate Change.

IPCC. 4 April 2022. *Climate Change 2022: mitigation of climate change*. Working Group III contribution to the Sixth Assessment Report of the Intergovernmental Panel on Climate Change, Intergovernmental Panel on Climate Change.

Jänicke, Martin. 2017. 'The Multi-level System of Global Climate Governance: the model and its current state'. *Environmental Policy and Governance* 27 (2): 108–21.

Jenkins, Jesse D. 2014. 'Political Economy Constraints on Carbon Pricing Policies: what are the implications for economic efficiency, environmental efficacy, and climate policy design?' *Energy Policy* 69: 467–77.

Jordaan, Sarah M., Adrienne Davidson, Jamal A. Nazari, and Irene M. Herremans. 2019. 'The Dynamics of Advancing Climate Policy in Federal Political Systems'. *Environmental Policy and Governance* 29 (3): 220–34.

Karapin, Roger, Andreas Balthasar, Miranda A. Schreurs, and Frédéric Varone. 2020. 'Federalism as a Double-Edged Sword: the slow energy transition in the United States'. *Journal of Environment & Development* 29 (1): 26–50.

Kingsbury, Damien. 2013. 'Indonesia: arbitrary polity, unitary state'. In *Routledge Handbook of Regionalism and Federalism*, eds. John Loughlin, John Kincaid, and Wilfried Swenden, 401–12. Abingdon: Routledge.

Landau, Martin. 1973. 'Federalism, Redundancy and System Reliability'. *Publius* 3 (2): 173–96.

May, Peter J. 1992. 'Policy Learning and Failure'. *Journal of Public Policy* 12 (4): 331–54.

Mildenberger, Matto. 2020. *Carbon Captured: how labor and business control climate politics*. Cambridge MA: MIT.

Mueller, Sean. 2020. 'Grundgesetz und deutscher Föderalismus von außen betrachtet: Inspirationen für die Schweiz?' In *Reformbaustelle Bundestaat*, eds. Felix Knüpling, Mario Kölling, Sabine Kropp, and Henrik Scheller, 247–66. Wiesbaden: Springer.

Mueller, Sean, and Alan Fenna. 2022. 'Dual Versus Administrative Federalism: origins and evolution of two models'. *Publius*. doi.org/10.1093/publius/pjac008

Nordhaus, William D. 2013. *The Climate Casino: risk, uncertainty, and economics for a warming world*. Yale University Press.

OECD. 2019. *Taxing Energy Use 2019: using taxes for climate action*. Paris: Organisation for Economic Co-operation and Development.

Ostrom, Elinor. 2012. 'Nested Externalities and Polycentric Institutions: must we wait for global solutions to climate change before taking action at other scales?' *Economic Theory* 49 (2): 353–69.

Poirier, Johanne, and Cheryl Saunders. 2015. 'Comparing Intergovernmental Relations in Federal Systems: an introduction'. In *Intergovernmental Relations in Federal Systems: comparative structures and dynamics*, eds. Johanne Poirier, Cheryl Saunders, and John Kincaid, 1–13. Don Mills ON: Oxford University Press.

Pralle, Sarah B. 2003. 'Venue Shopping, Political Strategy, and Policy Change: the internationalization of Canadian forest advocacy'. *Journal of Public Policy* 23 (3): 233–60.

Rabe, Barry G. 2018. *Can We Price Carbon?* Cambridge MA: MIT.

Rabe, Barry G., and Christopher P. Borick. 2012. 'Carbon Taxation and Policy Labeling: experience from American states and Canadian provinces'. *Review of Policy Research* 29 (3): 358–82.

Reid, Anthony. 2007. 'Indonesia's Post-Revolutionary Aversion to Federalism'. In *Federalism in Asia*, eds. Baogang He, Brian Galligan, and Takashi Inoguchi, 144–64. Cheltenham: Edward Elgar.

Schnabel, Johanna. 2020. *Managing Interdependencies in Federal Systems: intergovernmental councils and the making of public policy*. Cham: Palgrave Macmillan.

Steytler, Nico, ed. 2009. *Local Government and Metropolitan Regions in Federal Systems*. Kingston & Montreal: McGill-Queen's University Press.

Tapscott, Chris. 2015. 'Republic of South Africa: an uncertain path to federal democracy'. In *Political Parties and Civil Society in Federal Countries*, eds. Klaus Detterbeck, Wolfgang Renzsch, and John Kincaid, 199–226. Don Mills ON: Oxford University Press.

Weaver, R. Kent. 2020. 'Policy Dynamics in Federal Systems: a framework for analysis'. *Publius* 50 (2): 157–87.

Weibust, Inger. 2009. *Green Leviathan: the case for a federal role in environmental policy*. Farnham: Ashgate.

Woods, Neal D. 2021. 'An Environmental Race to the Bottom? "No more stringent" laws in the American States'. *Publius* 51 (2): 238–61.

2

Climate Governance and Federalism in Australia

ALAN FENNA

2.1 Introduction

Australia is one of the highest per capita producers of greenhouse gas (GHG) emissions and is regularly accused of dragging its heels on emissions reduction – as exemplified by the absence of a carbon pricing scheme (e.g., CAT 2020; CT 2020; Germanwatch 2020). What role has federalism played here?

As outlined in the Introduction (Chapter 1) to this book, federalism has a number of amphibolous qualities. On the one hand, it provides opportunities for locally tailored and experimental policymaking, a degree of 'fail safe' redundancy, and an opportunity for policy experimentation and inter-jurisdictional learning. On the other hand, it risks obstruction, patchy and counterproductive efforts, misalignment or discoordination; and collective action problems. Concern about coordination deficits has been particularly prominent in discussion of climate-change policy in federal systems such as Australia's.

> The politics of climate change mitigation are made even more difficult in Australia by a number of distinctive contingent factors. Firstly, Australia's federal system, and the difficulty of achieving national policy consensus, makes agreement challenging at the best of times; when the stakes are high and the issues complex and the consequences uncertain, it can be particularly difficult to achieve agreement, as the unresolved inter-state struggle over water allocation reminds us. (Beeson and McDonald 2013, 335)

The result is "fragmentation" (Jones 2009) – a problem, it is claimed, that can only be addressed by "vertical coordination" (Gordon 2015, 123; also D. M. Brown 2012, 331–2; Kallies 2021).

This chapter outlines a situation where a country with heavy reliance on carbon-intensive energy resources has faced substantial greenhouse gas dilemmas, where those dilemmas manifest themselves in strong ideological and partisan differences, and where both the central government (the Commonwealth) and the States have broad licence in climate change policymaking. It finds that the need for coordination can be exaggerated. Federalism has been a facilitating rather than a hindering factor in Australia, more consistent with Derthick's (2010) notion of *compensatory federalism* where 'governments at one level of the system are able to compensate for weaknesses or defects at another level', or Hollander's (2010) emphasis on the often-unrecognised benefits of 'overlap and duplication'.

2.2 The Australian Conundrum

High-emissions industries are one of the cornerstones of the Australian economy and this dependence explains the country's cautious approach to emissions reduction and ambivalence towards international commitments. At the same time, though, this historic dependence means there has been a good deal of low-hanging fruit to pick. Australia is also endowed with enormous potential for renewable energy development.

2.2.1 Contributions

Australia is one the highest high per capita emitters of $C0_2e$ in the world, and the highest per capita emitter among the industrialised democracies: 21 tonnes per person in the year to March 2020, down substantially from 36 tonnes per person in 1990 (DISER 2020b).[1] With a negligible share of the world's population, though, Australia's aggregate output of 530Mt is comparatively small. Producing scarcely more than 1 per cent of the global total, Australia's emissions are insignificant compared with the United States at 15 per cent, or, at the extreme, China at almost 30 per cent.[2] This is important not only in itself, but also in understanding the climate change debate within Australia, since it means that no matter how strenuous Australia's emissions-reduction efforts, in themselves they can have only the most trivial material effect on climate change.

The leading source of Australian emissions is electricity generation, dominated by coal-fired power stations. In turn, this reflects the now-inconvenient reality that Australia has coal in abundance, which is also the country's second most valuable export after iron ore. Even worse as far as greenhouse gas emissions are concerned, Australia is particularly rich in the dirtier *brown* coal (lignite), holding one-quarter of the world's recoverable resources, suitable only for *in situ* usage. 'At 2018 rates of extraction, the accessible resource base ... will support over 1000 years of production' (Geoscience Australia 2020, 29). These reserves are, for all intents and purposes, unlimited and have long been the main fuel for electricity generation in one of the two most populous and industrialised States, Victoria.

In turn, electricity generated by vast coal deposits has underwritten investment in energy-intensive resource processing such as the transformation of bauxite, of which Australia is the world's leading producer, into alumina, of which Australia is the world's leading exporter; and the transformation in turn of alumina into aluminium. The Tomago aluminium refinery, for instance, alone uses 10 per cent of the entire New South Wales electricity supply and the Portland aluminium refinery accounts for a similar share of electricity use in Victoria. The significance of this lies in the reality that minerals processing is a large part of what passes for manufacturing in Australia's resource-dominated export profile and thus plays an outsized economic role. For over a century, Australian public policy has sought to diversify the economy away from its comparative advantage in primary products, but with mixed results (Fenna 2016). Aluminium is Australia's leading 'manufactured' export, making up 8 per cent of the total, and ranks sixteenth in total exports. It is not hard to see why climate change policy in Australia is, if not all about, certainly very much about, electricity generation. 'The early and orderly movement to zero-emission electricity is the cornerstone of the decarbonisation of the Australian economy' (Garnaut 2019, 49).

At the same time, Australia is not short of the main climate-friendly alternatives: wind and solar. 'Australia's renewable energy resource endowment is both large and rare' (Wood, Dundas, and Ha 2020). With distinct advantages to exploit in converting to a low-carbon economy, 'Australia would prosper exceptionally from doing its fair share in a strong global effort to reduce the disruption from climate change' (Garnaut 2019, 15). Research suggests that '100% renewable electricity in Australia' is feasible (Blakers, Lu, and Stocks 2017; Lu et al. 2021) – though 'net-zero' emissions would be more practical (Wood and Ha 2021). It is also convenient that the country's fleet of large coal-fired power stations is an ageing one, 'and most are scheduled to be retired by 2040', particularly those in Victoria and New South Wales (Wood and Ha 2021, 6). There is thus not the problem of stranded assets there might be.

Obstacles to reaching majority reliance on renewables for electricity generation include challenges in balancing the grid to cope with their distributed and variable nature, and realigning the existing transmission systems to accommodate new sources of supply (AEMC 2019a). Battery storage has become increasingly important for the former, and 'renewable energy zones' for the latter. When a severe storm caused an alarming blackout across South Australia in 2016 – immediately following the closure of the State's last coal-fired baseload power station – energy security became a prominent concern and possible political obstacle to an enthusiastic embrace of renewables.

These obstacles have become an issue precisely because renewable generation has grown so rapidly – notwithstanding Australia's comparative advantage in fossil fuels. That expansion in turn means that other emissions sources will become more significant. The transport sector will soon replace electricity generation as the country's leading source of emissions and thus transitioning away from combustion-engine vehicles will become the next frontier (Saddler 2021). Meanwhile, in 2020, Australia became the world's largest exporter of liquified natural gas (LNG), production of which generates significant emissions as well.

2.2.2 Consequences

At the same time as being a large per capita contributor to global warming, Australia is also particularly vulnerable to its consequences. The continent is in many ways a fraught ecosystem characterised by droughts, fires, floods, and cyclones. Australia is 'a country defined by extremes: erratic climate influences virtually every aspect of our lives' (Gergis 2018, 8). Extreme weather events are normal, but at risk of being accentuated by climate change, and Australia has been described as 'the most vulnerable nation in the developed world' (Gergis 2018, 264; also Christoff 2014).

There are regularly warnings that this is bringing with it more frequent and more severe bushfires (e.g., Abram et al. 2021; BOM and CSIRO 2020; ELCA 2019; Lukas et al. 2007) – particularly after the devastating ones of 2019–20 (Hughes et al. 2020; Steffen et al. 2019). This finds some support in the research (e.g., van Oldenborth et al. 2021). In addition, because Australia's population lives and plays disproportionally in the littoral zone, the threat of rising sea levels to coastal infrastructure and amenities has been recognised as a significant risk for over a decade now (e.g., DCC 2009; also Arbinolo and Gamper 2021). Governments mapping out comprehensive emissions-reduction programs in Australia typically link the importance of such action to the prediction of such risks (e.g., DELWP 2021, 9).

2.2.3 International Commitments

Australia's climate change commitments go back to 1990 when it signed up to the Toronto target of a 20 per cent reduction in emissions from 1988 levels by 2005. This was a soft or 'no regrets' commitment in that it was to be implemented only if it could be done without economic cost. Australia signed the Kyoto protocol in 1998, but only ratified it in 2007, after a change of government (see below). Policy inaction under the Kyoto protocol was legitimised by the provision in that treaty allowing Australia to use land clearing emissions to inflate the 1990 baseline measure (CAA 2019; Crowley 2010). Australia ratified the more pragmatic Paris Agreement of the UNFCCC in 2016, committing to 26–28 per cent reduction by 2030, a target the government described in self-congratulatory terms as 'ambitious' (DEE 2017; Kellow 2018; also see Hale 2016).

The government's Climate Change Authority (2015) had, however, recommended a target twice as ambitious if meaningful reduction was to be made and the economic opportunities of new technologies seized. The existing targets would be insufficient to get Australia near the net-zero-by-2050 aim that was being widely adopted, and existing measures insufficient to ensure Australia meets even its modest 2030 Paris target without using so-called carryover credits (AATE 2020). By contrast, all but one of the States have adopted their own emissions-reduction targets without any obligation to do so, as discussed below.

2.3 Climate Governance and the Federal System

The nature of Australian federalism is such that, with some important exceptions, either or both orders of government can play a substantial role in emissions reduction. Responsibility for climate change adaptation, meanwhile, sits more naturally with State and local government – with the latter being 'on the frontline in dealing with the impacts of climate change' (SCCC 2012). Even there, though, the Commonwealth inevitably has a role to play.

2.3.1 The Division of Powers in Theory and Practice

The Commonwealth Constitution lays out a scheme for a classic 'coordinate' or 'dual' federal system where the States have full responsibility for the majority of domestic policy tasks. The Commonwealth was assigned a limiting list of powers, chiefly concerned with managing Australia's external relations and ensuring the national economy. Few of the Commonwealth's powers were made exclusive, but it enjoys primacy in all concurrent fields. Particularly since 1920, though, the Commonwealth has steadily expanded its remit, supported by expansive High Court interpretation of its assigned powers (Aroney 2017; Fenna 2019). With the States being denied access to sales taxes by the High Court, and the Commonwealth taking exclusive control of the personal and corporate income tax in 1942, Australian federalism has been characterised by a pronounced vertical fiscal imbalance (VFI). And with the States thus holding responsibilities far in excess of their tax revenues, and the Commonwealth enjoying tax revenues well in excess of its needs, there has been ample scope for exercise of the 'spending power' as sanctioned by section 96 of the Constitution (Fenna 2008). Through conditional, or 'tied', grants, the Commonwealth exercises policy influence virtually at will in areas of State jurisdiction.

In addition, broad interpretation of its enumerated powers has given the Commonwealth enormous clout via key clauses such 'trade and commerce with other countries, and among the States' (§ 51.i), 'corporations' (§ 51.xx), and 'external affairs' (§ 51.xxix). The latter makes it possible to over-ride the States in any respect of which Australia has signed an international treaty, and has provided the foundation since 1982 for the assumption by the Commonwealth of a substantial role in environmental policy (Fowler 2015; Saunders 1996; Twomey 2008).

The result of this expansion in Commonwealth power has been to create a great deal of overlapping and thus de facto concurrency. The States have generally maintained service delivery responsibility, but in various areas of their jurisdiction the Commonwealth exercises some degree of influence or control. At certain times it appears that the States are very much under the thumb of the Commonwealth, and there are periodically calls for them to be abolished. That traditionally came from the Labor side of politics, and now on the conservative side even the Liberal Party has largely abandoned its traditional defence of States' rights and shifted to a much more centralising view (Hollander 2008; Sharman 2001). The Covid-19 pandemic of 2020–2 showed, though, just how important the States remain, having assumed primary responsibility for protecting their citizens – even to the point of closing their respective borders – and on various fronts resisted Commonwealth pressure to be more relaxed in their approach (Fenna 2021).

Local government occupies a decidedly subordinate position in the Australian system (Grant and Drew 2017; Sansom 2009). In part, this is for straightforward constitutional reasons: local governments have no federal constitutional recognition and are entities of their respective State governments, exercising delegated powers with State government oversight.

2.3.2 Climate Change Governance and the Division of Powers

The States have primary jurisdiction over almost the full range of functions relevant to climate-change policymaking: criminal and civil law; land and resource management; transport and urban planning; infrastructure, including network utilities; public services; and the environment. Until privatisations in the 1990s, the States all owned and operated their respective electricity utilities – the chief emissions culprit. Some still do, and they all still regulate and control them.

At the same time, there is at least implicit recognition that the Commonwealth has a legitimate role in energy policy, in part because of the establishment of the national electricity market (NEM) linking the five eastern States (AEMC 2019b; COAG 2001). The NEM is a fairly recent phenomenon, and only connects the different systems rather than creating a single new one. It is managed by the Australian Energy Market Operator (AEMO), established in 2009.[3] The NEM's interconnexions become increasingly important as the reliance on variable renewable energy increases (Wood and Ha 2021, 45).

What the States no longer have is a broad power to tax. If we accept that a carbon tax of some form is the most economically efficient and administratively simple way of moderating emissions (Garnaut 2019, 117; OECD 2019; Rabe 2018), this represents a significant limitation on State action. Constitutionally, no Australian State is in the position to emulate British Columbia, 'a poster child of political courage and policy efficacy' (Rabe 2018, 204). Between its plenary power to tax, the external affairs power, the corporations power and the commerce and trade power, the Commonwealth has ample authority and resources to implement virtually any climate change mitigation policy it so chooses. The Commonwealth would have little difficulty imposing an aggressive emissions-reduction programme in Australia.

Local governments, meanwhile, have been taking steps to reduce emissions, not just in their own operations, but also more widely in their communities (Proudlove, Bravo, and Denis-Ryan 2020). A large part of their contribution can only be to the long-term, however, given the importance of the built environment, and hence urban planning and transport infrastructure, for sustainability (Lowe 2017). While climate change adaptation also involves all levels, local government is generally described as being 'at the forefront', particularly in respect to the coastal zone (Leitch 2017; Nalau, Preston, and Maloney 2015). After Labor won the 2007 federal election there was some expression of interest from Canberra in a greater Commonwealth role (e.g., HSCCCWEA 2009). However, subsequent intergovernmental consideration largely endorsed the status quo (SCCC 2012).

2.3.3 Cooperative Federalism in Australia

The large amount of de facto concurrency in the Australian system fuels a comprehensive network of intergovernmental relations – comprising numerous intergovernmental agreements, ministerial and specialist councils, and, at the apex, regular first ministers' meetings (Fenna and Phillimore 2015). From 1991 through until Covid-19 precipitated a change in 2020, the latter went by the name of COAG, the Council of Australian Governments (Fenna 2021).[4] Of particular relevance to climate change has been the COAG Energy Council.

Intergovernmental relations in Australia are overwhelmingly vertical rather than horizontal in nature, and top-down, given the Commonwealth's expanded constitutional authority and superior resources (Phillimore and Fenna 2017). A rare exception, noted below, was when Labor governments held office in all the States and Territories while the Liberal–National Party coalition ('the Coalition') held office at the Commonwealth level. This led to a flirtation with horizontal collaboration between the States. In general, collaboration has tended to occur when and insofar as it has been useful and attractive to the Commonwealth.

2.4 Climate Change Politics and Policy in Australian Federalism

A key factor in climate change policy in Australia has been the interaction between partisanship and federalism. In particular, the existence of two orders of government with relevant powers has allowed climate change action to be pursued through one channel when the other is blocked.

2.4.1 Ideology and Partisanship

Climate change has been a partisan issue in Australia, divided ideologically between Left and Right consistent with longstanding differences between the parties (Botterill and Fenna 2020). On the Left, the Australian Labor Party (ALP) and more so the Greens have favoured action. On the Right, the Liberal and (particularly) the National parties have resisted, sometimes staunchly so (Fielding et al. 2012; Tranter 2013). They regularly emphasise the 'immense' cost to Australia of turning its back on fossil fuels (e.g., Wild 2022). Underpinning this ideological and partisan divide has been the schism between the two-thirds of Australians who accept the notion of a scientific consensus on the proposition of anthropogenic climate change and the one-third who do not (Tranter 2017).

Epitomising the divide was the contrast between Labor prime minister Kevin Rudd's 2007 declaration that climate change is 'the great moral challenge of our generation', and then-Treasurer and subsequent Liberal prime minister Scott Morrison's appearance in parliament on 9 February 2017, prop in hand, announcing 'this is coal; don't be afraid, don't be scared; it won't hurt you'. Morrison went on to assert that coal has ensured Australia's prosperity for over a hundred years. There were calls from within the Coalition, not just for preservation of the coal-based status quo, but indeed for the government to subsidise construction of new coal-fired power stations (Coorey and McIlroy 2020).

The partisan divide was also evident in the contrast between Labor's commitment in the 2016 election to a 'net zero pollution' target for 2050 and the Liberal Party's absence of a long-term target altogether (Pearse 2018, 583). By 2020, the Liberal Party's position had shifted away from coal slightly but not away from hydrocarbons, with Australia's large reserves of natural gas being touted as the key transition fuel to sustain the country's processing and manufacturing industries for the foreseeable future (Fisher 2020; PM 2020). That position drew much criticism for being neither economically nor environmentally rational (e.g., Climate Council 2020; Ogge 2021; Stock et al. 2020; Wood and Dundas 2020). In general, the conservative side of politics has stuck to the 'no regrets' approach that had prevailed in 1990, rejecting measures that might impose costs on Australian industry and maintaining a commitment to the status quo through fossil fuel subsidies as well as funding for carbon capture and storage (Campbell, Littleton, and Armistead 2021). The suggestion after the 2019–20 bushfires that climate change is exacerbating Australia's natural disaster problems has led the conservative side of politics to emphasise 'preparation, resilience and adaptability' rather than change their stance on mitigation (Benson and Chambers 2020).

No party is ideologically monolithic, though, and there have been dissenting elements on both sides. The ALP is home to major unions representing workers in the extractive, minerals processing, and associated industries – unions that pressed the Party in late 2020 to take a more cautious approach to phasing out coal and gas (Brown 2020). The Liberal Party, meanwhile, has a progressive wing open to climate change action, and factional conflict around this issue has contributed to leadership turmoil and change. Their Coalition partner, the National Party, is more solidly attached to the status quo and often unabashedly pro-coal, arguing that 'Australia needs to build modern coal fired power stations to help manufacturing industries' (Nationals 2021, 18).

2.4.2 Regional Variation

While it important to note that 'carbon-intensive industries are often regionally concentrated, both in an international sense and in a subnational sense' (Brown 2012), this is less true of Australia than of some other federations such as Canada (Macdonald 2020). Hydrocarbons are widely distributed across the country, including in Victoria and New South Wales, the two metropolitan States. The three most populous States, those along the eastern seaboard, all rely on coal for electricity generation, and Queensland and NSW are the country's major coal exporters as well. There is one State with abundant hydroelectricity and all its electricity generated from renewables, but that is the minor – and offshore – State of Tasmania. Western Australia and South Australia have coal deposits, albeit of a smaller scale, and South Australia closed its coal mine and associated power stations in 2015–16. Meanwhile, Australia's abundant gas reserves, onshore and offshore, are also distributed around the country.

There is one jurisdiction – Western Australia – that is massively dependent on resource extraction, but unlike the Canadian province of Alberta, its dependence is first and foremost on iron ore and other minerals rather than hydrocarbons.[5] That said, it has a large and growing LNG industry, and indeed, produces over half of Australia's massive LNG exports. Queensland, the next most resource-based economy, is far more dependent on fossil fuel production, but is considerably more diversified than Western Australia. In its disproportionate contribution to emissions, Queensland is not entirely unlike Alberta. However, producing 32 per cent of Australia's emissions with 20 per cent of the country's population,

Queensland rather pales in comparison beside Alberta, which produces 38 per cent of Canada's total emissions with only 11.5 per cent of the country's population.[6] Moreover, while Queensland emissions came down somewhat from 1990, Alberta's increased 58 per cent over the same period (Macdonald 2020, 96).

2.4.3 Stasis: The Commonwealth

With the main parties being so strongly opposed on this issue, national policy directions have not just fluctuated wildly, but have been determined by the side of politics that has dominated over the past twenty-five years of climate change policymaking. At the national level, this has been the Coalition parties (1996–2007 and 2013–22). Thus, a combination of changes in government and Coalition dominance explains why 'Australia is the only country in the world to have adopted then abandoned carbon pricing', discussed below (Crowley 2017, 2).

A further factor has been Australia's strong bicameralism, most importantly in the federal parliament, but also in five of the six States. In the Senate, proportional representation ensures that the governing party at the national level only very rarely enjoys a majority and thus cannot be assured of getting its legislation passed. Enacting controversial new measures can thus be difficult, and, for Labor, support from the Greens has often been important. While working at times to hobble Labor's efforts to introduce emissions-reduction policies, the Senate has also worked to frustrate Coalition efforts at dismantling Labor policies.

The Coalition parties generally addressed emissions reduction through subsidy-based initiatives. However, there was one early programme that represented more of an imposition on industry: 'the first mandatory renewable energy target (MRET) in the world' (Kent and Mercer 2006, 1046). Introduced in 2001, its goals, though, were modest: requiring that 2 per cent of electricity (9,500 GWh) be generated from renewables by the end of the decade.

2.4.4 Enter the States

The partisan divide is equally evident at the State level, where the 'impact of parties' can be seen across a range of policy fields, not least of all climate change (Phillimore and Fenna 2020). While the Coalition parties were entrenched in Canberra from 1996 until 2007, the opposite was true at the State level. Those State and Territory Labor governments acted individually and collectively to fill the gap left by Commonwealth inaction. In general, the States picked up the baton when they were in Labor hands, and then dropped it when the other side of politics took over (Crowley 2013, 380). As in the United States (Berry, Laird, and Stefes 2015; Bromley-Trujillo and Holman 2020), partisanship has been an important variable in Australian climate change policy at the State level. However, this has diminished recently, and some convergence has occurred, with non-Labor governments in South Australia and New South Wales reconciling themselves to climate change action – sometimes to the consternation of their federal colleagues.

2.4.4.1 Individual State Action

Individually, there was a range of steps the States could – and did – take. The signature initiative from this period was early experimentation with an emissions trading scheme (ETS). 'In 2003, the New South Wales Labor (NSW) government introduced one of the first mandatory greenhouse gas ETSs in the world, followed by the Australian Capital Territory Labor government, which introduced a complementary scheme in 2005' (Crowley 2013, 371). The NSW Greenhouse Gas Abatement Scheme operated for almost a decade before being terminated following Labor's election defeat in 2011. While NSW Labor seemed to have 'lost its way' in many policy areas, climate change was a rare exception (Sartor 2011, 288 and *passim*).

More common at the State level were actions to support the conversion from coal to renewable sources for electricity generation. Complementing the Commonwealth's MRET, for instance, was *Victoria's Renewable Energy Act 2006*, introduced by the Labor government that had come to office in 1999 with upper house support of the Greens. It set a target of 10 per cent renewables by 2016. The chief mechanism by which States energised the renewables market was feed-in tariffs providing what was effectively a cross-subsidy for the uptake of rooftop solar.

2.4.4.2 Collective State Action

From 2002 until 2008, Labor held office in every State and Territory, laying the basis for collective action. An unprecedented level of horizontal intergovernmentalism followed, including the formation of CAF, the Council for the Australian Federation.[7] The States and Territories developed plans for a National Emissions Trading Scheme beginning with 'the establishment of a 'National Emissions Trading Taskforce' in 2004' (Twomey 2012, 108). The notion had been mooted by a Commonwealth government agency but not pursued (AGO 1999) and was consistent with the compromise approach that became internationally fashionable in this period (Meckling 2011). At the February 2007 meeting of COAG, the States and Territories pledged to go it alone if the Commonwealth failed to come onside.[8]

2.4.5 Commonwealth Takes the Lead: The Carbon Tax

By the time the Emission Trading Taskforce's Report was released, Labor had come to power in Canberra and the Commonwealth assumed leadership on the question after what 'has been described as the world's first climate change election' (Beeson and McDonald 2013, 331; Rootes 2008). Given the changed political landscape, the Taskforce Report emphasised that a 'collaborative arrangement through the Council of Australian Governments (COAG) is required. Such an example of cooperative federalism would build on more than three years of consistent work through the Taskforce' (NETF 2007, xiv). The States and Territories also commissioned the Climate Change Review (Garnaut 2008), which likewise only released its report once leadership had migrated to the Commonwealth.

'Australia's climate change policy changed dramatically in late 2007 with the ratification of Kyoto by the newly elected Labor government' led by Kevin Rudd (Crowley 2010). The centrepiece of the Rudd government's climate policy was the introduction of legislation for an emissions trading scheme as promoted by the Labor States. This was officially the Carbon Pollution Reduction Scheme, or CPRS (Macintosh, Wilkinson, and Deniss 2010). According to the Commonwealth government's key policy research body, the Productivity Commission, the CPRS would displace existing programmes, including the MRET. 'With an effective ETS, much of the current patchwork of climate change policies will become redundant and there will only be a residual role for state, territory and local government initiatives' (PC 2008). While there may well be State-based programmes that become redundant in such circumstances, this fails to acknowledge the many ways in which other policy instruments can reinforce, support, augment, or be otherwise complementary to an overarching national programme (Buzbee 2015).

To placate industry, the CPRS was substantially watered down – to the point where the Greens withdrew their support and the bill was defeated in the Senate in 2010. The Labor government did succeed, though, in renewing and substantially lifting the renewable energy target (RET) from the old MRET's 2 per cent to 20 per cent, or 45,000 GWh, by 2020 (St John 2014). Consistent with the highly collaborative approach the Commonwealth was taking with the States in the first two years after Labor had come to power (Fenna and Anderson 2012), this was developed through COAG. However, like intergovernmental relations in Australia more generally, it was a top-down process where the Commonwealth tended to dominate (Jones 2010).

Labor formed a minority government under new leader Julia Gillard in 2010 with the support of the Greens and other independents. Having made a formal agreement with the Greens, who held the balance of power in the Senate, Labor introduced a bill for a new carbon-pricing scheme, the *Clean Energy Future* plan in 2012 (Crowley 2013). The scheme came into effect on 1 July that year, imposing a carbon price and establishing the framework for an emissions-trading scheme.

2.4.6 Axing the Tax

Labor was defeated in elections the following year, and the adoption of a carbon tax despite promises in 2010 to the contrary contributed to that loss (Economou 2015, 348). 'Axe the tax' had been the Coalition's war cry (Talberg 2016, 145), and the new government promptly did just that (Crowley 2017). Indications are that over its two-year life, the carbon price mechanism did make a dent in emissions (Diesendorf 2019, 42; Grudnoff 2020; also Best, Burke, and Jotzo 2020). In its stead, the Coalition implemented their 'Direct Action Plan' which focused on voluntary measures and subsidies such as the Emissions Reduction Fund (DEE 2017). Thanks to the Senate, the renewable energy target survived, although it was scaled back to 33,000 GWh. For the entire duration of the Coalition's most recent nine years in office federally, from 2013 to 2022, it maintained this line. For a brief period, the faction within the Liberal Party favouring action held sway, and in conjunction with the States developed a compromise called the National Energy Guarantee (ESB 2018); however, a leadership change ended that foray. When, eventually, at the United Nations Climate Change Conference in late 2021, the prime minister announced that Australia would commit to net zero emissions by 2050, it was made clear that this would be achieved in what they called 'the Australian way' – meaning without jeopardising existing industries and comparative advantage (DPMC 2021).

2.4.7 Back to the States

By 2019, all States except Western Australia had emissions-reduction targets in place, as had one of the two Territories, the ACT. Net zero by 2050 was the standard, with ACT aiming for 2045 (CCA 2019). To help achieve these goals, most jurisdictions had renewable energy targets in place by 2019. The most ambitious were Victoria, 40 per cent by 2025; South Australia (SA) 50 per cent by 2025; ACT 100 per cent by 2020. By 2021, all jurisdictions had committed to net-zero by 2050 or sooner (Cleary and Graham 2021).

While federally the Coalition had promptly abolished the centrepiece of Labor's climate change policy once they returned to office in 2013, Labor almost as promptly introduced a suite of emissions-reduction measures when they returned to power in Victoria the following year (DELWP 2016). Victoria's *Climate Change Act 2017* legislated a net zero target for 2050 and mandated a strategy to reach those goals along with requirements for 'Adaptation Action Plans'. This was supported by the closure of 'Australia's "dirtiest" power station', which burned brown coal and had been single-handedly producing 3 per cent of the country's total greenhouse gas emissions (Environment Victoria 2020). The State's 2021 plan announced emissions-reduction targets of '28–33 per cent by 2025, and 45–50 per cent by 2030' together with a wide range of initiatives to achieve them (DELWP 2021; Malos 2021).

The renewables leader, though, has been South Australia, whose Labor government maximised opportunities provided by the Commonwealth's original MRET scheme to convert enthusiastically to renewables. With very limited coal supplies, ageing power stations, a considerable dependence on electricity imported from Victoria, and a transmission network and environment ideally suited to wind and solar generation, there was every incentive to do so (McGreevy et al. 2021).

If the results are anything to go by, these policies have been strikingly successful (Bourne et al. 2019). Twenty years ago, SA had 'no renewable energy production and imported around 30% of its electricity requirements from coal generators interstate. By 2018, it was generating 52% of its electricity from renewables and exported around 3% of its annual production interstate' (McGreevy et al. 2021). The ACT, meanwhile, was fully converted to renewable energy for its electricity generation by 2020. At the same time as being shamed for its slow progress on national emissions reduction policy, Australia has garnered more favourable international attention for making itself the world leader in rooftop solar as a result, in no small part, of these State-level initiatives (e.g., Albeck-Ripka and Penn 2020). Overall, they allowed Australia to reach its 2020 renewable energy target a year early (Stocks, Baldwin, and Blakers 2019). There is variation across the States, with resource-intensive Queensland and Coalition-controlled NSW being slower to act (Bourne et al. 2019). However, in 2020, even NSW broke with their Commonwealth counterparts and launched an ambitious renewables strategy that antagonised both the installed generators and their Canberra colleagues, particularly the National Party (Brown and Maddison 2020; Durie 2020; Williams 2020). By the time Labor lost office in South Australia, conversion to renewables was a fait accompli and fully accepted by the incoming Liberal government (McGreevy et al. 2021). The latest SA (2021, 18) action plan envisages 'a level of renewable energy that is more than 500% of current local grid demand by 2050'.

Somewhat slower to adopt emissions reduction targets was Western Australia – reflecting the fact that because of its large and growing natural gas industry it is the only jurisdiction whose emissions have maintained an upward trajectory (DISER 2021; Hare et al. 2018). Nonetheless, the sum total of State efforts – executed, in progress, and planned – is considerable. 'Current state and territory government 2030 targets combined are the equivalent of a national target of 37 per cent reduction below 2005 levels. This is well beyond the federal government's current 26–28 per cent national emissions reduction target' (Malos 2021).

2.5 Analysis and Assessment

In the first quarter of 2021, generation of electricity by black coal in Australia fell to its lowest level ever, while gas generation not only fell to its lowest level since 2005, but was exceeded for the first time by solar (AER 2021). Renewable energy sources were producing more than one-quarter of the country's electricity – substantially exceeding the original 20-per-cent-by-2020 target (CEC 2021). The leading source of GHG emissions in Australia is undergoing an accelerating transition from fossil fuels to renewables and as a result, according to one account, Australian emissions are peaking (Blakers and Stocks 2019).

How is this possible, when for all but six of the past twenty-five years Australia has been governed at the national level by political parties determined to do as little as possible to combat climate change? For many climate change activists, federalism is part of the answer, given the extent to which 'states and territories lead the way' (Climate Council 2021; also Edis 2019) – consistent with international experience (Schaffer and Bernauer 2014). The Commonwealth government's claim that 'Australia is on track to meet and beat its 2030 target' under the Paris Agreement (DISER 2020a) is in all likelihood disingenuous, it must be said – based on a sleight of hand around the inclusion of land-clearing changes and the effect of temporary events (Maraseni and Reardon-Smith 2019; Merzian and Hemming 2021). Emissions have been increasing across all the main sectors and, as a consequence, so have total emissions. The sole exception has been electricity generation, which reflects the scope for compensatory policymaking afforded by the country's federal system. Here we reflect how the three sets of federalism's strengths and weaknesses mooted in the introduction (Chapter 1) to this book have played out so far in Australian climate-change governance.

2.5.1 Federalism as Facilitator of Climate-Change Governance

Locally Tailored Response. With so much of climate change policymaking coming from the States, the result has certainly been a degree of policy diversity. However, this did not so much reflect the need to tailor policy to differing circumstances as much as the differing availability of resources and differing political pressures. There has been no particular advantage to federalism in this regard, since a uniform national policy such as that introduced by the Commonwealth in 2012 but soon shut down would have worked eminently well.

Compensatory Federalism. Much more evident in the Australian case has been the opportunity that divided jurisdiction provides for policy obstacles at one level to be circumvented by action at the other level, and for the States to play a catalytic role (Bernstein and Hoffman 2018). The dominance of the conservative parties at the national level has given Australia its reputation for being dilatory in regard to climate change. Labor governments in the States and Territories proved themselves ready, willing and able to initiate, enact, and implement emissions-reduction policies in compensation. There are obvious parallels here to the experience in the United States (Engel 2020; Thomson 2014).

We cannot pretend that this has occurred entirely in the absence of Commonwealth action: the introduction of the MRET, even if at very modest levels, prompted and facilitated State climate change policy; NSW then took the lead with its ETS and the States collectively then took up the baton and pushed the Commonwealth towards an emissions trading scheme (Jones 2014, 428–30) and towards a greatly increased renewable energy target. This might be seen as an instance of what Carlson (2009) calls 'iterative federalism', or Fisher (2013) 'boomerang federalism' – perhaps a more apposite term in the Australian context – whereby mutual reinforcement occurs between the central government and the constituent units. With the dominance of the conservative parties in Canberra since 2013 and the failure to replace the expired RET with a new regime in 2020, it has become a more straightforward case of State action compensating for Commonwealth government inaction. Has there been a downside to this? Whether it relieved pressure on the Commonwealth to take action is impossible to determine, but it seems unlikely.

Experimentation and Learning. It is difficult to find examples of genuine experimentation. The closest approximation was the introduction in NSW of an emissions-trading scheme, but that was only an experiment in the Australian context. As Engel (2015, 170–1) has concluded about climate change activism in the American States, there has been little need for policy innovation in emissions reduction; it is not coming up with new ways of doing things that has been the issue, but simply doing them.

Just as it is difficult to find evidence of policy innovation by the States, it is difficult to find evidence of genuine policy learning or the diffusion of good ideas. States typically rushed into similar policies together (such as feed-in tariffs), or took actions that reflected their own circumstances and politics, as appears to have been the case in other areas of environmental policy making (Hollander 2013, 142).

2.5.2 Federalism as Hindrance

Veto Points? Divided jurisdiction did not create jurisdictional obstacles to national action in Australia. It was the least of the problems facing the Rudd and Gillard federal Labor governments in the carbon tax years and has not prevented the States from implementing a range of mitigation policies. Strong bicameralism did play a role in constraining policymaking by the Commonwealth, but that was not a function of Australia's federal system, rather it was a function of what has been called Australia's 'semi-parliamentarism' (Ganghof, Eppner, and Pörschke, 2018).

Patchwork of Policies? This is in part the perennial complaint that federalism is messy: plagued with policy gaps, inconsistencies, and redundancies. It goes beyond that, however, to suggest that shirking or free riding will undo the efforts of others (Gordon 2015). As noted in the Introduction (2.1) above, one school of thought holds that climate change mitigation will be stymied unless governments work together. 'Without intergovernmental cooperation there will be no success' (Jones 2009, 17).

The fact that States demonstrated varying levels of passion for emissions-reduction means that the aggregate effect has not been as great as it might, but it did not create any perverse dynamics in Australia. To the extent that there were laggards this diminished the overall mitigation effort; contrary to Gordon's argument, though, it did not in any way negate the efforts of the leaders.

It is quite possible that practical mechanics makes coordination desirable or even necessary in some instances. As we saw, the States worked hard to develop a coordinated pan-Australian ETS before the Commonwealth government changed hands to Labor in 2007. Similarly, the incoming Labor government worked closely with the States to develop its ETS, although that operated only briefly. Much has been accomplished, though, via individual, non-coordinated, State action. Given that Australia's electricity networks are still predominantly State networks – notwithstanding the NEM – there is little reason why much emissions-reduction policymaking in that sphere cannot occur on a State-by-State basis and be effective. Indeed, right up until its defeat in May 2022, the incumbent Commonwealth government continued to work at cross-purposes with the States, while the shift to renewables only seemed to gain momentum.

Collective Action Problems? The fact that no individual jurisdiction could make a significant dent on the problem seems to have done little to discourage State governments from embarking on often-ambitious mitigation policies in Australia.

2.6 Conclusion

As Weaver (2020) notes, the way federalism affects policymaking is highly contingent – the consequence of a variety of potentially reinforcing or neutralising causes. In the case of Australian climate change governance, federalism provided opportunities for policymaking that would not have been available in a unitary system with a national government likewise unfavourably disposed. Moreover, it did so with the States acting, to a large extent, autonomously. This reflected the character of Australian federalism, Australia's political economy and economic geography, partisan alignments, and the nature of the issue itself.

To conclude that federalism has facilitated climate change governance in Australia is not to argue that the outcome has been 'better' than it might have been under a well-coordinated approach, or if the Commonwealth had taken strong unilateral action via a carbon tax. After all, some of the key State policies were undoubtedly far less efficient (e.g., VAGO 2021). Nor is it to argue that this occurred in the absence of Commonwealth action altogether – it didn't. It is to argue, though, that under the circumstances, federalism provided a context for climate change mitigation that would not have occurred otherwise.

This required that the States have sufficient jurisdictional capacity, which they demonstrably do. It also required that, as a whole, the States and Territories were disposed to take action. As the Canadian case shows, one jurisdiction whose economic welfare is tied to large and increasing emissions can swamp the mitigation efforts of all the others (Harrison 2013, S107; Macdonald 2020, 98). Australia's political economy is much less fraught. In addition, while burdened with enormous coal and gas reserves, the country is also blessed with extraordinary solar- and wind-power potential and able to reap the benefits of the rapidly declining costs of those technologies. Whether the States and Territories can play a similar role in reforming other high-emissions sectors is the next question.

Notes

1 More recent figures have been distorted by the subduing effect of the Covid-19 pandemic.

2 The coal and gas imported from Australia by other countries, China among them, produces three times as much emissions as Australia's entire domestic output (AATE 2020).

3 To confuse things, there is also the Australian Energy Regulator (AER), an agency of another agency, the Australian Competition and Consumer Commission (ACCC), which polices the rules; the Australian Energy Market Commission (AEMC), with makes the rules; and the Energy Security Board (ESB), established in 2017 to oversee strategic change pursuant to the report of the Independent Review into the Future Security of the National Electricity Market – the 'Finkel Review' – *Blueprint for the Future*.

4 In the middle of March 2020, first ministers' meetings were re-styled 'National Cabinet', escalated in frequency and regularity, and took on a more collegial character. This was welcomed by the States (e.g., Victoria 2020) and not much later the PM declared that the arrangement would supersede COAG altogether.

5 WA is extraordinarily dependent on resources, which contributed $135.3bn of the State's $316.3bn GSP in 2019. While a significant portion of that was natural gas, by far the largest part was iron ore, gold, and aluminium, and in total the State contributed fully half of Australia's goods exports (DJTSI 2021). Alberta's great resource, meanwhile, is its enormous body of tar sands – exploitation of which is not only difficult and expensive, but energy and emissions intensive.

6 2018 and 2017 figures respectively. Queensland's emissions have increased since then, driven in particular by the growth of LNG exports. WA has a similar share of the population as Alberta but 'only' produces 17 per cent of Australia's emissions.

7 Inspired by Canada's Council of the Federation, CAF was active for a couple of years, but fell into desuetude once Labor took office in Canberra and lost office in some of the States.

8 The PM did eventually commission a 'Task Group on Emissions Trading', but losing office made that moot (PMTGET 2007).

References

AATE (Australian Academy of Technology and Engineering). 2020. *Australia's Emissions Targets: Are we on track?* AATE.

Abram, Nerilie J., Benjamin J. Henley, Alex Sen Gupta, et al. 2021. 'Connections of Climate Change and Variability to Large and Extreme Forest Fires in Southeast Australia'. *Communications Earth and Environment* 2 (8): 1–17.

AEMC (Australian Energy Market Commission). 2019a. *Integrating Distributed Energy Resources for the Grid of the Future: Economic regulatory framework review.* Sydney: AEMC.

AEMC. 2019b. *National Energy Governance*. Sydney: AEMC. www.aemc.gov.au/regulation/national-governance.

AER (Australian Energy Regulator). 2021. *Wholesale Markets Quarterly: Q1 2021.* Canberra: Commonwealth of Australia.

AGO (Australian Greenhouse Office). 1999. *National Emissions Trading Discussion Paper 1: Establishing the boundaries.* Canberra: Commonwealth of Australia.

Aklin, Michaël, and Matto Mildenberger. 2020. 'Prisoners of the Wrong Dilemma: Why distributive conflict, not collective action, characterizes the politics of climate change'. *Global Environmental Politics* 20 (4): 4–27.

Albeck-Ripka, Livia, and Ivan Penn. 2020. 'How Coal-Loving Australia Became the Leader in Rooftop Solar'. *New York Times*, 29 September.

Arbinolo, Marta, and Catherine Gamper. 2021. 'Adapting to a Changing Climate in the Management of Coastal Zones'. *OECD Environment Policy Papers 24*. Paris: Éditions OCDE.

Aroney, Nicholas. 2017. 'The High Court of Australia: Textual unitarism vs structural federalism'. In *Courts in Federal Systems: Federalists or unitarists?*, eds. Nicholas Aroney and John Kincaid, 29–68. University of Toronto Press.

Beeson, Mark, and Matt McDonald. 2013. 'The Politics of Climate Change in Australia'. *Australian Journal of Politics and History* 59 (3): 331–48.

Benson, Simon, and Geoff Chambers. 2020. 'PM's Bid to Boost Disaster Powers'. *The Australian*, 28 January, 1.

Bernstein, Steven, and Matthew Hoffman. 2018. 'The Politics of Decarbonisation and the Catalytic Impact of Subnational Climate Experiments'. *Policy Sciences* 51 (2): 189–211.

Berry, Michael J., Frank N. Laird, and Christoph H. Stefes. 2015. 'Driving Energy: The enactment and ambitiousness of state renewable energy policy'. *Journal of Public Policy* 35 (2): 297–328.

Best, Rohan, Paul J. Burke, and Frank Jotzo. 2020. 'Carbon Pricing Efficacy: Cross-country evidence'. *Environmental and Resource Economics* 77 (1): 69–94.

Blakers, Andrew, Bin Lu, and Matthew Stocks. 2017. '100% Renewable Electricity in Australia'. *Energy* 133: 471–82.

Blakers, Andrew, and Matt Stocks. 2019. *Australia's Greenhouse Emissions are Peaking*, Policy Briefings, Canberra. http://re100.eng.anu.edu.au/resources/assets/191024Emissionsarepeaking.pdf.

BOM (Bureau of Meteorology) and CSIRO (Commonwealth Scientific and Industrial Research Organisation). 2020. *State of the Climate 2020.* Canberra: Commonwealth of Australia. www.bom.gov.au/state-of-the-climate.

Bromley-Trujillo, Rebecca, and Mirya R. Holman. 2020. 'Climate Change Policymaking in the States: A view at 2020'. *Publius: The Journal of Federalism* 50 (3): 446–72.

Brown, Douglas M. 2012. 'Comparative Climate Change Policy and Federalism: An overview'. *Review of Policy Research* 29 (3): 322–33.

Brown, Greg. 2020. 'Unions Urge Split Labor to Back Gas, Coal'. *The Australian*, 29 October. www.theaustralian.com.au/nation/politics/unions-push-for-split-labor-to-back-gas-coal/news-story/07794dc294b1fb9a5ca771543cb83c1a.

Brown, Greg, and Max Maddison. 2020. 'Canavan Reignites Coalition Energy Wars'. *The Australian*, 1 December.

Burck, Jan, Ursula Hagen, Christoph Bals, et al. 2020. 'Climate Change Performance Index'. Accessed 21 November. www.climate-change-performance-index.org.

Buzbee, William W. 2015. 'Climate Federalism, Regulatory Failure and Reversal Risks, and Entrenching Innovation Incentives'. In *The Law and Policy of Environmental Federalism: A Comparative analysis*, ed. Kalyani Robbins, 145–68. Cheltenham: Edward Elgar.

CAA (Climate Analytics Australia). 2019. *Australia's Proposed 'Kyoto Carryover': Nature, scale, legal issues and environmental integrity of the Paris Agreement.* Fremantle: CAA. https://climateanalytics.org/media/report_australia_kyoto_carry over_dec2019.pdf.

Campbell, Rod, Eliza Littleton, and Alia Armistead. 2021. *Fossil Fuel Subsidies in Australia: Federal and state government assistance to fossil fuel producers and major users 2020–21.* Canberra: Australia Institute.

Carlson, Ann E. 2009. 'Iterative Federalism and Climate Change'. *Northwestern University Law Review* 103 (3): 1097–1162.

CAT (Climate Action Tracker). 2020. *Australia*. CAT. https://climateactiontracker.org/countries/australia.

CCA (Climate Change Authority). 2015. *Final Report on Australia's Future Emissions Reduction Targets.* Canberra: Commonwealth of Australia. www.climatechangeauthority.gov.au/sites/default/files/2020-07/Final-report-Australias-future-emissions-reduction-targets.pdf.

CCA. 2019. *Australia's Climate Change Policies at the Australian and State and Territory Government Levels: A stocktake.* Canberra: Commonwealth of Australia (CCA). https://apo.org.au/node/224211.

CEC (Clean Energy Council). 2021. *Clean Energy Australia: Report 2021.* Melbourne: CEC. www.cleanenergycouncil.org.au/resources/resources-hub/clean-energy-austra lia-report.

Christoff, Peter, ed. 2014. *Four Degrees of Global Warming: Australia in a hot world.* London: Earthscan/Routledge.

Cleary, Alison, and Simon Graham. 2021. *State and Territory Climate Action: Leading policies and programs in Australia*. ClimateWorks Australia.

Climate Council. 2019. *'This is Not Normal': Climate change and escalating bushfire risk.* Potts Point: Climate Council of Australia. www.climatecouncil.org.au/resources/new-report-this-is-not-normal-climate-change-and-escalating-bushfire-risk.

Climate Council. 2021. *Hindsight is 2020: The year in climate news.* Potts Point: Climate Council of Australia. www.climatecouncil.org.au/climate-news-summary-2020.

COAG (Council of Australian Governments). 2001. *Communiqué*. 8 June. Canberra: Council of Australian Governments.

Coorey, Phillip, and Tom McIlroy. 2020. 'Rebel Nats Defy PM on Gas'. *Australian Financial Review*, 12 February.

Courtenay Botterill, Linda, and Alan Fenna. 2020. 'Initiative-Resistance and the Australian Party System'. *Australian Journal of Politics and History* 66 (1): 63–77.

Crowley, Kate. 2010. 'Climate Clever? Kyoto and Australia's decade of recalcitrance'. In *Global Commons, Domestic Decisions: The comparative politics of climate change*, eds. Kathryn Harrison and Lisa McIntosh Sundstrom, 201–28. Cambridge MA: MIT.

Crowley, Kate. 2013. 'Irresistible Force? Achieving carbon pricing in Australia'. *Australian Journal of Politics and History* 59 (3): 368–81.

Crowley, Kate. 2017. 'Up and Down with Climate Politics 2013–2016: The repeal of carbon pricing in Australia'. *Wiley Interdisciplinary Reviews: Climate Change* 8 (3): 1–13.

CT – Climate Transparency. 2020. *Climate Transparency Report: Comparing G20 climate action and responses to the COVID-19 crisis.* www.climate-transparency.org/wp-content/uploads/2020/11/Climate-Transparency-Report-2020.pdf.

DCC (Department of Climate Change). 2009. *Climate Change Risks to Australia's Coasts: A first pass national assessment.* Canberra: Commonwealth of Australia (DCC). www.environment.gov.au/system/files/resources/fa553e97-2ead-47bb-ac80-c12adffea944/files/cc-risks-full-report.pdf.

DEE (Department of the Environment and Energy). 2017. *Australia's 7th National Communication on Climate Change*. Canberra: Commonwealth of Australia (DEE). https://unfccc.int/files/national_reports/national_communications_and_biennial_reports/application/pdf/024851_australia-nc7-br3–1-aus_natcom_7_br_3_final.pdf>.

DELWP (Department of Environment, Land, Water and Planning). 2016. *Victoria's Climate Change Framework.* Melbourne: State of Victoria DELWP. www.climatechange.vic.gov.au/__data/assets/pdf_file/0021/55254/DELWPClimateChange_Framework.pdf.

DELWP. 2021. *Victoria's Climate Change Strategy.* Melbourne: State of Victoria DELWP. www.climatechange.vic.gov.au/victorias-climate-change-strategy/Victoria-Climate-Change-Strategy.pdf.

Derthick, Martha. 2010. 'Compensatory Federalism'. In *Greenhouse Governance: Addressing climate change in America*, ed. Barry G. Rabe, 58–72. Washington DC: Brookings Institution.

Diesendorf, Mark. 2019. 'Energy Futures for Australia'. In *Decarbonising the Built Environment: Charting the transition*, eds. Peter Newton, Deo Prasad, Alistair Sproul, and Stephen White, 35–52. Singapore: Palgrave Macmillan.

DISER (Department of Industry, Science, Energy and Resources). 2020a. *Australia's Emissions Projections 2020.* Canberra: Commonwealth of Australia (DISER). www.industry.gov.au/publications/australias-emissions-projections-2020.

DISER. 2020b. *Quarterly Update of Australia's National Greenhouse Gas Inventory: March 2020.* Canberra: Commonwealth of Australia (DISER). www.industry.gov.au/sites/default/files/2020-08/nggi-quarterly-update-march-2020.pdf.

DISER. 2021. *National Greenhouse Accounts 2019: State and Territory greenhouse gas inventories: Annual emissions.* Canberra: Commonwealth of Australia (DISER). www.industry.gov.au/data-and-publications/national-greenhouse-accounts-2019/state-and-territory-greenhouse-gas-inventories-annual-emissions.

DJTSI (Department of Jobs, Tourism, Science and Innovation). 2021. *Western Australia Economic Profile: March 2021.* Perth: Government of Western Australia (DJTSI).

DPMC (Department of the Prime Minister and Cabinet). 2021. *The Plan to Deliver Net Zero the Australian Way.* Canberra: Commonwealth of Australia.

Durie, John. 2020. 'NSW Energy Minister Matt Kean Fills the Vacuum Left by Federal Inaction'. *The Australian*, 9 November. www.theaustralian.com.au/business/mining-energy/nsw-energy-minister-matt-kean-fills-the-vacuum-left-by-federal-inaction/news-story/6a6c5f31e6594ce7f2e3d7715237cec6.

Economou, Nick. 2015. 'The Environment in the 2013 Election: Controversies over climate change, the carbon tax and conservation'. In *Abbott's Gambit: The 2013 Australian federal election*, eds. Carol Johnson, John Wanna and Hsu-Ann Lee, 341–58. Acton ACT: ANU Press.

Edis, Tristan. 2019. 'While the Government is in Denial, the States are Making Staggering Progress on Renewable Energy'. *Guardian*, 2 July. www.theguardian.com/commentisfree/2019/jul/02/while-the-federal-government-is-in-denial-the-states-are-making-staggering-progress-on-renewable-energy.

ELCA (Emergency Leaders for Climate Action). 2019. *Australia Unprepared for Worsening Extreme Weather*. www.climatecouncil.org.au/wp-content/uploads/2019/04/fire-chiefs-statement-pages.pdf.

Engel, Kirsten H. 2020. 'Climate Federalism in the Time of COVID-19: Can the States "save" American climate policy?' *Northern Kentucky Law Review* 47 (1): 116–38.

Engel, Kirsten H. 2015. 'The Enigma of State Climate Change Policy Innovation'. In *The Law and Policy of Environmental Federalism: A comparative analysis*, ed. Kalyani Robbins, 169–87. Cheltenham: Edward Elgar.

Environment Victoria. 2020. '*Hazelwood FAQs*'. Carlton: Environment Victoria. https://environmentvictoria.org.au/hazelwood-faqs.

ESB (Energy Security Board). 2018. *National Energy Guarantee: Final detailed design.* Canberra: Commonwealth of Australia (DEE). www.energyministers.gov.au/publications/national-energy-guarantee-final-detailed-design-commonwealth-elements.

Fenna, Alan. 2008. 'Commonwealth Fiscal Power and Australian Federalism'. *University of New South Wales Law Journal* 31 (2): 509–29.

Fenna, Alan. 2016. 'Shaping Comparative Advantage: The evolution of trade and industry policy in Australia'. *Australian Journal of Political Science* 51 (4): 618–35. https://doi.org/10.1080/10361146.2016.1239565.

Fenna, Alan. 2019. 'The Centralization of Australian Federalism 1901–2010: Measurement and interpretation'. *Publius* 49 (1): 30–56. https://doi.org/10.1093/publius/pjy042.

Fenna, Alan. 2021. 'Australian Federalism and the COVID-19 Crisis'. In *Federalism and the Response to COVID-19: A comparative analysis*, eds Rupak Chattopadhyay, Felix Knüpling, Diana Chebenova, Liam Whittington and Phillip Gonzalez. Abingdon: Routledge.

Fenna, Alan, and Geoff Anderson. 2012. 'The Rudd Reforms and the Future of Australian Federalism'. In *The Future of Australian Federalism: Comparative and interdisciplinary perspectives*, eds Gabrielle Appleby, Nicholas Aroney and Thomas John, 393–413. Cambridge: Cambridge University Press.

Fenna, Alan, and John Phillimore. 2015. 'Intergovernmental Relations in Australia: New modes, old realities'. In *Federalism as Decision-Making: Changes in structures, procedures and policies*, eds. Francesco Palermo and Elisabeth Alber, 192–212. Leiden: Brill Nijhoff.

Fielding, Kelly S., Brian W. Head, Warren Laffan, Mark Western, and Ove Hoegh-Guldberg. 2012. 'Australian Politicians' Beliefs about Climate Change: Political partisanship and political ideology'. *Environmental Politics* 21 (5): 712–33.

Fisher, Brian S. 2020. *Powering Out of Pandemic: Unleashing the potential of gas.* Barton ACT: Menzies Research Centre.

Fisher, Dana R. 2013. 'Understanding the Relationship between Subnational and National Climate Change Politics in the United States: Toward a theory of boomerang federalism'. *Environment and Planning C: Government and Policy* 31 (5): 769–84.

Fowler, Robert. 2015. 'The Australian Experience with Environmental Federalism: Constitutional and political perspectives'. In *The Law and Policy of Environmental Federalism: A comparative analysis*, ed. Kalyani Robbins, 271–303. Cheltenham: Edward Elgar.

Ganghof, Steffen, Sebastian Eppner, and Alexander Pörschke. 2018. 'Australian Bicameralism as Semi-Parliamentarism: Patterns of majority formation in 29 democracies'. *Australian Journal of Political Science* 53 (2): 211–33.

Garnaut, Ross. 2008. *Climate Change Review: Final report*. Port Melbourne: Cambridge University Press.

Garnaut, Ross. 2019. *Superpower: Australia's low-carbon opportunity*. Carlton VIC: La Trobe University Press with Black Inc.

Gergis, Joëlle. 2018. *Sunburnt Country: The history and future of climate change in Australia*. Carlton VIC: Melbourne University Press.

Germanwatch, NewClimate Institute, Climate Action Network. 2020. *Climate Change Performance Index*. www.climate-change-performance-index.org.

Godinho, Catrina (Summary Report Lead). 2020. *Climate Transparency Report: Comparing G20 Climate Action and Responses to the COVID-19 Crisis*. Climate Transparency. www.climate-transparency.org/wp-content/uploads/2020/11/Climate-Transparency-Report-2020.pdf.

Gordon, David J. 2015. 'An Uneasy Equilibrium: The coordination of climate governance in federated systems'. *Global Environmental Politics* 15 (2): 121–41.

Grant, Bligh, and Joseph Drew. 2017. *Local Government in Australia: History, theory, and public policy*. Singapore: Springer.

Grudnoff, Matt. 2020. *The Carbon Pricing Mechanism under the Gillard Government*. Canberra: Australia Institute. https://australiainstitute.org.au/report/the-carbon-pricing-mechanism-under-the-gillard-government.

Hale, Thomas. 2016. '"All Hands on Deck": The Paris Agreement and nonstate climate action'. *Global Environmental Politics* 16 (3): 12–22.

Hare, Bill, Niklas Roming, Ursula Fuentes Hutfilter, Michiel Shaeffer, and Matt Beer. 2018. *Western Australia's Gas Gamble: Implications of exploiting Canning Basin and other unconventional gas resources for achieving climate targets*. Berlin: Climate Analytics. https://climateanalytics.org/media/climateanalytics-report-westernaustraliasgasgamble-2018.pdf.

Harrison, Kathryn. 2013. 'Federalism and Climate Policy Innovation: A critical reassessment'. *Canadian Public Policy* 39 (2): S95–S108.

Hollander, Robyn. 2008. 'John Howard, Economic Liberalism, Social Conservatism and Australian Federalism'. *Australian Journal of Politics and History* 54 (1): 85–103.

Hollander, Robyn. 2010. 'Rethinking Overlap and Duplication: Federalism and environmental assessment in Australia'. *Publius* 40 (1): 136–70.

Hollander, Robyn. 2013. 'Borrowing from Neighbours: Policy transfer to tackle climate change in the Australian Federation'. In *Policy Transfer and Learning in Public Policy and Management: International contexts, content and development*, eds. Peter Carroll and Richard Common, 128–45. Abingdon: Routledge.

Hollander, Robyn. 2015. 'ESD, Federalism and Intergovernmental Relations in Australia'. *Australasian Journal of Environmental Management* 22 (1): 21–32.

HSCCCWEA (House Standing Committee on Climate Change, Water, Environment and the Arts). 2009. *Managing our Coastal Zone in a Changing Climate: The time to act is now*. Canberra: Commonwealth of Australia.

Hughes, Lesley, Will Steffen, Greg Mullins, Annika Dean, Ella Weisbrot, and Martin Rice. 2020. *Summer of Crisis.* Potts Point: Climate Council of Australia. www.climatecouncil.org.au/resources/summer-of-crisis.

Jones, Stephen. 2009. 'The Future of Renewable Energy in Australia: A test for cooperative federalism?' *Australian Journal of Public Administration* 68 (1): 1–20.

Jones, Stephen. 2010. 'The National Renewable Energy Target: An example of post-Machiavellian policy-making?' *Australian Journal of Public Administration* 69 (2): 165–77.

Jones, Stephen. 2014. 'Flirting with Climate Change: A comparative policy analysis of subnational governments in Canada and Australia'. *Journal of Comparative Policy Analysis: Research and Practice* 16 (5): 424–40.

Kallies, Anne. 2021. 'The Australian Energy Transition as a Federalism Challenge: (un) cooperative energy federalism?' *Transnational Environmental Law* 10 (2): 211–35.

Kellow, Aynsley. 2018. *Negotiating Climate Change: A forensic analysis.* Cheltenham: Edward Elgar.

Kent, Anthony, and David Mercer. 2006. 'Australia's Mandatory Renewable Energy Target: An assessment'. *Energy Policy* 34 (9): 1046–62.

Keywood, Melita D., Mark F. Hibberd, and Kathryn M. Emmerson. 2017. *Australia State of the Environment 2016: Atmosphere.* Canberra: Commonwealth of Australia (DEE). https://soe.environment.gov.au/sites/default/files/soe2016-atmosphere-final-web-v3.pdf?v=1499655757.

Leitch, Anne. 2017. 'Climate Change: Local governments adapting to the wicked problem of sea-level rise'. In *The Routledge Handbook of Australian Urban and Regional Planning*, eds. Neil Sipe and Karen Vella, 209–30. New York: Routledge.

Lowe, Ian. 2005. *Living in the Hothouse: How global warming affects Australia.* Carlton VIC: Scribe.

Lowe, Ian. 2017. 'How Planning can Address the Challenge of Transitioning to Low-Carbon Urban Economics'. In *The Routledge Handbook of Australian Urban and Regional Planning*, eds. Neil Sipe and Karen Vella, 199–208. New York: Routledge.

Lu, Bin, Andrew Blakers, Matthew Stocks, Cheng Cheng, and Anna Nadolny. 2021. 'A Zero-Carbon, Reliable and Affordable Energy Future in Australia'. *Energy* 220: 1–10.

Lukas, C., K. Hennessy, G. Mills, and J. Bathols. 2007. *Bushfire Weather in Southeast Australia: Recent trends and projected climate change impacts.* Melbourne: Bushfire Cooperative Research Centre, Australian Bureau of Meteorology and CSIRO Marine and Atmospheric Research. https://doi.org/10.25919/5e31c82ee0a4c.

Macdonald, Douglas. 2020. *Carbon Province, Hydro Province: The challenge of Canadian energy and climate federalism.* Toronto ON: University of Toronto Press.

Macintosh, Andrew, Debra Wilkinson, and Richard Deniss. 2010. 'Climate Change'. In *The Rudd Government: Australian Commonwealth Administration 2007–2010*, eds. Chris Aulich and Mark Evans, 199–220. Acton ACT: ANU Press.

Malos, Anna. 2021. 'Victorian Government Takes Important Steps on Emissions Reduction'. *ClimateWorks Australia*, 20 May. www.climateworksaustralia.org/news/victorian-government-takes-important-steps-on-emissions-reduction.

Maraseni, Tek, and Kathryn Reardon-Smith. 2019. 'Meeting National Emissions Reduction Obligations: A case study of Australia'. *Energies* 12 (3): 438.

McGreevy, Michael, Colin MacDougall, Matt Fisher, Mark Henley, and Fran Baum. 2021. 'Expediting a Renewable Energy Transition in a Privatized Market via Public Policy: The case of South Australia 2004-18'. *Energy Policy* 148 (A): 1–14.

Meckling, Jonas. 2011. *Carbon Coalitions: Business, climate politics, and the rise of emissions trading*. Cambridge MA: MIT.

Merzian, Richie and Polly Hemming. 2021. *Banking on Australia's Emissions: Why creative accounting will not get us to net zero emissions.* Canberra: Australia Institute. https://australiainstitute.org.au/wp-content/uploads/2021/05/P1076-Banking-on-Australias-Emissions-Commitments-.pdf.

Nalau, Johanna, Benjamin L. Preston, and Megan C. Maloney. 2015. 'Is Adaptation a Local Responsibility?' *Environmental Science and Policy* 48: 89–98.

The Nationals. 2021. '*Manufacturing 2035: 800,000 New Jobs*'. Barton ACT: The Nationals. www.bridgetmckenzie.com.au/media-releases/the-nationals-strategic-plan-to-double-australian-manufacturing-employment.

NETF (National Emissions Trading Taskforce). 2007. *Possible Design for a National Greenhouse Gas Emissions Trading Scheme: Final framework report on scheme design.* City East QLD: Council for the Australian Federation. www.caf.gov.au/Documents/nett-final-report.pdf.

OECD (Organisation for Economic Co-operation and Development). 2019. *Taxing Energy Use 2019: Using taxes for climate action*. Paris: OECD. www.oecd.org/tax/taxing-energy-use-efde7a25-en.htm.

Ogge, Mark. 2021. *Wrong Way, Go Back: Why the gas-fired recovery plan will fail to reduce energy prices or create jobs but will increase emissions.* Canberra: Australia Institute. https://australiainstitute.org.au/report/wrong-way-go-back.

PC (Productivity Commission). 2008. *What Role for Policies to Supplement an Emissions Trading Scheme?* Canberra: Commonwealth of Australia (PC). www.pc.gov.au/research/supporting/garnaut-emission-trading.

Pearse, Rebecca. 2018. '"Continuity and Change": Environmental policy and the coming energy transition'. In *Double Disillusion: The 2016 Australian federal election*, eds. Anika Gauja, Peter Chen, Jennifer Curtin, and Juliet Pietsch, 571–91. Canberra: ANU Press.

Phillimore, John, and Alan Fenna. 2017. 'Intergovernmental Councils and Centralization in Australian Federalism'. *Regional and Federal Studies* 27 (5): 597–621. https://doi.org/10.1080/13597566.2017.1389723.

Phillimore, John, and Alan Fenna. 2020. 'Road or Rail? The partisan politics of urban transport policy in the Australian States'. *Australian Journal of Politics and History* 66 (1): 50–62.

PM (Prime Minister), MEER (Minister for Energy and Emissions Reduction), and MRWNA (Minister for Resources, Water and Northern Australia). 2020. 'Gas-Fired Recovery: Media Release'. *Prime Minister of Australia*, 15 September. www.pm.gov.au/media/gas-fired-recovery.

PMTGET (Prime Ministerial Task Group on Emissions Trading). 2007. *Report of the Task Group on Emissions Trading.* Barton ACT: Department of the Prime Minister and Cabinet. https://apo.org.au/sites/default/files/resource-files/2007-06/apo-nid968.pdf.

Proudlove, Richard, Coral Bravo, and Amandine Denis-Ryan. 2020. *Net Zero Momentum Tracker: Local government report.* Melbourne: ClimateWorks Australia. www.climateworksaustralia.org/resource/net-zero-momentum-tracker-local-government-report.

Rabe, Barry G. 2018. *Can We Price Carbon?* Cambridge MA: MIT.

Rootes, Christopher. 2008. 'The First Climate Change Election? The Australian general election of 24 November 2007'. *Environmental Politics* 17 (3): 473–80.

SA (Government of South Australia). 2021. *South Australian Government Climate Change Action Plan 2021–2025*. Government of South Australia (Department for

Environment and Water). www.environment.sa.gov.au/topics/climate-change/climate-change-action-plan-2021-2025.

Saddler, Hugh. 2021. *National Energy Emissions Audit Report: February/March 2021.* Canberra: Australia Institute. https://australiainstitute.org.au/wp-content/uploads/2021/05/NEEA-Report-Feb-March-2021-WEB.pdf.

Sansom, Graham. 2009. 'Commonwealth of Australia'. In *Local Government and Metropolitan Regions in Federal Systems*, ed. Nico Steytler, 7–36. Montreal & Kingston: McGill-Queen's University Press.

Sartor, Frank. 2011. *The Fog on the Hill: How NSW Labor lost its way*. Melbourne: Melbourne University.

Saunders, Cheryl. 1996. 'The Constitutional Division of Powers with Respect to the Environment in Australia'. In *Federalism and the Environment: Environmental policymaking in Australia, Canada, and the United States*, eds. Kenneth M. Holland, Frederick L. Morton and Brian Galligan, 55–76. Westport CT: Greenwood.

SCCC (Select Council on Climate Change). 2012. *Roles and Responsibilities for Climate Change Adaptation in Australia.* Canberra: Council of Australian Governments. www.environment.gov.au/climate-change/adaptation/strategy.

Schaffer, Lena Maria, and Thomas Bernauer. 2014. 'Explaining Government Choices for Promoting Renewable Energy'. *Energy Policy* 68: 15–27.

Senior, A., A. F. Britt, D. Summerfield, et al. 2020. *Australia's Identified Mineral Resources 2019.* Canberra: Commonwealth of Australia (Geoscience Australia). http://dx.doi.org/10.11636/1327-1466.2019.

Sharman, Campbell. 2001. 'Federalism and the Liberal Party'. In *Liberalism and the Australian Federation*, ed. John R. Nethercote, 287–302. Annandale NSW: Federation Press.

St John, Alexander. 2014. *The Renewable Energy Target: A quick guide*. Canberra: Parliament of Australia (Department of Parliamentary Services). www.aph.gov.au/About_Parliament/Parliamentary_Departments/Parliamentary_Library/pubs/rp/rp1314/QG/RenewableEnergy.

Steffen, Will, Lesley Hughes, Greg Mullins, Hilary Bambrick, Annika Dean, and Martin Rice. 2019. *Dangerous Summer: Escalating bushfire, heat and drought risk.* Potts Point: Climate Council of Australia.

Stock, Andrew, Greg Bourne, Will Steffen, and Tim Baxter. 2020. *Passing Gas: Why renewables are the future*. Potts Point: Climate Council of Australia. www.climatecouncil.org.au/resources/passing-gas-renewables-are-future.

Stocks, Matthew, Ken Baldwin, and Andrew Blakers. 2019. *Powering Ahead: Australia leading the world in renewable energy build rates.* Canberra: Energy Change Institute, Australian National University. https://energy.anu.edu.au/files/Renewable%20energy%20target%20report%20September%202019_1_0.pdf.

Stokes, Leah Cardamore. 2020. *Short-Circuiting Policy: Interest groups and the battle over clean energy and climate policy in the American states*. New York: Oxford University Press.

Talberg, Anita. 2016. 'The Climate Policy Carousel'. In *From Abbott to Turnbull: A new direction? Australian Commonwealth Administration 2013–2016*, ed. Chris Aulich, 145–70. West Geelong VIC: Barrallier Books.

Thomson, Vivian. 2014. *Sophisticated Interdependence in Climate Policy: Federalism in the United States, Brazil, and Germany*. London: Anthem.

Tranter, Bruce. 2013. 'The Great Divide: Political candidate and voter polarisation over global warming in Australia'. *Australian Journal of Politics and History* 59 (3): 397–413.

Tranter, Bruce. 2017. 'It's Only Natural: Conservatives and climate change in Australia'. *Environmental Sociology* 3 (3): 274–85.

Twomey, Anne. 2008. 'Commonwealth of Australia.' In *Foreign Relations in Federal Countries*, ed. Hans Michaelman, 37–64, Vol. 5 of *A Global Dialogue on Federalism*, ed. John Kincaid. Kingston & Montreal: McGill-Queen's University Press.

Twomey, Anne. 2012. 'NSW and Federalism'. In *From Carr to Keneally: Labor in office in NSW 1995–2011*, eds. David Clune and Rodney Smith, 100–9. Crows Nest NSW: Allen & Unwin.

VAGO (Victorian Auditor-General's Office). 2021. *Delivering the Solar Homes Program.* Melbourne: VAGO. www.audit.vic.gov.au/report/delivering-solar-homes-program.

van Oldenborgh, Geert Jan, Folmer Krikken, Sophie Lewis, et al. 2021. 'Attribution of the Australian Bushfire Risk to Anthropogenic Climate Change'. *Natural Hazards and Earth System Sciences* 21 (3): 941–60.

Victoria, Government of. 2020. *Victorian Government Submission to the Senate Select Committee on COVID-19.* Melbourne: Government of Victoria.

Weaver, R. Kent. 2020. 'Policy Dynamics in Federal Systems: A framework for analysis'. *Publius: The Journal of Federalism* 50 (2): 157–87.

Weisbrot, Ella, Tim Baxter, Greg Bourne, Andrew Stock, and Fiona Ivits. 2019. *State of Play: Renewable energy leaders and losers.* Potts Point: Climate Council of Australia. www.climatecouncil.org.au/resources/states-renewables-2019.

Wild, Daniel. 2022. *The Economic and Employment Consequences of Net Zero Emissions by 2050 in Australia.* Melbourne: Institute of Public Affairs.

Williams, Perry. 2020. 'NSW Underwriting Plan Sparks Power Struggle'. *The Australian,* 10 November. www.theaustralian.com.au/business/mining-energy/nsw-underwriting-plan-sparks-power-struggle/news-story/eb442b85a5a5eb21d042c03af6b4e5a5.

Wood, Tony. 2019. *Australia's Energy Transition: A blueprint for success.* Carlton: Grattan Institute.

Wood, Tony, and Guy Dundas. 2020. *Flame Out: The future of natural gas.* Carlton: Grattan Institute.

Wood, Tony, Guy Dundas, and James Ha. 2020. *Start with Steel: A practical plan to support carbon workers and cut emissions.* Carlton: Grattan Institute.

Wood, Tony, and James Ha. 2021. *Go For Net Zero: A practical plan for reliable, affordable, low-emissions electricity.* Carlton: Grattan Institute.

3

Climate Governance and Federalism in Brazil

FABIANA BARBI SELEGUIM AND FERNANDO REI

3.1 Introduction

Greenhouse gas (GHG) emissions in Brazil are largely connected to changes in land use and land management practices (SEEG 2018).[1] In 2018 land use and land-cover change contributed to 44 per cent of Brazil's emissions, followed by agriculture, which accounted for 25 per cent. Deforestation has been the main source of land use emissions, representing 93 per cent of the sector's total for the period 1990–2018. In terms of vulnerability to climate change, Brazil is ranked 92 out of 181 countries in the 2020 ND-GAIN Index.[2] Extreme temperatures, rising seas, as well as the complex challenges of different regions across the country experiencing significant water scarcity and heavy rainfall are predicted to place significant pressure on vulnerable groups, urban infrastructure, the economy and the country's unique ecosystems (World Bank 2021).

While the literature on climate change governance in Brazil has been centred on the multi-level governance framework (Inoue 2012; de Macedo and Jacobi 2019; Setzer 2017), this chapter concerns itself specifically with federalism. It identifies and analyses the main climate change strategies at federal, state and municipal levels in Brazil, focusing on two processes: (i) the favourable context for decentralized policymaking; and (ii) the scope for experimental policymaking and associated learning process among the constituent units.

Two features of Brazilian federalism are highlighted. One is that all levels of government have constitutional responsibilities for climate change policy, with a distinct range of policymaking powers. That is, state and municipal governments can step in and compensate for inaction by the federal government. States and municipalities have been taking the lead in climate governance, especially to compensate for the refusal of federal government to act on climate change. Brazilian states and municipalities have developed climate change policies relating to both mitigation and adaptation. The second is the availability of multiple forums for 'experimental' policymaking triggering processes of policy diffusion among the constituent units. Such forums constitute institutional arrangements that encompass multi-level units as well as agents from different segments of society, such as transnational cooperation networks and agencies, research groups, among others.

3.2 The Practice of Federalism in Brazil

The return of democracy in 1985 (after two decades of military rule) also meant the return of federalism in Brazil. The country became a federal republic under the Constitution of 1988, which established three levels of government: the central government or Union; state governments and the Federal District government; and municipal governments. States and municipalities have autonomous administrations that collect their own taxes and receive a share of the taxes collected by the Federal government. States are headed by a governor and municipalities by a mayor. Both entities have elected legislative bodies. The twenty-six states and the Federal District have their own constitution. The 5,570 municipalities are not governed by states as is the case in most federations (Viswanathan 2014). They are granted the status of federal entities – at the same level as the states and are governed by an organic law, which must comply with federal and state constitutions. There is great discrepancy in the size (geographical area and population), and social and economic indicators among the subnational jurisdictions, but all Brazilian municipalities enjoy the same legal status. In 2015, seventeen municipalities had more than 1 million inhabitants, representing 22 per cent of the population; 44 per cent of municipalities had fewer than 10,000 inhabitants, representing 6.3 per cent of the population (OECD 2016).

The country's size (over 8.5 million km^2) and territorial diversity (physical, social and economic) have justified the choice for a system of government that allows for the decentralization of policies and a focused management of territories. In the context of climate change policymaking, federalism should help the State accounting for the diversity of effects experienced throughout the country, as well as the distinct types of actions that are needed to address the problem (Arretche 2000).

The Constitution assigns the federal government authority to act in foreign policy and international relations; propose and execute the national security and defence policy; conduct the country's economy and finances, including issuing currency; organize, regulate and provide services in the area of communication; explore nuclear services and facilities. State powers are those outside the federal government's area of activity and that were not expressly prohibited by the Constitution. Municipalities can legislate on matters of local interest (article 30, I), in addition to complementing, when possible, federal and state legislation (article 30, II). Metropolitan regions can be created by the States (article 25, paragraph 3).[3]

The Union occupies a central position in environmental protection. It is responsible for implementing the general environmental policy, as established by Act 6,938 of 1981, which was enacted prior to the Constitution. The Union is also responsible for designing and executing national and regional planning (article 21, IX), which form the basis of environmental protection and climate change policies. But the Union shares authority with the constituent units over several themes related to environmental protection: forests, hunting, fishing, fauna, nature conservation; defence of soil and natural resources, protection of historical, cultural, tourist and landscape heritage; liability for damage to the environment, consumers, goods and rights of artistic, aesthetic, historical, tourist and landscape value; and health (article 24 of the Constitution). Control of sources of GHG emissions is shared between the environmental agencies (Complementary Act 140 of 2011) and subnational governments are allowed to engage in the global climate governance agenda.

3.3 Climate Change in Brazil

3.3.1 Contributions to Climate Change and Its Impacts

Brazilian GHG emissions reached 2,175 bn tonnes of carbon dioxide equivalent (CO_2e) in 2019 (SEEG 2020), placing Brazil as number seven in the ranking of the world's largest emitters. Emissions from the energy sector grew 1 per cent in 2019 compared to the previous year to 413 million tons of CO_2e. Meanwhile, emissions from deforestation increased 19 per cent, to 968 million tons of CO_2e – making this the main contributor to the increase in emissions, responsible for 44 per cent of the country's emissions.

Due to deforestation, Brazil is still far from being a low carbon economy. Emissions per capita exceed 10t CO_2e/inhabitant (2018) and are still higher than the global average of 7t CO_2/inhabitant (SEEG 2020). The agricultural states of Pará and Mato Grosso are responsible for most of the country's emissions. Livestock activity has contributed to the increase of emissions, in addition to deforestation. On the other hand, the most industrialized state in the country, São Paulo, which represents one-third of the national GDP and has one-fifth of the country's population, occupies fourth place (SEEG 2020).

Despite the economic downturn resulting from the Covid-19 pandemic, GHG emissions in Brazil in 2020 increased by 8 per cent compared to 2019 (SEEG 2020). This was due to the lack of government command and control policy in tackling illegal deforestation and forest fire prevention.

The relevance of Conservation Units and Indigenous Lands for the protection of the Amazon has been extensively documented by numerous studies (Adeney et al. 2009; Barber et al. 2014; Noltea et al. 2013; Soares-Filho et al. 2010), in the sense of reaffirming the fundamental role that these areas play in curbing illegal deforestation and, therefore, in reducing Brazilian GHG emissions (Guetta et al. 2019).

Regarding climate change impact, extreme weather events are predicted to intensify and become more frequent, causing severe impacts on the six Brazilian biomes, coastal areas, the food system and security and water availability. The country's fishing potential may be reduced by 6 per cent over the next forty years, and by 2030 the country could lose about 11 million hectares suitable for agriculture. In turn, food insecurity could increase due to the decrease in subsistence agricultural production, with a consequent lack of food for populations directly exposed to climatic adversity. The effect of climate change will be concentrated mainly in the poorest regions of Brazil and will accentuate social inequalities (PBMC 2013).

3.4 Climate Change and Federalism in Brazil

3.4.1 Climate Change Commitments

Brazil signed and ratified the United Nations Framework Convention on Climate Change (UNFCCC), the Kyoto Protocol and the Paris Agreement. Brazil adopted its first voluntary commitment to cut GHG emissions in 2009, as part of the pioneering National Act on Climate Change Policies (PNMC). The legislation committed the country to a deviation in emissions between 36.1 per cent and 38.9 per cent by 2020, compared to projections from a business-as-usual scenario.

The adoption of PNMC meant a significant evolution of the institutional and legal framework on climate change. It was no longer an international agenda alone, but part of the country's development agenda, involving economic sectors, civil society and all levels of government in the policy formulation process (Hale et al. 2018; Senado 2019). The PNMC functioned as a guide for the implementation of decentralized climate policies developed by states and municipalities, based on their exclusive constitutional powers, whether within the scope of command and control of activities that generate GHG emissions (article 23, VI), or in concurrent legislative authority on the environment and pollution control (article 24, VI). Based on this shared power, some states and municipalities, such as the state and the city of São Paulo, had already enacted their climate laws before the Federal government.[4]

The 2010 decree stated that gross Brazilian emissions should be between 2,068 Mt CO_2e and 1,977 Mt CO_2e by 2020, including sectorial plans to cut emissions economy-wide. The most significant was the Plan for Deforestation Prevention and Control in Amazonia, in which the target was to slash deforestation rates 80 per cent by 2020 compared to the 1996–2005 average. The agriculture sector plan established that Brazil should recover 15 million hectares of degraded pastures – a figure that would be double in 2015 by the NDC for 2030 (Angelo and Rittl 2019).[5]

Brazil achieved impressive results by reducing the deforestation rate in the Amazon by 83.5 per cent from 2004 to 2012. During this period, forest destruction fell from 27,772 km^2 to 4,571 km^2 a year. In 2010, the Plan for the Prevention and Control of Deforestation and Burning Practices in the Cerrado Region was also created, resulting in a 33 per cent reduction of deforestation in that biome by 2018. During this period, action was taken to improve land use and land tenure regularization; create more conservation units; create and improve environmental monitoring systems; strengthen environmental surveillance; promote sustainable productive activities; and create economic incentives for forest conservation. Several federal agencies were mobilized to implement and monitor such plans, in addition to the creation of a high-level governance structure (SEEG 2018).

In 2015, Brazil submitted a reduction target under the Paris Agreement, becoming the first major developing country to commit to reduce its emissions in relation to a base year, as opposed to reductions based on projected emissions or per unit of GDP. Its NDC committed the country to a 37 per cent reduction in emissions by 2025 and 43 per cent in 2030 compared to 2005 levels and the publication of a National Adaptation Plan, in 2016 (Brasil 2020).

However, since the election of President Jair Bolsonaro in 2019, this structure has been collapsing. Climate action plans were paralysed and their governance structure extinguished. Likewise, the Amazon Fund, specially created to obtain international financing for mitigation and adaptation to climate change, has been threatened (SEEG 2018). Deforestation in the Amazon increased by 29.5 per cent in 2019, the worst rate in the last eleven years and the third highest in the historical series that began in 1988 (INPE 2020).

3.4.2 Climate Change at State Level

Nineteen out of twenty-seven states have passed a legislation establishing a climate change policy. Most of them, sixteen, were approved between 2007 and 2012. Not all state policies have clear mitigation and adaptation strategies, though. Seven states with a climate policy have defined neither mitigation nor adaptation strategies. In these cases, the climate initiative exists only 'on paper'. Twelve states have defined their mitigation strategies, with specific foci depending on the emissions profile of each state. In the Amazonian states such as Acre, Amazonas and Rondônia where most emissions come from deforestation, mitigation efforts focus on environmental services and deforestation prevention and control. In São Paulo and Minas Gerais, mitigation action is more centred in the energy and transportation sectors. Only eight of these states have developed their inventories, which are central to mitigation planning. Regarding adaptation, less than half of the policies (only eight) have defined actions. They are most related to environmental disaster risk management, centred in areas vulnerable to climate related events, such as flooding and landslides. Such strategies do not incorporate future climate change projections for their territories, which is essential to plan the urban space and coastal areas in order to guarantee the best use and occupation of these spaces, the safety of people and economic and social development.

In terms of institutional mechanisms for policy implementation, fifteen states have created a Climate Forum or Climate System, which count on the participation of state secretaries and agencies, municipalities, academia, private sector and civil society organizations. In most cases, these institutional arrangements played a key role in the policy elaboration and approval phases. They were created with the aim of developing a climate change policy in the first place, but after the policy approval not all forums kept active. In some cases, a specific governmental body related to the climate issue has absorbed the policy implementation. Fifteen states have created a climate change board, management, department, superintendence or coordination within their governmental structure. Table 3.1 summarizes states' climate change policies.

3.4.3 Climate Change at Municipal Level

Twelve out of 5,570 municipalities have a specific law establishing a climate policy, corresponding to a population of over 30 million people (IBGE 2010). Between 2003 and 2011, six cities approved their climate change laws (Belo Horizonte, Curitiba, Feira de Santana, Manaus, Palmas, Rio de Janeiro and São Paulo) and five municipalities (Fortaleza, Porto Alegre, Recife, Santos and Sorocaba) approved theirs after 2014, with greater attention to adaptation. Not all municipal policies have clear mitigation or adaptation strategies. Seven out of the twelve municipalities have mitigation strategies and six of them have adaptation actions. Three municipalities have defined neither mitigation nor adaptation actions. Most mitigation strategies include setting or planning to set GHG emission reduction targets. Other actions involve green areas conservation and energy efficiency. The adaptation strategies mainly involve the civil defence and urban planning sectors.

Coastal cities represent an important gap in Brazil's local climate policies. They are considered even more vulnerable to climate change for their geographical specificity, their interface between continent, atmosphere and ocean, and because

Table 3.1 *Climate change policies in Brazilian States*

State	Climate act / policy	Year of strategy	Mitigation strategies	Adaptation strategies	Institutional mechanisms for implementation
Acre	Act no. 2,308	2010	State System of Incentives for Environmental Services (Carbon) State Plan of Deforestation Prevention and Control (2010)	Environmental Disaster Risk Management Plan (2012)	Institute for Climate Change and Regulation of Environmental Services (2011)
Amazonas	Acts no. 3,135 and 4,266	2007 / 2015	Environmental Services Management System; State Plan of Deforestation Prevention and Control	Non-existent	Forum on Global Climate Change, Biodiversity and Environmental Services (2009)
Bahia	Act no. 12,050	2011	State Plan on Climate Change to be defined	State Plan on Climate Change to be defined	Forum on Global Climate Change and Biodiversity (2005)
Ceará	Act no. 16,146	2016	State Plan on Climate Change in elaboration	Adaptation Plan in elaboration	Forum on Climate Change and Biodiversity (2008)
Distrito Federal	Act no. 4,797	2012	Mitigation Plan to be elaborated	Adaptation Plan to be elaborated	not defined
Espírito Santo	Act no. 9,531	2010	GHG emission reduction targets to be set for 2025	non-existent	Forum on Global Climate Change, Rational Use of Water and Biodiversity
Goiás	Act no. 16,497	2009	State Plan on Climate Change Mitigation and Adaptation (2012) focused on low carbon agriculture	State Plan on Climate Change Mitigation and Adaptation (2012)	Forum on Climate Change (2016)
Mato Grosso	Complementary Act no. 582	2017	State Plan on Climate Change to be elaborated	State Plan on Climate Change to be elaborated	State Climate Change System (2017)
Minas Gerais	Act no. 45,229	2009	Energy and Climate Change Plan (2015)	to be elaborated	Climate Change Forum (2005)

Table 3.1 (*cont.*)

State	Climate act / policy	Year of strategy	Mitigation strategies	Adaptation strategies	Institutional mechanisms for implementation
Paraíba	Act no. 9,336	2011	GHG emission reduction target of 36.1–38.9% until 2020	Not defined	State Secretary of Environment and Water Resources
Paraná	Act no. 17,133	2012	State Plan on Climate Change in elaboration	State Plan on Climate Change in elaboration	Forum on Global Climate Change (2008)
Pernambuco	Act no. 14,090	2010	State Plan on Climate Change (2011)	State Plan on Climate Change (2011)	State System for Combating Climate Change (2010)
Piauí	Act no. 6,140	2011	State Plan on Climate Change to be elaborated	State Plan on Climate Change to be elaborated	State Secretary of Environment and Water Resources
Rio de Janeiro	Act no. 5,690	2010	State Plan on Climate Change (2012)	State Plan on Climate Change (2012)	Forum on Global Climate Change (2007)
Rio Grande do Sul	Act no. 13,594	2010	Sectorial Plan for Mitigation and Adaptation to Climate Change for the Consolidation of a Low Carbon Economy in Agriculture (2013)	Sectorial Plan for Mitigation and Adaptation to Climate Change for the Consolidation of a Low Carbon Economy in Agriculture (2013)	Forum on Climate Change (2007)
Rondônia	Act no. 4,437	2018	State System of Climate Governance and Environmental Services (2018)	State System of Climate Governance and Environmental Services (2018)	State System of Climate Governance and Environmental Services Management Council (2018)

Santa Catarina	Act no. 14,829	2009	Not defined	Not defined	Forum on Global Climate Change (2009)
São Paulo	Act no. 13,798	2009	Biogas Program (2012); Transportation Program (2014)	Program for the Prevention of Natural Disasters and the Reduction of Geological Risks (2011)	Forum on Global Climate Change and Biodiversity (2005)
Tocantins	Act no. 1,917	2008	Sectorial Plan for Mitigation and Adaptation to Climate Change for the Consolidation of a Low Carbon Economy in Agriculture (2013)	Sectorial Plan for Mitigation and Adaptation to Climate Change for the Consolidation of a Low Carbon Economy in Agriculture (2013)	Forum on Global Climate Change and Biodiversity (2007)

Source: the authors.

they are places with a high concentration of people and structures – which changes these events to the status of disasters, since people and structures can be severely affected. Brazil has a coastline of almost 7,500 km, where many – and some of the most important – cities in the country are located and where most of the population is concentrated. Only five coastal cities (Fortaleza, Recife, Rio de Janeiro, Salvador and Santos) have an adaptation strategy.

In terms of institutional mechanisms for policy implementation, ten out of the twelve cities established a Climate Forum or Committee, with the participation of municipal secretaries and agencies, universities and research institutes, private sector and civil society organizations. A summary of these municipal climate policies is shown in Table 3.2.

3.4.4 Climate Change Policy Construction Process

The development and implementation of climate mitigation and climate adaptation policies at the three levels of government in Brazil relies on several institutions that establish dialogue and coordination, information sharing, capacity development, planning, implementation, monitoring and evaluation across the different levels of government. These institutions and their roles are described in Table 3.3.

However, in recent years, federal institutions at the federal level – the Interministerial Committee on Climate Change, the Interministerial Commission on Global Climate Change and the National Climate Change Fund – have been affected by the climate change denialist position of the current government (dos Santos Estevo 2021). The work of the Brazilian Panel on Climate Change has also been affected by funding reductions, as has the Brazilian Forum of Climate Change.[6]

3.5 Climate Policy and Federalism in Brazil: The Role of the Subnational Level

This section analyses the ways subnational level of government has been taking a leadership in climate governance in Brazil through two processes: (i) the favourable context for decentralized policymaking; and (ii) the scope for experimental policymaking and associated learning among the constituent units.

3.5.1 The Favourable Context for Decentralized Policymaking

As in other countries (e.g., Rabe 2008), decentralized and experimental climate policymaking has emerged in Brazil in a context of bottom-up climate policymaking seeking to fill a void left by federal inaction. The city of Santos,

Table 3.2 *Climate change policies in Brazilian municipalities*

City / State	Climate act/ policy	Year	Mitigation strategies	Adaptation strategies	Institutional mechanisms for implementation
Belo Horizonte (MG)	Act no. 10,175	2011	30% GHG emissions reduction until 2015	Adaptation plan in elaboration	Municipal Committee of Climate Change and Eco economy (2006)
Curitiba (PR)	Decree no. 1,186	2009	Mitigation plan in elaboration	Adaptation plan in elaboration	Curitiba Forum of Climate Change (2009)
Feira de Santana (BA)	Act no. 3,169	2011	Objective to reduce GHG emissions but no target set	To be defined	Municipal Forum of Global Climate Change and Biodiversity (2011)
Fortaleza (CE)	Act no. 10,586	2017	15.5% GHG emissions reduction until 2020 and 20% until 2030	Adaptation plan in elaboration	Fortaleza Forum of Climate Change (2015)
Manaus (AM)	Act no. 254	2010	Mandatory use of equipment aimed at the rational use of energy and water in buildings and tax incentives for sustainable practices	Mandatory use of equipment aimed at the rational use of energy and water in buildings and tax incentives for sustainable practices	Municipal government
Palmas (TO)	Act no. 1,182	2003	Green areas conservation and energy efficiency plan	Not defined	Municipal Department of Environment
Porto Alegre (RS)	Complementary Act no. 872	2020	GHG emission reduction targets to be defined after inventory execution	Resilience Plan (2016)	Municipal Committee of Climate Change and Energy Efficiency (2016)

Table 3.2 (*cont.*)

City / State	Climate act/ policy	Year	Mitigation strategies	Adaptation strategies	Institutional mechanisms for implementation
Recife (PE)	Act no. 18,011	2014	GHG emission reduction plan with targets by sector of activity (2016)	Adaptation Plan (2019)	Recife Committee of Sustainability and Climate Change (Comclima) (2013) Executive Group of Sustainability and Climate Change (Geclima) (2013)
Rio de Janeiro (RJ)	Act no. 5,248	2011	GHG emission reduction targets: 8% in 2012; 16% in 2016; 20% in 2020	Climate Change Adaptation Strategy (2016)	Forum Carioca of Climate Change and Sustainable Development (2009)
Santos	Adaptation Plan	2016	Not defined	Adaptation Plan (2016)	Municipal Commission of Climate Change Adaptation (2015)
São Paulo (SP)	Act no. 14,933	2009	Guidelines for the City of São Paulo Action Plan for Mitigation and Adaptation to Climate Changes (2011)	Guidelines for the City of São Paulo Action Plan for Mitigation and Adaptation to Climate Changes (2011)	Municipal Committee of Climate Change and Eco economy (2005)
Sorocaba (SP)	Act no. 11,477	2016	GHG emission reduction targets to be defined after inventory execution	Adaptation Plan (in elaboration?)	Local Committee of Climate Change and Working Group on Climate Change (2019)

Source: the authors.

Table 3.3 *Institutions responsible for the development and implementation of climate change policies in Brazil*

Institutional arrangements at federal level	**Interministerial Committee on Climate Change:** articulates federal government internally among ministries to guide the implementation, monitoring and evaluation of the National Plan on Climate Change; supports the international articulation within the scope of the Climate Convention. **Brazilian Forum of Climate Change:** main institutional arrangement that enables dialogue, coordination and information sharing. **Brazilian Panel on Climate Change:** established in the IPCC model, as the technical-scientific extension of the National Policy; interfaces directly with the Brazilian Forum; supported by the **Brazilian Network for Research on Global Climate Change**.
Institutional arrangements at subnational level	**State and municipal forums of climate change**, stimulated by the Brazilian Forum. **Brazilian Association of State Environmental Entities (ABEMA):** promotes political articulation and coordination in Brazilian states. **National Association of Municipal Environmental Entities (ANAMMA)** and the **Forum of Environmental Secretaries of Brazilian Capital Cities (CB27)**: similar role of Abema at city level.
Capacity development	**Transnational networks:** mainly ICLEI, 100 Resilient Cities and C-40 play a key role in capacity building opportunities and development of technical knowledge (e.g., elaboration of GHG emissions inventories and resilience plans).
Implementation	**National Climate Change Fund:** instrument of the National Climate Change Policy to finance projects, studies and enterprises that aim to mitigate and to adapt to the effects of climate change; linked to the Ministry of Environment.
Monitoring & evaluation	**Interministerial Committee on Climate Change (CIM)**: guides and prepares the implementation, monitoring and evaluation of the National Plan on Climate Change.

Source: the authors.

in the state of São Paulo, illustrates this case. In the absence of a national and state adaptation strategy and already feeling the effect of climate change, the local government of Santos started developing its adaptation plan in 2015 and published it a few months after the National Adaptation Plan (PNA) in 2016. As a coastal city, Santos is highly vulnerable to climate change, primarily due to the risks related to relative sea-level rise; the occurrence of extreme events of rain, storm surges, and storm tides; and the socio-environmental consequences of these events (Marengo et al. 2017; Souza et al. 2019).

The PNA was approved seven years after the approval of the National Policy on Climate Change, which suggests that, at the federal level, adaptation has taken a longer time to be internalized. Furthermore, for Brazilian specialists (Di Giulio et al., 2016a), the PNA falls short of what is needed in relation to adaptation policies. Its implementation has been interrupted by the current federal administration. At state level, only the Federal District developed an adaptation plan, which is currently under public consultation.[7]

In Brazil, the metropolitan level is essential for environmental protection and climate governance. Although climate change effects are localized, they are often linked to transformations and disruptions in ecosystems and ecological processes that include multiple jurisdictions. Therefore, a coordinated response by governments at multiple scales is more efficient in responding to climate impacts (Keskitalo et al. 2016; Nalau et al. 2015). For example, measures involving water management and flood protection systems, such as warning systems that require effective communication and coordination mechanisms, go beyond municipal boundaries. Public transportation also needs to be thought of at the level of the metropolitan region, with the collaboration of municipalities. However, climate change is not considered in any existing political–institutional structures and public policies at metropolitan level (Sathler et al. 2019; Torres et al. 2019). Metropolitan regions along with Integrated Development Regions (RIDEs) concentrate 54.3 per cent of the country's population (IBGE 2010). Therefore, it is essential that the federal and state governments promote action at the metropolitan scale, acting as interlocutors and encouraging the dissemination of climate policies and the creation of institutional arrangements that enable the construction of an integrated climate agenda at the inter-municipal level.

One example of articulation at municipal and metropolitan scales is the Reconecta Program at Campinas Metropolitan Region, in the state of São Paulo, headed by the municipality of Campinas. While not explicitly a climate change policy, the programme focuses on ecosystem services and supports the integration of conservation and recovery measures for fauna and flora in the twenty municipalities that are part of the metropolitan region. The programme has INTERACT-Bio Project as a partner, it is coordinated and implemented by ICLEI – Local Governments for Sustainability (Interact-Bio 2021).

Due to the localized nature of many climate change effects, it is important that regional and local governments can design their own adaptation policies (Biesbroek et al. 2014; Termeer et al. 2011). For instance, while coastal regions need to enhance their resilience to floods, mountainous regions may need to focus on landslides. Following tragic landslides in the mountainous region of Rio de Janeiro,[8] the National Civil Defence and Protection Policy (Act No. 12,608) was approved in 2012, which instituted the National Civil Defence and Protection System. The focus of public policies at the federal level shifted from post-disaster response and reconstruction to preventive actions, which mainly seek to safeguard human life and have been influencing state and municipal policies. This has had a significant impact at the local level, where climate change adaptation planning is strongly related to the civil defence department (Di Giulio et al. 2016b). Decentralized action in Brazil involving risk and disaster management has contributed to the development and implementation of actions with the potential to adapt to the effects of climate change at local level.

However, decentralized and uncoordinated climate action may lead to profound disparities between different regions in a federation, resulting from distinct capacities, resources and assets that local and regional governments have to adapt to climate change (Gordon 2015). That is the case of the municipalities and regions within the state of São Paulo, which have different levels of resilience and capacities. To reduce such disparities, the State launched a programme ('Resilient Municipalities') to support thirteen selected municipalities and the Metropolitan Region of Santos in the design of their adaptation and resilience plans; the pilot phase started in 2021 (São Paulo Governo do Estado 2021).

Moreover, Brazilian cities have faced difficulties when putting such plans into practice. Many climate policies fail to get implemented due to insufficient financial resources (Barbi and de Macedo 2019). Participation in transnational municipal networks (TMNs) is a path taken by most Brazilian cities which have climate policies. Several cities have joined the ICLEI network, whose methodology in the Green Climate Cities Programme includes helping cities find financing for their climate actions. In the case of Santos, the city engaged in the ProAdapta Project, supported by GIZ. Another possibility is to consider climate action budget in the multi-year municipal plan.

The emphasis on decentralization in the cooperative Brazilian federalism model without defined distribution of responsibilities may be a source of tension (Viswanathan 2014). When states or municipalities do not carry out their responsibilities, there is no adequate mechanism for the federal authorities to remedy the situation.

Municipalities with flexibility and capacity to establish their own climate policies can also experiment with innovative solutions for combating climate change (Biesbroek and Lesnikowski 2018) and become a model for other municipalities. This is the case of Santos and Campinas, which are leading the way to climate adaptation in their metropolitan regions, engaging the neighbouring municipalities in climate planning.

3.5.2 The Scope for Experimental Policymaking and Associated Learning

The main mechanism of policy diffusion and social influence in Brazil is policy learning – using established definitions of policy learning as the generation of knowledge on resolving a policy problem and of lessons on best strategies to secure policy adoption (May 1992). Informal institutions such as networks involving officials and experts within and between governments can facilitate policy diffusion processes in a federation, particularly the development and sharing of policy lessons (Butler et al. 2016; Vinke-De Kruijf and Pahl-Wostl 2016). Climate policy diffusion in Brazil is supported by TMNs, mainly ICLEI, C-40 and 100 Resilient Cities, and other institutional arrangements, such as the Brazilian Association of State Environmental Entities (ABEMA) at state level and the National Association of Municipal Environmental Entities (ANAMMA) and the Forum of Environmental Secretaries of Brazilian Capital Cities (CB27), both at local level. By engaging in TMNs and in these other institutional arrangements, it is expected that governments will learn from one another's experiences in designing and implementing climate policies.

The availability of multiple forums for policymaking offers some advantages for combating climate change. Federalism, in particular, makes it possible for state and municipal level to adopt climate policies to compensate for the void left by another level of government's inability or refusal to deal with climate change (Derthick 2010). In the context of federal inaction on climate change by the Bolsonaro administration, multiple Brazilian states and municipalities have engaged in institutional arrangements through which they commit to reduce their carbon emissions. One example is the Brazilian Alliance for Climate Action (ACA), established in 2021 to mobilize state and local authorities, business leaders, investors, academics, the press, religious bodies and civil society organizations to increase climate action (ACA Brasil 2021). Six months after its creation, four states and nineteen municipalities had signed the declaration, assuming the responsibility of meeting the Brazilian NDC and collaborating to make it even more ambitious.

Another example is the 'Governors for Climate' movement, within the framework of the Brazil Center on Climate, which sent a letter to US President Joe Biden signed by twenty-four state governors proposing strategic partnerships between the USA and their states on the eve of the 2021 Leaders' Climate Summit (CBC 2021). With this, they seek to reactivate and create state forums on climate change, attract investments and establish connections between states and international funding agencies, and overall, find opportunities and compensate for the void generated by the national government's refusal to address climate change.

At the same time, at municipal level, the Forum of Environmental Secretaries of Brazilian Capital Cities (CB27) published the Letter in Defence of Biomes, a document that reinforced the idea that the defence of Brazilian biomes is a necessary condition for the preservation of biodiversity and essential for the climate emergency mitigation and adaptation (CB27 2021). Time will tell whether these mobilizations at state and local levels will be translated into consistent policy mitigation and adaptation strategies.

A well-functioning federal state should facilitate an effective multi-level climate governance system in which all levels collaborate to develop and implement synergistic climate policies based on their experience and resources, thus achieving the balance between centralization and decentralization (Carlson 2009). However, the current administration is hindering multi-level climate governance, essential when it comes to sectors such as land use management, energy, water resources and others. President Bolsonaro was elected and supported by 'ruralists', the large rural landowners bloc in Congress, who are interested in agriculture expansion, reduction of conservation areas and authorization for increased use of pesticides (dos Santos Estevo 2021). This situation illustrates what Jordaan et al. (2019) describe as political conflicts and ideological divisions arising from political polarization between different governments in a federation undermining the emergence of effective forms of multi-level climate governance. Overcoming this current political polarization in Brazil is one of the country's biggest challenges in order to promote multi-level climate governance.

Furthermore, federal structures may enhance climate change resilience through collaborative projects, networks and co-funding arrangements that enable the opening of paths to meet the needs of affected areas, to build capacity at local level for managing climate risks and to address underlying differences between jurisdictions regarding exposure to climate hazards and their adaptive capacity (Pahl-Wostl et al. 2012). In Brazil, some federal funding such as the Climate Fund have been compromised by the federal government, making it difficult to build resilience to climate change.

Federated entities can also act in concert through their own networks, horizontal coordinated climate governance, where they can collaborate and share knowledge with one another (Hanssen et al. 2013). At local level, one example is the participation of 109 Brazilian cities in the Global Covenant of Mayors for Climate and Energy, a global alliance of local governments voluntarily committed to climate change mitigation and adaptation (GCMCE nd). ICLEI has mobilized Brazilian cities and supported the consolidation of a strategic network of institutions committed to the implementation of long-term national strategies, comprising the Brazilian Association of Municipalities (ABM), the National Front of Mayors (FNP) and the National Confederation of Municipalities (CNM) (ICLEI 2021).

Within the multiplicity of forums for the formulation of climate policies, regulatory overlaps and coordination deficits within a federation can hinder collaboration and generate inefficiencies in the adoption and implementation of climate policies (Jordaan et al. 2019). In Brazil, a sector related to climate governance, such as land use management, is a source of tension between different levels of government.

Brazilian federalism has provided a favourable context for decentralized and experimental policymaking in climate change governance in the country, especially at local level, where municipalities are leading the way to adaptation strategies. Cooperation between municipalities through transnational municipal networks and international cooperation has collaborated to the policy diffusion of climate strategies in the country. The availability of multiple forums for climate policymaking at the subnational level has played an important role, especially at a time when the federal government is actively dismantling the national climate agenda.

3.6 Conclusion

In recent years Brazilian constituent units have been able to move forward with climate change policies in the absence of effective action by the central government. A favourable legislative framework and the political will of local leaders provided constituent units with the necessary authority and conditions to develop experimental climate policies. State and municipal governments with environmental bodies, financial resources and technical staff have been able to drive local and regional agendas more effectively than the central administration.

However, for key sectors such as land use change and forestry, federal government action is essential. Subnational governments efforts alone are simply uncapable of containing deforestation in the Cerrado and the Amazon region and therefore cannot address the main sources of the country's emissions. Given the key role of the land use and forestry sector in Brazil's emissions and the huge global importance of its forests for environmental services, biodiversity and carbon sequestration, the Brazilian government urgently needs to strengthen mitigation action in this sector, as well as to coordinate such action with the constituent units. Our analysis also highlighted an important gap regarding the metropolitan level. This is due to the lack of institutional arrangements that provide metropolitan regions with governance capacity.

When it comes to conflicting interests, the executive and legislative powers have constitutional authority over issues related to climate change mitigation and adaptation at the three levels. However, with the Congress currently dominated by interests that oppose such policies, and the national government adopting a denialist position towards climate change, the Senate has played the role of moderator, often 'locking up' agendas that are not aligned with climate action.

While subnational action in Brazilian federalism has been an important facilitator of climate strategies and policies, it is necessary to monitor to what extent the commitments made and plans designed at subnational level will move forward in a post-pandemic political and economic context.

Notes

1 When considering historical emissions, Brazil is the fourth-largest global GHG emitter, responsible for 5 per cent of historical emissions worldwide (Carbon Brief, 2021).

2 The ND-GAIN Index11 ranks 181 countries using a score which calculates a country's vulnerability to climate change and other global challenges as well as their readiness to improve resilience (University of Notre Dame, 2022).

3 From the second half of the 1990s onwards, a large number of metropolitan regions were created by state governments. There are now more than sixty metropolitan regions in the country (Fernandes, Araújo, 2015).

4 Lei nº 12.187, de 29 de dezembro de 2009. *Institui a Política Nacional sobre Mudança do Clima – PNMC.* Diário Oficial da União. Brasília, DF, nº 248, Seção 1, p. 109, 29 dez.

5 Decreto nº 7.390, de 9 de dezembro de 2010. *Regulamenta os arts. 6o, 11 e 12 da Lei no 12.187, de 29 de dezembro de 2009, que institui a Política Nacional sobre Mudança do Clima – PNMC.*

6 The Brazilian Forum of Climate Change has changed its name to Forum Climate Brazil (Azevedo 2019).

7 www.sema.df.gov.br/wp-conteudo/uploads/2021/02/Texto-Consulta-Publica-2021-Plano-de-Adaptacao-Distrito-Federal_publicado.pdf, accessed on 23/06/2021.

8 The disaster occurred between 11 and 12 January, 2011, affecting seven cities in the mountainous region of the State of Rio de Janeiro, when heavy rains caused floods and landslides, leaving more than 900 dead, around 350 missing and thousands of people unsheltered, as well as serious damage to the region's infrastructure, economy and geography.

References

ACA Brasil. 2021. 'Signatários'. https://acabrasil.org.br/signatarios, last accessed on 30/06/2021.

Adeney, J. Marion, Norman L. Christensen Jr., and Stuart L. Pimm. 2009. 'Reserves Protect against Deforestation Fires in the Amazon'. *PLoS ONE* 4 (4): e5014. doi: 10.1371/journal.pone.0005014.

Albuquerque, Igor (ICLEI), Ane Alencar (Ipam), Claudio Angelo (Observatório Clima) et al. 2020. 'Análise das Emissões Brasileiras de Gases de Efeito Estufa e suas implicações para as metas do Brasil – 1970–2019'. SEEG (Sistema de Estimativas de Emissões de Gases de Efeito Estufa). https://seeg-br.s3.amazonaws.com/2019-v7.0/documentos-analiticos/SEEG-Relatorio-Analitico-2019.pdf, accessed on 04/06/2020.

Arretche, Marta. 2000. *Estado Federativo e Políticas Sociais: Determinantes da Descentralização*. Rio de Janeiro: Revan; São Paulo: FAPESP.

Azevedo, Ana Lucia. 2019. 'Exonerado por Bolsonaro, Alfredo Sirkis diz que "Brasil precisa de uma Lava-Jato do desmatamento"'. *O Globo Brasil*, 10 May. https://oglobo.globo.com/sociedade/exonerado-por-bolsonaro-alfredo-sirkis-diz-que-brasil-precisa-de-uma-lava-jato-do-desmatamento-23656390, accessed on 26/05/2021.

Barber, Christopher P., Mark A. Cochrane, Carlos M. Souza Jr., and William F. Laurance. 2014. 'Roads, Deforestation, and the Mitigating Effect of Protected Areas in the Amazon'. *Biological Conservation* 177: 203–9. doi: 10.1016/j.biocon.2014.07.004.

Barbi, Fabiana, and Laura Valente de Macedo. 2019. 'Transnational Municipal Networks and Cities in Climate Governance: Experiments in Brazil'. In *Urban Climate Politics: Agency and empowerment*, eds. Jeroen van der Heijden, Harriet Bulkeley, and Chiara Certomà, 59–79. Cambridge: Cambridge University Press. doi: 10.1017/9781108632157.

Biesbroek, Robbert, and Alexandra Lesnikowski, A. 2018. 'Adaptation'. In *Governing Climate Change: Polycentricity in action?*, eds. Andrew Jordan, Dave Huitema, Harro van Asselt, and Johanna Forster, 303–19. Cambridge: Cambridge University Press. doi: 10.1017/9781108284646.018.

Biesbroek, Robbert, Katrien Termeer, Judith Klostermann, and Pavel Kabat. 2014. 'Rethinking Barriers to Adaptation: Mechanism-based explanation of impasses in the governance of an innovative adaptation measure'. *Global Environmental Change* 26 (1): 108–18. doi: 10.1016/j.gloenvcha.2014.04.004.

Brasil: MMA (Ministério do Meio Ambiente). 2016. *Plano Nacional de Adaptação à Mudança do Clima: Volume II: Estratégias Setoriais e Temáticas*. Portaria MMA nº 150 de 10 de maio de 2016. Brasília: MMA. https://antigo.mma.gov.br/clima/adaptacao/plano-nacional-de-adaptacao.html.

Brasil: MRE (Ministério das Relações Exteriores). 2020. Press release n. 157.

Carbon Brief. 2021. 'Historical Responsibility for Climate Change is at the Heart of Debates over Climate Justice'. www.carbonbrief.org/analysis-which-countries-are-historically-responsible-for-climate-change.

Carlson, A. E. 2009. 'Iterative Federalism and Climate Change'. *Northwestern University Law Review* 103 (3): 1097–162.

CBC. 2021. 'Carta governadores brasileiros a Joe Biden'. *CBC*. www.centrobrasilnoclima.org/acoes/Carta-governadores-brasileiros-a-Joe-Biden, last accessed on 30/06/2021.

CB27. 2021. 'Regenera o Verde: Capitais Pela Defesa dos Biomas'. *CB27*. https://cb27.s3.amazonaws.com/port-cb27-carta-regeneracao-verde-port.pdf, last accessed on 30/06/2021.

Derthick, M. 2010. 'Compensatory Federalism'. In *Greenhouse Governance: Addressing climate change in America*, ed. Barry G. Rabe, 58–72. Washington DC: Brookings Institution Press.

Di Giulio, Gabriela Marques, David Lapola, Roger Torres, et al. 2016a. 'Plano Nacional de Adaptação à Mudança do Clima: possibilidades e desafios'. *Jornal da Ciência 24 (1):1*. www.jornaldaciencia.org.br/edicoes/?url=http://jcnoticias.jornaldaciencia.org.br/24-plano-nacional-de-adaptacao-a-mudanca-do-clima-possibilidades-e-desafios, accessed on 16/06/2021.

Di Giulio, Gabriela Marques, Ana Maria Bedran Martins, and Maria Carmen Lemos. 2016b. 'Adaptação climática: Fronteiras do conhecimento para pensar o contexto brasileiro'. *Estudos Avançados* 30 (88): 25–41. doi: 10.1590/S0103-40142016.30880004.

dos Santos Estevo, Jefferson. 2021. 'Riscos e Mudanças Climáticas: Os Casos de Brasil e China (2011– 2019)'. PhD diss. Unicamp. www.unicamp.br/unicamp/teses/2021/05/04/riscos-e-mudancas-climaticas-os-casos-de-brasil-e-china-2011-2019.

Fernandes, Antônio Sérgio Araújo, and Suely Mara Vaz Guimarães de Araújo. 2015. 'A Criação de Municípios e a Formalização de Regiões Metropolitanas: Os Desafios da Coordenação Federativa'. *Revista Brasileira de Gestão Urbana* 7 (3): 295–309.

GCMCE (Global Covenant of Mayors for Climate and Energy). nd. http://pactodealcaldes-la.eu/en, last accessed on 30/06/2021.

Guetta, Mauricio, Antonio Francisco Perrone Oviedo, and Nurit Bensusan. 2019. 'Litigância climática em busca da efetividade da tutela constitucional da Amazônia'. In *Litigância climática: Novas fronteiras para o Direito Ambiental no Brasil*, eds. Joana Setzer, Kamyla Cunha, and Amália Botter Fabbri. São Paulo, SP: Thomson Reuters Brasil.

Gordon, D. 2015. 'An Uneasy Equilibrium: The coordination of climate governance in federated systems'. *Global Environmental Politics* 15 (2): 121–41.

Hale, Thomas. 2018. *The Role of Sub-state and Nonstate Actors in International Climate Processes*. London: Chatham House: The Royal Institute of International Affairs. www.chathamhouse.org/sites/default/files/publications/research/2018-11-28-non-state-sctors-climate-synthesis-hale-final.pdf, accessed on 07/07/2020. 2018.

Hanssen, Gro Sandkjaer, Per Kristen Mydske, and Elisabeth Dahle. 2013. 'Multi-Level Coordination of Climate Change Adaptation: By national hierarchical steering or by regional network governance?' *Local Environment* 18 (8): 869–87. doi: 10.1080/13549839.2012.738657.

IBGE (Instituto Brasileiro de Geografia e Estatística). Censo 2010. https://cidades.ibge.gov.br.

ICLEI (Governos Locais pela Sustentabilidade). 2021. 'Pacto Global de Prefeitos pelo Clima e a Energia'. *ICLEI*. https://americadosul.iclei.org/projetos/pacto-global-de-prefeitos, last accessed on 30/06/2021.

INPE (INSTITUTO NACIONAL DE PESQUISAS ESPACIAIS). 2020. 'Terra Brasilis Taxa de Desmatamento Amazônia Legal 1988–2020'. http://terrabrasilis.dpi.inpe.br/app/dashboard/deforestation/biomes/legal, accessed on 26/10/2022.

Inoue, C. 2012. 'Governance of Global Climate Change in the Brazilian Amazon: The case of Amazonian municipalities of Brazil'. *Revista Brasileira de Política Internacional*, 55 (special edition): 170–89.

INPE (Instituto Nacional de Pesquisas Espaciais). 2021. 'Monitoramento do Desmatamento da Floresta Amazônica Brasileira por Satélite'. www.obt.inpe.br/OBT/assuntos/programas/amazonia/prodes, accessed on 04/06/2020.

Interact-Bio. 2021. Plano de Ação para Implementação da Área de Conectividade da Região Metropolitana de Campinas. https://americadosul.iclei.org/wp-content/uploads/sites/78/2021/04/60-ly-plano-de-acao-campinas-digital-3.pdf, last accessed 30/05/2021.

IPCC (Intergovernmental Panel on Climate Change). 2014. 'Summary for Policymakers'. In *Climate Change 2014: Impacts, adaptation, and vulnerability: Part A: Global and sectoral aspects: Working Group II contribution to the IPCC fifth assessment report*, eds. Christopher B. Field, Vicente R. Barros, David Jon Dokken, et al., 1–34. Cambridge: Cambridge University Press. doi: 10.1017/CBO9781107415379.

Jordaan, Sarah M., Adrienne Davidson, Jamal A. Nazari, and Irene M. Herremans. 2019. 'The Dynamics of Advancing Climate Policy in Federal Political Systems'. *Environmental Policy and Governance* 29 (3): 220–34. doi: 10.1002/eet.1849.

Keskitalo, E. Carina H., Sirkku Juhola, Nina Baron, Håkon Fyhn, and Johannes Klein. 2016. 'Implementing Local Climate Change Adaptation and Mitigation Actions: The role of various policy instruments in a multi-level governance context'. *Climate* 4 (1): 7. doi: 10.3390/cli4010007.

Marengo, Jose A., and Fabio Rubio Scarano, eds. 2017. *Impacto, Vulnerabilidade e Adaptação das Cidades Costeiras Brasileiras às Mudanças Climáticas: Relatório Especial do Painel Brasileiro de Mudanças Climáticas*. Rio de Janeiro: PBMC. doi: 10.13140/RG.2.2.36042.16329.

de Macedo, Laura Silvia Valente, and Pedro Roberto Jacobi. 2019. 'Subnational Politics of the Urban Age: Evidence from Brazil on integrating global climate goals in the municipal agenda'. *Palgrave Communications* 5, 18. doi: 10.1057/s41599-019-0225-x.

May, P. J. 1992. 'Policy Learning and Failure'. *Journal of Public Policy* 12 (4): 331–54.

Nalau, Johanna, Benjamin L. Preston, and Megan C. Maloney. 2015. 'Is Adaptation a Local Responsibility?' *Environmental Science and Policy* 48: 89–98.

Noltea, Christoph, Arun Agrawal, Kirsten M. Silvius, and Britaldo S. Soares-Filho. 2013. 'Governance Regime and Location Influence Avoided Deforestation Success of Protected Areas in the Brazilian Amazon'. *PNAS* 110 (13): 4956–61. doi: 10.1073/pnas.1214786110.

Observatório do Clima. 'The Worst is Yet to Come: A year of environmental havoc under Brazil's far-right President Jair Bolsonaro'. www.oc.eco.br/wp-content/uploads/2019/12/Relato%CC%81rio-COP25-v3.pdf, accessed on 17/06/2020.

OECD. 2016. 'Country Profile: Brazil'. www.oecd.org/regional/regional-policy/profile-Brazil.pdf.

Pahl-Wostl, Claudia, Louis Lebel, Christian Knieper, and Elena Nikitina. 2012. 'From Applying Panaceas to Mastering Complexity: Toward adaptive water governance in river basins'. *Environmental Science and Policy* 23: 24–34. doi: 10.1016/j.envsci.2012.07.014.

PBMC (Painel Brasileiro de Mudanças Climáticas). 2013. '*Executive Summary: Impacts, vulnerability and adaptation to climate change: Contribution from Working Group 2 to the Primeiro Relatório de Avaliação Nacional sobre Mudança Climáticas of the Painel Brasileiro de Mudanças Climáticas (GT2 RAN1 PBMC)*'. [Eduardo Delgado Assad, and Antônio Rocha Magalhães, eds.]. COPPE. Rio de Janeiro: Universidade Federal do Rio de Janeiro.

Rabe, B. 2008. 'States on Steroids: The intergovernmental Odyssey of American climate policy'. *Review of Policy Research* 25 (2): 105–28.

São Paulo Governo do Estado. 2021. 'Municípios Paulistas Resilientes'. *Infrastructura e Meio Ambiente*. www.infraestruturameioambiente.sp.gov.br/municipiosresilientes≥, last accessed on 14/07/2021.

Sathler, Douglas, Julio Cesar Paiva, and Sandra Baptista. 2019. 'Mudanças Climáticas: Planejamento urbano e governança ambiental nas sedes das principais regiões metropolitanas e integradas de desenvolvimento do Brasil'. *Caderno de Geografia* 29 (56): 262–86. doi: 10.5752/P.2318-2962.2019v29n56p262.

SEEG. 2018. 'Emissões de GEE no Brasil e suas implicações para políticas públicas e a contribuição brasileira para o Acordo de Paris'. https://seeg-br.s3.amazonaws.com/Documentos%20Analiticos/SEEG_5/SEEG5_DOC_Analiticos_1990-2016.zip., accessed on 26/10/2022.

SEEG. 2020. 'Nota Técnica: Impacto da pandemia de Covid-19 nas emissões de gases de efeito estufa no Brasil'. https://seeg-br.s3.amazonaws.com/OC_nota_tecnica_FINAL.pdf, accessed on 04/06/2020.

SEEG. 2022. 'Emissões Totais'.http://plataforma.seeg.eco.br/total_emission# accessed on 22/04/2021.

Setzer, J. 2017. 'How Subnational Governments are Rescaling Environmental Governance: The case of the Brazilian State of São Paulo'. *Journal of Environmental Policy & Planning* 19 (5): 503–19.

Soares-Filho, Britaldo, Paulo Moutinho, Daniel Nepstad, et al. 2010. 'Role of Brazilian Amazon Protected Areas in Climate Change Mitigation'. *PNAS* 107 (24): 10821–26. doi: 10.1073/pnas.0913048107.

Souza, C. R. G., Souza, A. P., and Harari, J. 2019. 'Long Term Analysis of Meteorological-Oceanographic Extreme Events for the Baixada Santista Region'. In *Climate Change in Santos, Brazil: Projections, impacts and adaptation options*, eds. Lucí Hidalgo

Nunes, Roberto Greco, and José A. Marengo, 97–134. Cham: Springer Nature Switzerland. doi: 10.1007/978-3-319-96535-2.

Termeer, Catrien, Art Dewulf, Helena van Rijswick, et al. 2011. 'The Regional Governance of Climate Adaptation: A framework for developing legitimate, effective, and resilient governance arrangements'. *Climate Law* 2 (2): 159–79. doi: 10.3233/CL-2011-032.

University of Notre Dame. 2022. 'ND-GAIN: Notre Dame global adaptation initiative: Country index'. https://gain.nd.edu/our-work/country-index.

Vinke-De Kruijf, Joanne, and Claudia Pahl-Wostl. 2016. 'A Multi-Level Perspective on Learning about Climate Change Adaptation through International Cooperation'. *Environmental Science & Policy* 66: 242–49. doi: 10.1016/j.envsci.2016.07.004.

Viswanathan, R. 2014, 14 August. 'Federalism, Swaraj and Brazil'. *Gateway House*. www.gatewayhouse.in/federalism-swaraj-and-brazil, accessed on 24/04/2021.

Volden, Craig, Michael M. Ting, and Daniel P. Carpenter. 2008. 'A Formal Model of Learning and Policy Diffusion'. *American Political Science Review* 102 (3): 319–32. www.jstor.org/stable/27644523.

World Bank. 2021. 'Climate Risk Profile: Brazil'. *The World Bank Group*. https://climateknowledgeportal.worldbank.org/sites/default/files/2021-07/15915-WB_Brazil%20Country%20Profile-WEB.pdf.

WWF-Brasil. 2017. 'Guia de Adaptação às Mudanças do Clima para entes federativos. Brasília'. www.wwf.org.br/?62222/Guia-de-adaptao-s-mudanas-do-clima-para-entes-federativos.

4

Climate Governance and Federalism in Canada

KATHRYN HARRISON

4.1 Introduction

Endowed with abundant fossil fuels, Canada has built one of the most carbon-intensive economies in the world. The implication is not only that Canada has a long way to go to net zero, but that in doing so it faces entrenched resistance from industry and citizens alike. Most Canadians commute to work from relatively large homes in personal vehicles. Although voters support climate action in theory, they are less enthusiastic about policies likely to increase their cost of living. There is similar resistance from energy-intensive industries, most notably the export-oriented oil and gas industry that accounts for the largest share of Canada's emissions. Political opposition has been successful to date. Despite a succession of ambitious climate targets since 1990, Canada's emissions increased by 21 per cent from 1990 to 2018 (ECCC 2020b).

Fossil fuels are not only integral to Canada's economy but also to its federation. Provincial governments control publicly held 'Crown resources', on which they have relied both for government revenues and economic development. However, the distribution of resources varies greatly in such a geographically vast country. Some provinces have abundant hydro-electric potential, others fossil fuels. The resulting variation in the carbon intensity of provincial economies has yielded equally great variation in provincial governments' climate ambitions and support for federal policies.

As provinces have responded to cycles in public attention to climate change, intergovernmental relations in the Canadian federation have varied over time. In this chapter, I identify three phases in Canadian climate federalism. From 1990 to 2006, a 'joint decision trap' prevailed in which the least ambitious (and most fossil fuel-dependent) provinces vetoed national solutions. From 2007 to 2015, a truncated innovation and diffusion dynamic emerged in a vacuum of federal inaction. Provincial leaders adopted more ambitious and, in some cases, innovative climate policies. However, action by the least carbon-intensive provinces did not prompt fossil fuel-dependent provinces to follow. Reductions hard won by provincial leaders were overwhelmed by emissions growth in recalcitrant provinces. The third phase, since 2016, is characterized by federal unilateralism. While the mere threat of federal action initially yielded provincial collaboration in an ambitious pan-Canadian climate plan, successful implementation of that plan ultimately turned on the federal government's willingness to follow on its threat.

The history of Canadian climate policy underscores the finding of a broad scholarly literature that federalism has different effects under different conditions (Weaver 2020). In addition, one cannot attribute the success or (more often) failure of Canadian climate policy to any one factor, including federalism. Still, on balance I argue that from 1990 to 2015 federalism exacerbated the challenge of climate action in Canada. The combination of provincial governments' defence of the fossil fuel industry and an informal norm of intergovernmental consensus yielded weak policies in fossil fuel-rich provinces and constrained both interprovincial and federal action. Evaluation of policy developments since 2016 is more complicated, however. Leadership by a subset of provinces facilitated a stronger federal role – though a change in the governing party at the federal level was also critical. At the same time, continued deference to the provinces resulted in a patchwork of inconsistent policies that diminished both the efficacy and cost-effectiveness of climate change mitigation, and achieved climate action conditional on increased fossil fuel exports.

4.2 Climate Change in Canada

Canada ranks tenth globally in absolute greenhouse gas emissions, contributing 1.6 per cent of the global total, and eleventh in per capita emissions (Ritchie and Roser 2020). As evident from Figure 4.1, however, there is significant variation in emissions across the ten provinces. In part this reflects the uneven distribution of the population: at one extreme, 39 per cent of Canadians live in Ontario, at the other just 0.4 per cent on Prince Edward Island. However, it also reflects significant differences in per capita emissions, as illustrated by Figure 4.2. With only 12 per cent of the population, Alberta contributes over 40 per cent of Canada's emissions. Per capita emissions in 2018 ranged from 63 to 66 tonnes CO_2e/yr, respectively, in the oil-producing provinces of Alberta and Saskatchewan to 10 tonnes/yr in hydro-rich Quebec. Figure 4.1 also reveals divergent emissions trends across the provinces. Although most provinces experienced emissions growth from 1990 to 2005, followed by stable or declining emissions thereafter, emissions in the oil-producing provinces of Alberta and Saskatchewan continued

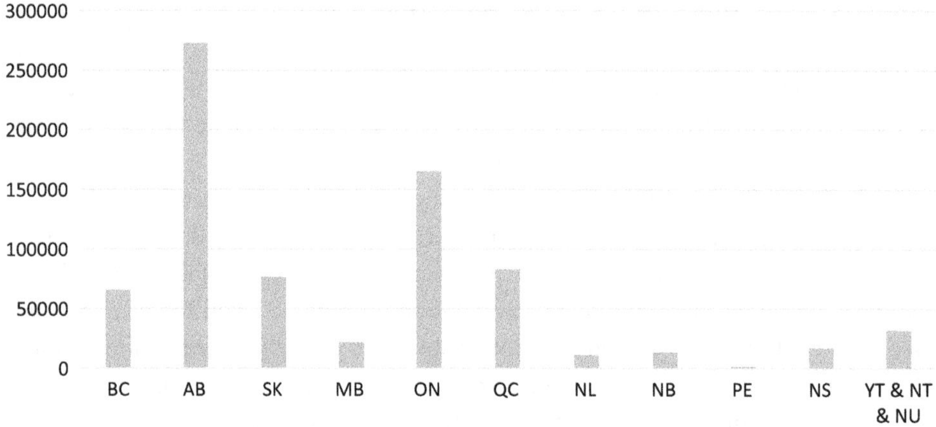

Figure 4.1 Total GHG emissions (kt CO_2eq) by province or territory.
Source: ECCC 2020b.

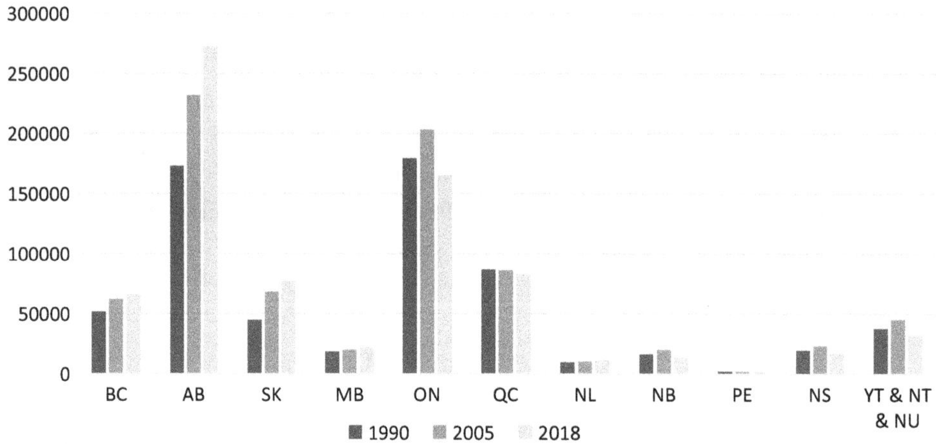

Figure 4.2 GHG Emissions per person (tonnes CO_2eq/yr) by province or territory.
Source: ECCC 2020b.

to grow. Oil and gas production now contributes the largest share of national emissions at 26 per cent, followed closely by transportation (ECCC 2020b).

As a polar-adjacent country, Canada is experiencing twice the global rate of warming (NRC 2019). Despite this, until recently Canada has been less affected by worsening heat waves than many countries by virtue of cold or temperate weather for much of the year and air conditioning enjoyed by 60 per cent of Canadian households (Statistics Canada [2015] 2021). A critical exception is Canada's far north, which has already experienced more than 2°C of warming, with consequences for the welfare and culture of remote Indigenous communities, wildlife and infrastructure (including roads that rest on ice and permafrost). The impact of climate change has become more apparent in western Canada as well, with regular summer wildfires and resulting unsafe air quality. A 'heat dome' in 2021 is believed to have resulted in roughly 600 deaths (Woo 2021). The village of Lytton in British Columbia shattered previous Canadian temperature records at 49.6C, before burning to the ground the next day.

Mainstream political parties have embraced the scientific consensus of the IPCC (Intergovernmental Panel on Climate Change) (though pronounced differences remain among their supporters). A Progressive Conservative government ratified the UN Framework Convention on Climate Change in 1992, in so doing embracing the non-binding target to return emissions to 1990 levels by 2000. A Liberal government signed the Kyoto Protocol in 1997, committing Canada to a 6 per cent reduction below 1990 levels by 2008 to 2012. With failure looming, a successor Conservative government withdrew Canada from the Kyoto Protocol in 2011, but signed on to a Copenhagen Accord target to reduce Canada's emissions 17 per cent below 2005 levels by 2020. In 2015, a Liberal government committed Canada to a 30 per cent reduction below 2005 levels by 2030. Time and again, Canada has committed to significant reductions, only to move the goal posts into the future.

In the absence of policy constraints, emissions continued to increase, most notably from oil production and motor vehicles. Throughout the 1990s and early 2000s, Canadian governments released many climate plans, but in practice implemented only ineffective voluntary programmes and modest subsidies (Harrison 2010; Simpson and Rivers 2008). After 2005, national regulation of motor vehicle emissions constrained emissions growth, but a $15/tonne carbon price in Alberta had little impact on emissions from oil extraction, which continued to climb as a result of both increased production and a shift from conventional to heavy oil from the tar sands (also known as oil sands). Since 2016, however, national measures – including a proposed low carbon fuel standard, phase-out of coal-fired electricity, anticipated harmonization with forthcoming US motor vehicle standards, and a schedule to increase the national carbon price to $170/tonne by 2030 – held the promise that Canada may for the first time meet an international climate target (ECCC 2020a, 62–3). In April 2021, the federal government raised the bar, committing to a new target of a 40–45 per cent reduction below 2005 levels by 2030, which will require even stronger policy measures.

4.2.1 Varieties of Climate Federalism

Canada's constitution, drafted in 1867, merged British parliamentary institutions with the American innovation of federalism. Adoption of a federal system reflected, among other factors, a distinctive francophone majority in what would become the province of Quebec. However, in the immediate wake of the US civil war, the drafters of Canada's constitution intended a more centralized federation, exemplified by the federal government's unlimited taxation powers and residual power to make laws for the 'Peace, Order and Good Government of Canada'. In compensation for more limited tax powers, the provinces were granted control of 'Crown lands' as a source of income. In practice, however, decades of broad judicial interpretation of provincial authority with respect to 'property and civil rights' and the courts' reluctance to resort to the federal residual power has yielded one of the most decentralized federations in the world (Dardanelli et al. 2019).

As the scope of government activity has grown, overlapping federal and provincial powers has become the norm, including with respect to environment and climate. The provinces' ownership of roughly 80 per cent of the land within their borders, and in most cases retention of rights to minerals beneath the remaining private 20 per cent, entails extensive proprietary power to protect or exploit those resources.[1] Legislative jurisdiction with respect to 'property and civil rights' also provides clear provincial authority to regulate pollution sources within their borders, whether public or private. The federal government has jurisdiction with respect to interprovincial 'works and undertakings', such as pipelines and electricity transmission lines, and products sold in interprovincial or international commerce, such as motor vehicles. Noteworthy for global climate change, however, there is no federal power to implement international treaties in areas of provincial jurisdiction. Two other federal powers do offer broader scope for climate change mitigation, though (Hogg 2008). The federal government has relied on its criminal law power, previously found to support federal regulation of toxic substances, to set sectoral standards for fuels, methane emissions and power plants.[2] Finally, legal scholars have long argued that the federal power to make laws for the peace, order and good government of Canada could support regulation or pricing of greenhouse gases (Chalifour and Wormington 2020), a question settled affirmatively by the Supreme Court only in 2021 (as discussed below). Persistent uncertainty with respect to its constitutional authority reflects the fact that until recently the federal government has introduced few climate change policies that could give rise to constitutional challenges.

Institutions, ideas in the form of intergovernmental norms and material interests (and resulting political incentives) in federal–provincial relations help to explain federal inaction. With respect to institutions, with only ten provinces and three territories it is feasible for representatives of all members of the Canadian federation to convene around a single table. That the provinces and federal government all have single member plurality elections, which tend to yield parliamentary majorities, also means that when 'first ministers' meet they usually are in a position to deliver on any agreements they reach. These institutional features have given rise to a practice known as 'executive federalism', which is exemplified by federal–provincial relations concerning the environment. Federal and provincial environment ministers meet at least annually under the auspices of the Canadian Council of Ministers of the Environment. The federal minister sits as one among equals with their provincial counterparts, the chair rotating annually among them. By convention, discussions take place behind closed doors, with a further norm – though not a constitutional requirement – of consensus decision-making. Indigenous governments, who in many cases contest Crown ownership of their unceded, ancestral lands, are not invited.

With respect to interests, political incentives that flow from regional economic diversity and provincial ownership of natural resources present a critical backdrop to federal–provincial relations concerning climate change. Provincial governments historically have sought to create jobs and raise revenue by exploiting Crown resources, and to defend vulnerable local industries. However, climate policy disproportionately threatens the economies of 'petro-provinces' that are dependent on oil and gas extraction (Carter 2020). Variation in provincial dependence on fossil fuel production also coincides with variation in public opinion and partisanship. Two provinces that account for 91 per cent of Canada's oil production (NRC 2020), Alberta and Saskatchewan, are the only provinces where less than half of voters accept that climate change is caused by human activity (Mildenberger et al. 2016). As in other countries, right-of-centre parties are more closely aligned with business. In the context of a major fossil fuel exporting country, that means stronger opposition to climate change mitigation from the Conservative Party of Canada and its provincial counterparts, which typically govern in Alberta and Saskatchewan.

Turning to ideas, cultural identity is also a factor in Canadian climate policy. Sensitivity to federal paternalism often prompts opposition from larger provinces, but that is especially true of Quebec, which fiercely guards its autonomy on behalf of a distinct francophone nation within Canada.

4.2.2 The 'Joint Decision Trap'

The first phase of climate federalism, from 1990 to 2007, was characterized by a 'joint decision trap' (Scharpf 1988), in which the norm of consensus decision-making enabled provinces with the most carbon-intensive economies to block joint provincial or federal action. The federal environment minister from 1993 to 1996 later recalled that 'it became clear that the rule of "consensus" in the environmental agenda would mean moving to the lowest common denominator. There was no way that Alberta would agree to *any* reduction in fossil-fuel emissions' (Copps 2004).

Provincial resistance to climate action was reinforced by economic competition, though consistent with a 'stuck at the status quo' dynamic rather than an all-out 'race to the bottom' (Harrison 1996a; Olewiler 2006). Moreover, with greater international than interprovincial trade (Statistics Canada [2012] 2021), the status quo in question was set not by other provinces so much as the United States, Canada's main trading partner. In advance of international climate negotiations in Kyoto in 1997, the federal and provincial governments agreed that Canada would commit to return its emissions to 1990 levels by 2010, matching the US target. When Canada subsequently agreed in Kyoto to a 6 per cent cut below 1990 levels, comparable to the US target of 7 per cent below, the provinces were outraged that the federal commitment was made unilaterally. Cooperation was restored, however, at a first ministers conference the next day. The prime minister reassured the premiers that Canada's implementation plan would be developed in a joint process co-chaired by the federal government and Alberta, and that only therafter would a decision be made on ratification (Harrison 2010). The National Climate Process sponsored hundreds of meetings over the next four-and a-half years without reaching agreement on a plan to meet Canada's Kyoto target.

Resistance from Canadian business and more carbon-intensive provinces was strengthened by US President George W. Bush's announcement in 2001 that the USA would not ratifiy the Kyoto Protocol. Federal–provincial conflict came to a head in the spring of 2002, by which time Alberta was considering legal action to block federal ratification (Macdonald 2020). When the federal government unilaterally released its own implementation plan, even the two provinces that supported ratification, Quebec and Manitoba, signed a unanimous statement by the provinces condemning federal unilateralism (Harrison 2010). Alberta withdrew from the joint process, and released a provincial plan that welcomed continued emissions growth.

The federal government had rejected the premise of the joint decision trap in finalizing a unilateral federal plan and ratifying the Kyoto Protocol in December 2002. However, as with previous moments of environmental assertiveness, the federal government failed to follow through. Implementation of the federal plan stalled in anticipation of Prime Minister Jean Chretien's retirement. Chretien's successor, Paul Martin, produced a new plan in 2005, which sought to restore federal–provincial harmony with the promise of billions of dollars of federal funding for the provinces. However, the Martin plan also failed to get off the ground before the Liberals lost the 2006 election.

The Conservative government led by Stephen Harper was still less threatening to carbon-intensive provinces by virtue of weaker climate goals (Harrison 2010). The new government immediately abandoned Canada's Kyoto target, and later withdrew from the treaty. Although it promised a strategy of sector-specific regulations, the Harper government only regulated two sectors. Motor vehicle emission standards matched those adopted by the USA, an essentially costless strategy since Canada manufactures vehicles and parts for an integrated North American market. Regulations adopted for electricity generators would not affect existing facilities for two decades (McCarthy 2012). The Conservative government declined to regulate the oil industry, though it accounted for most of Canada's emission growth.

4.2.3 Leaders without Followers

As federal inaction persisted, a new federal–provincial dynamic emerged as some provincial governments took climate policy into their own hands. Provincial actions are potentially important for two reasons. First, provincial 'laboratories of democracy' can produce innovative approaches that inform policies of other jurisdictions (Boyd and Olive 2021). Second, and arguably more important, is leadership in the context of economic competition. The challenge of climate policy is less that governments don't know how to reduce emissions (ideas), than that they are reluctant to impose costs on local actors should their actions not be matched by other jurisdictions (interests). Provincial leadership thus held the promise of reassuring provinces that were reluctant to act lest industry relocate, but not actively seeking to attract investment with lax standards (Harrison 1996b).

With respect to innovation, it was Alberta, ironically, that took the lead in adopting output-based carbon pricing for tar sands producers in 2007. British Columbia followed in 2008 with a revenue-neutral carbon tax (Harrison 2013). In 2014, Quebec joined California in the first emissions trading scheme to extend coverage to small sources (Houle, Lachapelle, and Purdon 2015). Less innovative but potentially important in reassuring other provinces was Ontario's leadership in closing its five remaining coal-fired power plants between 2003 and 2015 (Harris, Beck, and Gerasimchuk 2015).

Two features of this second phase of Canadian climate federalism bear emphasizing. With the exception of Alberta's generous pricing of tar sands emissions – which, as discussed below, was motivated by defending the oil industry's reputation and pre-empting federal action – provincial leaders were those that already had the lowest per capita emissions in Canada. Moreover, provinces with more carbon-intensive economies did not match the leaders' actions. Provinces dependent on fossil fuel extraction continued to resist both federal actions and interprovincial collaboration. Alberta and Saskatchewan thus blocked consensus on a national cap and trade scheme in 2008 (Howlett, Laghi, and Séguin 2008; White and Greenberg 2008). Even Ontario was selective in its climate leadership. As the home of Canada's automobile manufacturing industry,

Ontario vetoed calls for stricter exhaust standards until US regulations forced Canada's hand (Howlett and Keenan 2008).

In this second period, Canadian federalism generated innovation without diffusion and leadership without followers. From 2005 to 2018, British Columbia, Quebec and Ontario collectively reduced their emissions by 38 million tonnes/yr CO_2e,[3] but Alberta and Saskatchewan increased theirs by 49 million tonnes/yr.

4.2.4 The Federal Backstop

The third phase of Canadian climate federalism was characterized by unprecedented federal unilateralism. Following election of a Liberal government in 2015, the federal government extended provincial leaders' actions Canada-wide, in so doing rejecting the norm of granting all provinces, including those dependent on oil production, a veto over Canadian climate policy. Carbon pricing, which was central to this third phase, is explored in the next section.

4.3 Carbon Pricing in the Canadian Federation

Carbon pricing holds promise to achieve reductions targets more cost-effectively than traditional regulation but with a tradeoff of strong public opposition, especially in the case of carbon taxes. Indeed, the strength of public opposition has led some scholars to suggest that carbon pricing is not politically feasible at the level required to drive deep reductions (Green 2021; Jaccard 2020). This section considers how the three phases of Canadian climate federalism affected carbon pricing, with both negative and positive consequences for effective climate policy.

4.3.1 Federal Retreat, Provincial Laboratories of Democracy

A proposal for a carbon tax was floated in the early 1990s in the course of developing a national 'Green Plan', but the idea was quickly withdrawn in the face of opposition from Alberta and the oil industry (Hoberg and Harrison 1994). Interest in carbon pricing in the form of emissions trading emerged following the launch of the EU Emissions Trading Scheme (ETS) in 2005. The Martin government's 2005 Kyoto plan, Project Green, proposed to create a national emissions trading scheme, but would only have required industrial emitters to cut their emissions by 12 per cent below business as usual, a far cry from the more than 30 per cent reduction below projected emissions across all sectors Canada needed to meet its Kyoto target (Harrison 2010).

The promised emissions trading scheme was soon abandoned by the Harper government. In the meantime, however, provincial governments began to adopt their own carbon pricing policies. In 2007, Alberta adopted its Specified Gas Emitters Regulation (SGER), which gave large industry sources an option either to pay a carbon levy of $15/tonne or purchase offsets for emissions in excess of a 12 per cent intensity reduction. While innovative, the weaknesses of the system included a price too low to drive significant reductions from the oil industry, authorization of offsets that had already happened up to six years earlier, and the absence of any plan to increase either the price or intensity target. All in all, this suggests that the motive was less to deliver emissions reductions than to defend the reputation of Alberta's oil industry and pre-empt any federal regulatory or trading scheme. A further advantage was to keep the oil industry's payments in the province, including by authorizing offsets only within Alberta. It is telling that Alberta's 2008 Climate Change Strategy projected emissions that would still be above the province's 1990 Kyoto Protocol baseline in *2050.*

Next up, British Columbia (BC) announced a revenue-neutral carbon tax in 2008, increasing to $30/tonne in 2012 (Harrison 2012, 2013). The BC tax was celebrated internationally for its application of a consistent price to both industry and household emissions; and corresponding cuts in income taxes that promised to stimulate the economy (Partington 2013). Pundits speculated that BC's novel policy would rapidly spread to other provinces (Simpson 2009). However, the emergence of public opposition, the onset of a global recession and rejection of a similar carbon tax proposed by the Liberals in the 2008 federal election later quickly laid that optimism to rest (Harrison 2012). When no other provinces matched BC's price, the province's carbon tax was frozen at $30/tonne in 2012.

During this period, Canadian provinces also proposed to collaborate with US states to limit impacts on cross-border competitiveness. BC, Manitoba, Ontario and Quebec committed to join seven US states in the Western Climate Initiative (WCI) emissions trading plan. However, as it became clear that there would be no federal trading scheme to extend carbon pricing to non-WCI members in either country, all states and provinces but California and Quebec withdr (Houle, Lachapelle, and Purdon 2015). The two remaining members launched their own schemes in 2013, initiated cross-border trading in 2014 and extended coverage to fuel distributors in 2016.

During this period a handful of provinces adopted novel policies. Of particular note are Quebec and California's extension of emissions trading to transportation and heating fuels, and BC's revenue-neutral carbon tax. However, other provinces' failure to embrace those innovations suggests that the underlying challenge was less how to design carbon pricing policies than how to protect local economic interests. Although Ontario's phase out of coal-fired power reflected political leadership rather than innovation, it too was not matched by other coal-dependent provinces. The lack of policy diffusion ultimately undermined the ambition of provincial leaders. BC froze its tax at $30/tonne, and all but one province that had committed to WCI withdrew when it became clear that a federal backstop would not be forthcoming to protect their economies from laggard states and provinces. Ontario belatedly joined WCI trading in 2018, but withdrew the following year following election of a populist Conservative government (Raymond 2016), which simultaneously relaxed the province's emissions target.

The third phase in Canadian carbon pricing was triggered by two critical elections in 2015. The October federal election yielded a return to government for the Liberal Party, now led by Justin Trudeau. The stage for the Trudeau government had been set, however, by the May election of the first left-of-centre government in Alberta in eighty years. The Alberta New Democratic Party (NDP) government ran on a promise of stronger climate policies, but like its predecessors was keen for Alberta to set its own agenda in advance of an anticipated change in the federal government and an expected international agreement at COP21 in Paris. A Climate Change Advisory Panel was struck to devise a new provincial climate plan before the federal election. In parallel, secret negotiations between industry leaders and environmentalists had been ongoing since 2014. That group's agreement to cap oil sands emissions at 100 MT/yr, roughly 50 per cent higher than current emissions, was subsequently built into the Alberta climate plan (World Bank 2017). The government also committed to phasing out coal-fired power by 2030 (an NDP election promise) and on the recommendation of the Advisory Panel, to revising SGER to eliminate benefits for more carbon-intensive facilities, raising the carbon price to $20/tonne in 2017 and $30/tonne in 2018 and, most controversially, extending carbon pricing to households via a carbon tax on transportation and heating fuels. In a remarkable though brief moment of consensus, Alberta Premier Rachel Notley was joined on stage by leaders from the oil industry, environmental groups, labour and First Nations when she announced Alberta's Climate Leadership Plan in November 2015 (Boyd 2019).

A critical feature of the Alberta carbon pricing plan little noted in all the attention to the 100 MT oil-sands cap and carbon tax on households was a decision to match the price rather than targets of other jurisdictions. Ontario and Quebec had previously announced targets to reduce their emissions by 37 per cent and 37.5 per cent, respectively, below 1990 levels by 2030. Even the Harper government had announced a pre-Paris target for Canada of a 30 per cent reduction below 2005 by 2030. However, the Alberta Climate Leadership Plan rejected a province-wide emissions target in favour of an alternative approach of matching the highest extant carbon price in Canada: BC's $30/tonne. Since a carbon price would not have a significant impact on oil-sands production below roughly $70/tonne, the effect was to allow continued growth in Alberta's emissions to about 2030, consistent with the 100 MT cap.[4]

4.3.2 The Pan-Canadian Framework

The federal Liberals promised in anticipation of a fall 2015 election that they would build on the carbon pricing policies of British Columbia, Alberta (at that point SGER only), Quebec and Ontario (which had promised to join WCI) to establish a consistent, but unspecified, national carbon price. While that was encouraging for climate-concerned voters, others were simultaneously reassured by two other aspects of the Liberal platform. The promise to build on provincial actions suggested deference to provincial jurisdiction and sensitivity to regional interests. Moreover, the Liberal platform stressed the need to get 'Canada's resources to market' – a thinly-veiled promise to approve at least one additional export pipeline among several under review at the time.

The Liberals formed a majority government a month before COP21 and quickly convened a first ministers conference. A few months later, federal and provincial governments unanimously announced the 'Vancouver Declaration', through which all provinces and territories agreed to collaborate on a climate plan that would 'build on the leadership shown and actions taken by the provinces and territories' in four areas, one of which was 'carbon pricing mechanisms adapted to each province's and territory's specific circumstances' (Prime Minister of Canada, Justin Trudeau 2016). In late 2016, the first ministers unveiled the new pan-Canadian Framework on Clean Growth and Climate Change (PCF). Although the USA had just elected a president expected to withdraw from the Paris Agreement, Canada's first ministers forged ahead. It was a stark contrast to 2002, when the USA's withdrawal from the Kyoto Protocol strengthened provincial opposition. All provinces and territories but Saskatchewan signed onto the agreement.[5] In so doing, they endorsed Canada's Paris Agreement target of 30 per cent below 2005 by 2030, and committed to undertaking complementary actions specific to federal and provincial jurisdiction. Carbon pricing was highlighted as the 'central component' of the plan. The PCF recognized provincial leadership on carbon pricing, but incorporated the federal government's recent announcement of a carbon pricing benchmark, which included several elements:

- An option for provinces to devise their own carbon pricing plans in the first instance, with the expectation that the federal government would enact its own backstop should a province fail to meet all elements of the federal benchmark.
- A condition of broad coverage, consistent with the application of BC's carbon tax to both industry and households.
- Acknowledgement that provinces could embrace either a carbon tax or a cap-and-trade scheme. Provinces that opted for the former were expected to meet or surpass \$10/tonne CO_2e in 2018, rising steadily to \$50/tonne in 2022. Those opting for cap-and-trade would be expected to set a 2030 provincial reduction

target at least as ambitious as Canada's Paris Agreement target, and caps before 2022 consistent with reductions that would be expected within the province under a tax at the federal benchmark price.

- A commitment by the federal government to return any carbon pricing revenues it might collect under the federal backstop to the province of origin.

Federal–provincial agreement on the PCF carbon pricing plan rested on three factors. The first was Alberta's new-found climate ambition. Although the province had long been an impediment to Canada-wide action on climate, the 2016 Alberta Climate Leadership Plan became a template for the PCF, including a price- rather than target-based federal benchmark. Therein lay a poison pill, however. While Alberta could embrace price competitiveness as an *alternative* to a provincial emissions target, Canada's price benchmark ultimately would need to meet its international target under the Paris Agreement. In fact, the federal benchmark in the PCF was inconsistent: cap-and-trade provinces were expected to commit to a cap at least as ambitious as Canada's Paris target, while carbon tax provinces only needed to commit to a price of $50/tonne in 2022. The PCF's admission – consistent with two decades of modelling efforts (Simpson, Jaccard, and Rivers 2008) – that the first phase of the plan would not be sufficient to meet Canada's Paris target implied that a price well above $50/tonne would be required. However that was never agreed to by first ministers nor made explicit in the pan-Canadian plan.

A second factor critical to federal–provincial agreement was unprecedented federal government resolve. The Liberals set out their broad approach in the 2015 election, but had not specified either a price level nor timing. The prime minister announced details of the federal benchmark, subsequently embedded in the PCF, in the House of Commons in October 2015, even as federal and provincial environment ministers were meeting (Harris 2016). Taken by surprise, ministers from Saskatchewan, Nova Scotia and Newfoundland left the meeting in protest, while other provinces, including Alberta, withheld their agreement (Macdonald 2020).

The third condition was brokerage deals with individual provinces, each of which entailed relaxing the federal benchmark. Nova Scotia got a special deal to extend coal-fired power generation, while BC got federal approval of an LNG export project (Macdonald 2020). However, the PCF keystone was Alberta, which consented to match the federal benchmark of $50/tonne only in exchange for federal approval of a pipeline to gain access to new markets for Alberta oil.[6] Access to 'tidewater' was viewed as critical by the Canadian oil industry, which historically had shipped all of its exports to the USA, which was now less eager for Canada's oil amid booming production of its own shale oil. The federal government approved the Trans Mountain Expansion Project, a new pipeline to carry 590,000 barrels per day of bitumen from Alberta to the BC coast, days before federal–provincial agreement on the pan-Canadian Framework was announced.

4.3.3 The Federal Backstop

In 2018, the federal government passed the *Greenhouse Gas Pollution Pricing Act*, which set out a two-pronged approach: a fuel surcharge (carbon tax) for households and other small sources, and an output-based pricing system for large industrial sources. The federal government would apply one or both only if a province or territory failed to meet the federal benchmark or requested federal administration of carbon pricing. The output-based pricing system applies the scheduled carbon price to emissions in excess of a sector-specific baseline, a design intended to protect competitiveness of trade-exposed industries. Where the federal policy was necessitated by provincial non-compliance with the federal benchmark, revenues are returned to the province of origin via programmes to support industry transition, rather than to provincial governments themselves.

For households, the choice of a carbon tax in the form of a fuel surcharge, rather than emissions trading, reflected the federal government's expectation in early 2018 that the backstop would apply only in one province, Saskatchewan.[7] For one fairly small province, a federal tax was simply easier. Ninety per cent of fuel surcharge revenues are returned to households in the form of equal dividends to households of equal size, with the remaining 10 per cent set aside to support small business and public sector institutions such as schools and hospitals.[8] The commitment to return revenues to the province of origin raised the prospect of very different per capita dividends given dramatic variation in provincial emissions intensity.

The federal government called on provinces to submit their carbon pricing plans for review in mid-2018. Neither Saskatchewan nor Ontario submitted plans. Manitoba committed only to maintain a $25/tonne carbon price through 2022, while New Brunswick proposed to create a provincial carbon tax by reducing an existing fuel tax by the same amount. The federal government responded by implementing its carbon pricing backstop for both industry and households over the objections of those four provinces in April 2019, despite a federal election looming later that year. When the Alberta NDP government was defeated in May 2019 by the United Conservative Party, the new government immediately rescinded that province's carbon tax. The federal government applied its carbon tax in that province as well in January 2020.

The federal government's unilateralism met with vehement provincial opposition. Two of the provinces where carbon pricing was imposed by the federal government, Alberta and Saskatchewan, had a long history of opposition to federal climate initiatives. All five provinces were governed by conservative parties. Federal and provincial conservatives railed against a federal 'tax grab', posting images on social media of themselves refuelling large vehicles the day before the federal carbon tax took effect (Maclean's 2019).

Alberta, Saskatchewan and Ontario all challenged the constitutionality of the federal greenhouse gas pricing law, arguing that the subject matter was exclusive provincial jurisdiction, and that the balance of powers in the federation would be imperilled should the federal government be authorized to regulate any release of greenhouse gases (Chalifour, Oliver, and Wormington 2020). Quebec intervened in support of provincial challenges of the federal Act. Although Quebec is one of the most ambitious provinces on climate, it is invariably among the most defensive of provincial jurisdiction. British Columbia was the only province to intervene in support of the federal government.

Following mixed decisions in lower courts, the Supreme Court of Canada upheld the federal carbon pricing law in March 2021.[9] The court found that the federal government has authority to establish a minimum national carbon price backstop within its power to make laws for the 'Peace, Order, and Good Government' of Canada. Central to the Court's ruling was recognition that provincial governments face a collective action problem: inaction by one or more provinces can undo (and has undone) hard-won gains by others, an outcome that cannot be overcome through cooperation among provinces that have no authority over each other. The court thus incisively diagnosed and handed the federal government a delimited power to overcome the economic competition and provincial vetoes that have long undermined Canadian climate policy.

4.3.4 A Pan-Canadian Patchwork

Beyond the courts, the federal policy also was tested in a national election in October 2019, in which the opposition Conservatives promised to eliminate the federal carbon pricing backstop. The Liberals lost seats but held on to a minority government. Alberta, Ontario and New Brunswick all subsequently received federal approval of their pricing schemes for industry, but only New Brunswick sought and received approval to implement its own carbon tax. As of 2021, the resulting carbon pricing landscape in Canada is complicated indeed. Two provinces (Quebec and Nova Scotia) have unlinked emissions trading schemes, four have provincial carbon taxes (British Columbia, Newfoundland, New Brunswick and Prince Edward Island), and four are subject to the federal carbon tax on households (Alberta, Saskatchewan, Manitoba and Ontario). Federal pricing for industry applies in three carbon-tax provinces (Prince Edward Island,

Saskatchewan and Manitoba) while all others have their own industry pricing schemes (British Columbia, Alberta, Ontario, New Brunswick, Newfoundland.)

Although the federal government took a hard line with provinces that openly defied federal expectations, it interpreted other provinces' carbon pricing proposals with considerable flexibility (Dobson, Winter, and Boyd 2019). The Nova Scotia premier boasted that his province's stand-alone trading scheme would result in a gasoline price increase of $0.01 rather than the federal carbon tax increase of $0.11 per litre (Laroche 2018). While that presumably reflects the ease of meeting a cap in a province experiencing business-as-usual emissions decline, it is hard to see how a lower Nova Scotia carbon price met the federal requirement that provincial emissions trading schemes must deliver the same reductions as would be achieved by the federal carbon tax. (In contrast, Quebec could achieve comparable reductions at a lower price by purchasing credits from California.) The carbon taxes of Prince Edward Island, New Brunswick and Newfoundland all waived application to home heating and partially reduced other provincial fuel taxes in compensation. Both New Brunswick and Ontario technically met the federal benchmark for coverage and pricing for industry, but undermined the intended stringency via generous baselines for individual facilities (Rabson 2020). The result was a patchwork of policy instruments, coverage and prices (Sawyer et al. 2021), in which provincial variation did not merely tailor solutions to local circumstances, but rather undermined climate policy ambition.

The 2016 pan-Canadian Framework acknowledged that additional actions would be needed to close the gap to Canada's 2030 Paris Agreement target. The federal government signalled that it would release a plan to close the gap by the end of 2020. After decades of failed plans, expectations were muted. However, judicial and electoral survival of its core climate policy, carbon pricing, appears to have emboldened the federal government. In its December 2020 plan, the federal government unilaterally committed to increasing the backstop price to $170/tonne in 2030 (ECCC 2020a; Harrison 2020). In contrast to 2016, the federal government does not appear to have consulted provinces (at least not all of them) in advance. Ontario, Alberta and Saskatchewan expressed shock and outrage. While the 2020 plan was expected to meet Canada's original Paris Agreement target of a 30 per cent reduction between 2005 levels by 2030, in the spring of 2021 the federal government upped the ante, unilaterally announcing a new Paris target of a 40–45 per cent reduction over the same period. Within months, the minority Liberal government (with NDP support) passed the Net Zero Climate Accountability Act, which reinforces unilateralism by mandating *federal* accountability for meeting national emission targets.

The federal government's 2020 carbon pricing announcement, 2021 Paris Agreement target, and new net zero accountability law all reject the expectation of federal–provincial consensus that historically prevailed in Canadian climate and environmental policy. The Liberals' subsequent re-election in the fall of 2021 locks in a more assertive federal approach for the foreseeable future.[10]

4.4 Conclusion

Per capita, Canada has contributed more than its share of global emissions. Yet for that very reason climate change mitigation represents a significant political challenge. Voters with carbon-intensive lifestyles call for governments to 'do something', but not to them. Carbon-intensive industries resist the imposition of costs that threaten a long-standing comparative advantage, inexpensive energy. The export-oriented fossil fuel industry, which contributes the largest share of Canada's emissions, has resisted both domestic and global action. Mitigation costs unevenly distributed across regions further concentrate costs and reinforce political opposition. Against that backdrop, care is warranted not to attribute policy success or failure only or even primarily to federalism.

Still, the question remains whether federalism has exacerbated or moderated the challenge of climate action in Canada. Two features of Canadian federalism clearly reinforce the challenges noted above: constitutional empowerment of defenders of regional economies, and provincial ownership of fossil fuels.

On each of the three dimensions introduced in the introductory chapter – tailoring, backup and innovation – the disadvantages prevailed over the benefits in the Canadian federation from 1990 to 2015. The period from 1990 to 2006 reveals the perils of reliance on provinces tailoring solutions to their own circumstances in a context of environmental spillovers and economic competition. Fossil fuel-dependent provinces' 'local preferences' were for continued expansion of fossil fuel production, which resulted in unchecked growth in carbon emissions. The most carbon-intensive provinces not only failed to act on their own or in concert with other provinces, but also blocked federal action.

From 2007 to 2015, provincial innovation and leadership presented the possibility of both cross-provincial learning and provincial leadership to fill a void of federal government inaction. Carbon pricing policies adopted by British Columbia and Quebec and a coal-power phaseout by Ontario were possible only through decentralized authority. The benefit of those provinces' actions was muted, however, by a lack of uptake by other provinces. The Canadian federation gave rise to innovation without diffusion and leaders without followers. Reductions achieved by provincial leaders were overwhelmed by emissions growth from fossil-fuel producing provinces. As observed by Weaver (2020), a laboratory of democracy dynamic rests on shared values and/or common political incentives, neither of which was present as Canadian provinces confronted climate change. The diversity of regional economic interests ensured sustained provincial opposition to climate mitigation among the provinces expecting the greatest costs. That opposition was amplified by an informal institution of federal–provincial consensus that allowed carbon-intensive provinces to veto federal action.

More ambitious climate policies since 2015 were the result of a strong federal commitment to climate action and willingness to reject the norm of intergovernmental consensus. Anticipation of federal action contributed to stronger action by a newly elected Alberta government in 2015. Thereafter, a unilateral federal benchmark was critical to federal–provincial agreement on a collaborative pan-Canadian plan in 2016. The federal government's follow through on that threat was necessary in 2019 when a subset of provincial governments reneged on prior commitments. Contrary to McDonald's characterization of federal unilateralism as 'inept diplomacy' that 'needlessly antagonized provinces' (2020), belated progress in Canadian climate policy has turned on what Gordon has called the 'firm hand' of federal coercion, rather than the 'handshake' of federal–provincial cooperation (2015). The benefits of the federal government's rejection of a long-standing norm of intergovernmental consensus are a reminder that cooperation in a federation is not the end in and of itself and, indeed, can undermine policymaking in the national interest.

If federalism exacerbated already significant obstacles to effective climate policy in Canada from 1990 to 2015, what about the period since 2016? One can argue that the foundation laid by carbon pricing policies already adopted by the four most populous provinces made it easier for the federal Liberals to propose a national carbon price in 2015. However, a decentralized approach yielded a patchwork of inconsistent policies and an ambition gap in carbon tax provinces, predicated on approval of a new pipeline. One of the challenges of studying the impact of federalism is determining the counterfactual (Weaver 2020): what might the Trudeau government have done in 2015 in a unitary system? Perhaps the incremental process of regional buy-in was a politically necessary step towards stronger federal commitments in 2020 and 2021. Then again, in a unitary state the government of Canada might have adopted more ambitious and consistent climate policies decades earlier.

Notes

1 Provincial Crown lands constitute 48 per cent of Canada's total territory (Neimanis [2011] 2013).
2 *R. v Hydro Quebec*, 3 SCR 213 (1997).
3 Despite its carbon tax, BC's emission increased by 3 million tonnes/yr or about 5 per cent.
4 Personal communication, industry executive and provincial public servants, June 2018.
5 Manitoba did not sign when the pan-Canadian Framework was announced in December 2006, but did so a few months later.
6 Interviews with federal officials and a former Alberta Cabinet member.
7 Personal communication, senior federal officials.

8 The dividend is 10 per cent higher in rural areas.
9 References re: Greenhouse Gas Pollution Pricing Act, 2021 SCC 11 (2021). https://scc-csc.lexum.com/scc-csc/scc-csc/en/item/18781/index.do.
10 Although the Liberals again formed a minority government, climate demands of NDP and Green members suggest a parliamentary majority for climate action.

References

Boyd, Brendan. 2019. 'A Province under Pressure: Climate change policy in Alberta'. *Canadian Journal of Political Science* 52 (1): 183–99.

Boyd, Brendan, and Andrea Olive, eds. 2021. *Provincial Policy Laboratories: Policy diffusion and transfer in Canada's federal system*. University of Toronto Press.

Carter, Angela V, 2020. *Fossilized: Environmental Policy in Canada's Petro-Provinces*. Vancouver: UBC Press.

Chalifour, Nathalie, Peter Oliver, and Taylor Wormington. 2020. 'Clarifying the Matter: Modernizing peace, order, and good government in the greenhouse gas pollution pricing act appeals'. *National Journal of Constitutional Law* 40 (2): 153–211. http://dx.doi.org/10.2139/ssrn.3651556.

Copps, Sheila. 2004. *Worth Fighting For*. Toronto: McClelland & Stewart.

Dardanelli, Paolo, John Kincaid, Alan Fenna, André Kaiser, André Lecours, and Ajay Kumar Singh. 2019. 'Conceptualizing, Measuring, and Theorizing Dynamic De/Centralization in Federations'. *Publius: The Journal of Federalism* 49 (1): 1–29. https://doi.org/10.1093/publius/pjy036.

Dobson, Sarah, Jennifer Winter, and Brendan Boyd. 2019. 'The Greenhouse Gas Emissions Coverage of Carbon Pricing Instruments for Canadian Provinces'. *The School of Public Policy Publications* 12 (6): 1–56. http://dx.doi.org/10.11575/sppp.v12i0.53155.

ECCC (Environment and Climate Change Canada). 2020a. *A Healthy Environment and a Healthy Economy*. Canada (ECCC). www.canada.ca/content/dam/eccc/documents/pdf/climate-change/climate-plan/healthy_environment_healthy_economy_plan.pdf.

ECCC. 2020b. *National Inventory Report 1990–2018: Greenhouse gas sources and sinks in Canada: Executive summary*. Canada (ECCC). www.canada.ca/en/environment-climate-change/services/climate-change/greenhouse-gas-emissions/sources-sinks-executive-summary-2020.html.

ECCC. nd. 'Modelling and Analysis of: A Healthy Environment and a Healthy Economy'. Canada (ECCC). www.canada.ca/content/dam/eccc/documents/pdf/climate-change/climate-plan/annex_modelling_analysis_healthy_environment_healthy_economy.pdf.

Gordon, David J. 2015. 'An Uneasy Equilibrium: The coordination of climate governance in federated systems'. *Global Environmental Politics* 15 (2): 121–41. https://doi.org/10.1162/GLEP_a_00301.

Green, Jessica F. 2021. 'Does Carbon Pricing Reduce Emissions? A review of ex-post analyses'. *Environmental Research Letters* 16 (4): 1–18. https://doi.org/10.1088/1748-9326/abdae9.

Harris, Kathleen. 2016. 'Justin Trudeau Gives Provinces Until 2018 to Adopt Carbon Price Plan'. *Canadian Broadcasting Corporation*, 3 October. www.cbc.ca/news/politics/canada-trudeau-climate-change-1.3788825.

Harris, Melissa, Marisa Beck, and Ivetta Gerasimchuk. 2015. *The End of Coal: Ontario's coal phase-out.* Winnipeg (MB): International Institute for Sustainable Development. www.iisd.org/system/files/publications/end-of-coal-ontario-coal-phase-out.pdf.

Harrison, Kathryn. 2010. 'The Struggle of Ideas and Self-Interest in Canadian Climate Policy'. In *Global Commons Domestic Decisions: The comparative politics of climate change*, eds. Kathryn Harrison and Lisa McIntosh Sundstrom, 169–200. Cambridge (MA): MIT. 10.7551/mitpress/9780262014267.001.0001.

Harrison, Kathryn. 1996a. *Passing the Buck: Federalism and Canadian environmental policy*. Vancouver: UBC Press.

Harrison, Kathryn. 1996b. 'The Regulator's Dilemma: Regulation of pulp mill effluents in the Canadian federal state'. *Canadian Journal of Political Science* 29 (3): 469–96. https://doi.org/10.1017/S0008423900008209.

Harrison, Kathryn. 2010. 'The Comparative Politics of Carbon Taxation'. *Annual Review of Law and Social Science* 6 (1): 507–29. https://doi.org/10.1146/annurev.lawsocsci.093008.131545.

Harrison, Kathryn. 2013. 'The Political Economy of British Columbia's Carbon Tax'. *OECD Environment Working Papers* 63. Paris: OECD. https://dx.doi.org/10.1787/5k3z04gkkhkg-en.

Harrison, Kathryn. 2020. 'Canada Finally Has a Climate Plan That Will Let It Meet Its Carbon Targets by 2030'. Canada: *Conversation*, 16 December. https://theconversation.com/canada-finally-has-a-climate-plan-that-will-let-it-meet-its-carbon-targets-by-2030-152133 (accessed 1 February 2021).

Hoberg, George, and Kathryn Harrison. 1994. 'It's Not Easy Being Green: The politics of Canada's Green Plan'. *Canadian Public Policy* 20 (2): 119–37. <www.jstor.org/stable/3552101>.

Hogg, Peter W. 2008. 'Constitutional Authority over Greenhouse Gas Emissions'. *Alberta Law Review* 46 (2): 507–20. https://doi.org/10.29173/alr234.

Houle, David, Erick Lachapelle, and Mark Purdon. 2015. 'Comparative Politics of Sub-Federal Cap-and-Trade: Implementing the western climate initiative'. *Global Environmental Politics* 15 (3): 49–73. https://doi.org/10.1162/GLEP_a_00311.

Howlett, Karen, and Greg Keenan. 2008. 'Deal Lets Ontario Join Climate-Change Drive'. *Globe and Mail*, 4 August.

Howlett, Karen, Brian Laghi, and Rhéal Séguin. 2008. 'Premiers Seize Climate-Change Initiative'. *Globe and Mail*, 19 July. www.theglobeandmail.com/news/national/premiers-seize-climate-change-initiative/article4221727.

Jaccard, Mark. 2020. *The Citizen's Guide to Climate Success*. Cambridge University Press. https://doi.org/10.1017/9781108783453.

Laroche, Jean. 2018. 'N.S. Unveils How Cap-and-Trade System will Affect Cost of Gas, Electricity'. *Canadian Broadcasting Corporation*, 23 October. www.cbc.ca/news/canada/nova-scotia/cap-and-trade-climate-change-environment-mcneil-1.4874444.

Macdonald, Douglas. 2020. *Carbon Province, Hydro Province: The challenge of Canadian energy and climate federalism*. University of Toronto Press.

Maclean's. 2019. 'Conservative Politicians Fill Up their Gas-Guzzlers'. *Macleans*, 1 April. www.macleans.ca/news/conservative-mpps-fill-up-their-massive-gas-guzzling-tanks-in-protest-of-the-carbon-tax.

McCarthy, Shawn. 2012. 'Ottawa Unveils New Coal-Fired Plant Emissions Rules'. *Globe and Mail*, 5 September. www.theglobeandmail.com/report-on-business/industry-news/energy-and-resources/ottawa-unveils-new-coal-fired-plant-emissions-rules/article4522237.

Mildenberger, Matto, Peter Howe, Erick Lachapelle, Leah Stokes, Jennifer R. Marlon, and Timothy Gravelle. 2016. 'The Distribution of Climate Change Public Opinion in Canada'. *PLoS ONE* 11 (8): 1–14. www.umontreal.ca/climat/engl/index.html.

Neimanis, V. P. [2011] 2013. '*Crown Land*'. In *The Canadian Encyclopedia*. Historica Canada. www.thecanadianencyclopedia.ca/en/article/crown-land.

NRC (Natural Resources Canada). 2019. *Canada's Changing Climate Report.* Canasa (NRC).
https://changingclimate.ca/CCCR2019.

NRC (Natural Resources Canada). 2020. *Crude Oil Facts*. Canada (NRC). www.nrcan.gc.ca/science-data/data-analysis/energy-data-analysis/energy-facts/crude-oil-facts/20064.

Olewiler, Nancy. 2006. 'Environmental Policy in Canada: Harmonized at the bottom?' In *Racing to the Bottom? Provincial interdependence in the Canadian Federation*, ed. Kathryn Harrison, 113–56. Vancouver: University of British Columbia Press.

Partington, P. J. 2013. 'A "Textbook Example" of Good Climate Policy, OECD Praises B.C.'s Carbon Tax'. *Pembina Institute* [blog], 21 October. www.pembina.org/blog/a-textbook-example-of-good-climate-policy-oecd-praises-b-c-s-carbon-tax.

Poitras, Jacques. 2020. 'A Made-in-New Brunswick Carbon Tax Is Here, but Will It Actually Curb Consumption?' *Canadian Broadcasting Corporation*, 1 April. www.cbc.ca/news/canada/new-brunswick/new-brunswick-carbon-tax-1.5516992.

Prime Minister of Canada, Justin Trudeau. 2016. 'Communiqué of Canada's First Ministers'. *Prime Minister of Canada, Justin Trudeau: News*, 3 March. https://pm.gc.ca/en/news/statements/2016/03/03/communique-canadas-first-ministers.

Rabson, Mia. 2020. 'Feds Reluctantly Approve Ontario and New Brunswick Carbon Prices for Big Emitters'. *CTV News*, 21 September. www.ctvnews.ca/politics/feds-reluctantly-approve-ontario-and-new-brunswick-carbon-prices-for-big-emitters-1.5113029.

Raymond, Leigh. 2016. *Reclaiming the Atmospheric Commons: The regional greenhouse gas initiative and a new model of emissions trading*. Cambridge, MA: MIT. https://doi.org/10.7551/mitpress/9780262034746.001.0001.

Ritchie, Hannah, and Max Roser. 2020. 'CO_2 and Greenhouse Gas Emissions'. *Our World in Data*. https://ourworldindata.org/greenhouse-gas-emissions.

Sawyer, D., S. Stiebert, R. Gignac, A. Campney, and D. Beugin. 2021. *2020 Expert Assessment of Carbon Pricing Systems*. Ottawa: Canadian Institute for Climate Choices.

Scharpf, Fritz W. 1988. 'The Joint-Decision Trap: Lessons from German federalism and European integration'. *Public Administration* 66 (3): 239–78. https://doi.org/10.1111/j.1467-9299.1988.tb00694.x.

Simpson, Jeffrey. 2009. 'Time for a Little Mea Culpa (Don't Blame the Editors)'. *Globe and Mail*, 2 January. www.theglobeandmail.com/opinion/time-for-a-little-mea-culpa-dont-blame-the-editors/article20439592.

Simpson, Jeffrey, Mark Jaccard, and Nic Rivers. 2008. *Hot Air : Meeting Canada's climate change challenge*. Toronto: McClelland & Stewart.

Statistics Canada. [2012] 2021. *Interprovincial and International Trade Flows, Basic Prices, Summary Level (x 1,000,000) [Data set]*. Canada (Statistics Canada). https://doi.org/10.25318/1210008801-eng.

Statistics Canada. [2015] 2021. *Air Conditioners [Data set]*. Canada (Statistics Canada). https://doi.org/10.25318/3810001901-eng.

Weaver, R Kent. 2020. 'Policy Dynamics in Federal Systems: A framework for analysis'. *Publius: The Journal of Federalism* 50 (2): 157–87. https://doi.org/10.1093/publius/pjaa003.

White, Marianne, and Lee Greenberg. 2008. 'Premiers Split on Cap-Trade System: Meeting Wraps Up'. *National Post*, 9 July, A4.

Woo, Andrea. 2021. 'Nearly 600 People Died in B.C. Summer Heat Wave, Vast Majority Seniors: Coroner'. *Globe and Mail.* 1 November, www.theglobeandmail.com/canada/article-nearly-600-people-died-in-bc-summer-heat-wave-vast-majority-seniors.

World Bank. 2017. 'Oil-Rich Alberta Finds Common Ground on Climate Action'. *World Bank: News*, 17 April. www.worldbank.org/en/news/feature/2017/04/17/how-oil-rich-alberta-became-a-global-climate-leader.

5

Climate Governance and Quasi-Federalism in China

HONGTAO YI AND SHUAI CAO

5.1 Introduction

Sufficient evidence has been accumulated in the past few decades to show that the global temperature is rising, and the climate is changing rapidly to an alarming degree (Hoegh-Guldberg 2018; Thuiller 2007). The practice of federalism has stood out, but with ambivalent effects, in climate governance (Karapin 2020). Despite being an authoritarian state, China has adopted a quasi-federal system to combat climate change, comprising five mechanisms: a target-responsibility system; the inclusion of environmental performance in local officials' promotion assessment; fiscal incentives; the central inspection system for ecological and environmental protection; and public participation. These are explained in detail in the case analysis of China's low-carbon pilot policy.

China's quasi-federalist system in climate governance features centralized decision-making, evaluation, and supervision, together with decentralized implementation. This has shown evidence of environmental federalism in experimenting with innovative solutions to climate crisis and promoting policy diffusion, but it has also experienced some challenges as demonstrated in fragmented patchwork of policies at the local level of government.

Overall, China's environmental quasi-federalism of devolved implementation under centralized policymaking has generally been effective in climate governance even though it has some drawbacks.

5.2 Climate Change in China

China has achieved phenomenal economic growth in the past few decades, rising from being an impoverished country to becoming the second-largest economy in the world (Ross 2019). However, China's economic achievement was accomplished at the cost of damaging the environment to an alarming degree (Chow 2015).

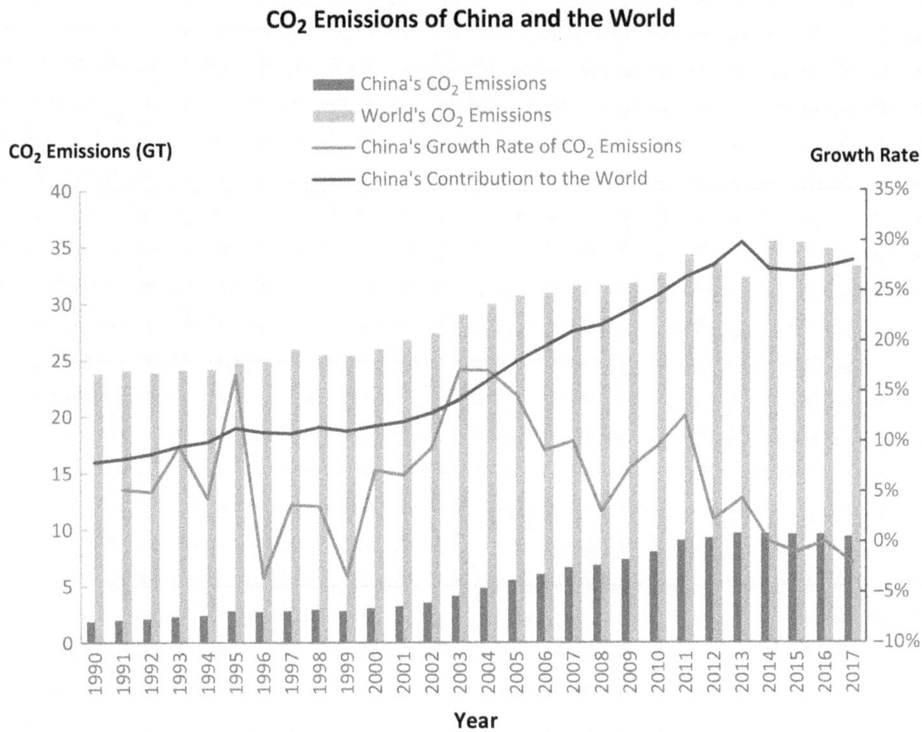

Figure 5.1 Annual CO_2 emissions of China and the world.
Source: CAIT Climate Data Explorer & International Energy Agency.

In 2006, China replaced the United States as the largest emitter of annual carbon dioxide, and it has remained at the top since then. According to the latest available data compiled by the International Energy Agency (IEA 2019) and CAIT Climate Data Explorer,[1] China's annual CO_2 emissions have been on the rise during the period of 1990 to 2017, starting from 1.9 gigatons (GT) annually which accounted for approximately 8 per cent of world's total in 1990 to 9.3 GT annually which contributed to 28 per cent of global emissions in 2017 (see Figure 5.1). Emissions growth continues, but at a slower rate.

There are three main GHG sources: energy, industry, and agriculture. In general, energy-related fossil fuel combustion is the primary driver of anthropogenic CO_2 emissions in most countries (IPCC 2014). In China, this is especially the case. China's energy-related CO_2 emissions (emissions from coal, oil, and natural gas) have long been the dominant portion of its total CO_2 emissions and have remained at over 90 per cent level from 1990 to 2016 (see Figure 5.2).

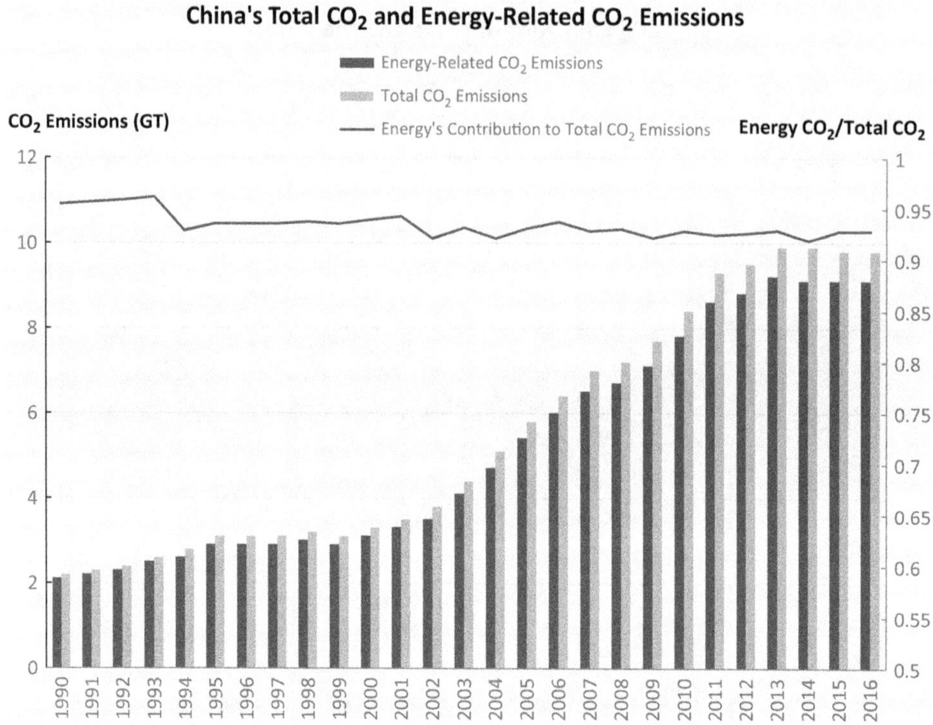

Figure 5.2 China's total CO_2 and energy-related CO_2 emissions.
Source: CAIT Climate Data Explorer.

This high percentage is inextricably linked to China's energy structure, where coal has long occupied over 60 per cent of all energy consumption compared to the global average of 27.8 per cent, and contributed even more to energy-related CO_2 emissions (Korsbakken, Andrew, and Peters 2019) As Figure 5.3 shows, coal-induced CO_2 has long been contributing close to or even more than 70 per cent of total fossil-fuel CO_2 emissions in China.

Constantly growing GHG emissions will continue to exacerbate global climate change, the effects of which are far-reaching in China. Annual mean air temperature over the past thirty years has risen by over 1.0°C, which is higher than the synchronous global average (Fang et al. 2018). To be more specific, northern China is warming faster than southern China (Ding et al. 2007). The most evident effect of global warming is melting glaciers. The glacier volume in the Qilian Mountains of north-western China decreased by 30 per cent ± 8 per cent from 1956 to 2010 (Tian et al. 2014). The glacier retreat has led to rising sea levels, which is alarming because China has a heavily populated 18,000-kilometre eastern coastline where many of the most economically prosperous cities are located. The occurrence of climate-related extreme weather events, such as droughts and floods,

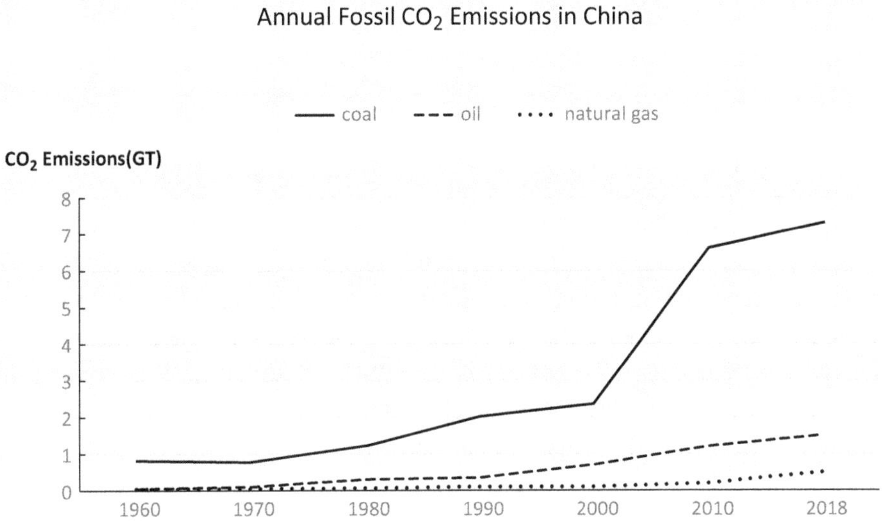

Figure 5.3 Annual fossil CO_2 emissions in China.
Source: CDIAC/UNFCCC/BP/USGS.

has increased unevenly in the north and south of China, resulting in severe damage to the stability and quality of food production, undermining China's ability to feed its people[2] (Piao et al. 2010). Socio-economic conditions including public health and sustainable economic development have also been affected by adverse climate change (Watts et al. 2015).

China has long been confronted with a series of stubborn climate adaptation challenges, among which a varied climate, scarce natural resources, an extensive economic growth model , and a large population are the main stumbling blocks (Nan and Jingyang 2014). With a vast territory, China has complex climate patterns and interactions, making climate adaptation more difficult. Insufficient natural resources coupled with a large population are discouraging adaptation action. The coal-dominant energy structure and low energy efficiency stand in the way as roadblocks to sustainable economic development (Dai and Finkelman 2020).

Despite the aforementioned challenges, China has been cooperating with the international community to combat climate change. In the COP26 summit in Glasgow in 2021, China promised to achieve carbon peak before 2030 and carbon neutrality before 2060. Under the Paris Agreement reached in 2015, China committed to peak its CO_2 emissions around 2030 and lift non-fossil energy to 20 per cent by 2030. According to a study, China is likely to achieve this goal if all current policies are effectively implemented (Gallagher et al. 2019). China's most recent pledge was to 'endeavour to reach carbon neutrality by 2060', which was considered a shockingly ambitious goal especially when the Covid-19 is still wreaking havoc on people's lives and economy in China and worldwide.[3]

5.3 Climate Change and Quasi-Federalism in China

The impact of federalism on climate governance has been widely studied (e.g., Austin et al. 2018; Balthasar, Schreurs, and Varone 2020; Jordaan et al. 2019). Despite being an authoritarian state, China adopted a quasi-federal system to mitigate and adapt to climate change with mixed measures and mechanisms. What underpins the quasi-federalist system is a multi-level structure with vertical and horizontal power dynamics. Five mechanisms embedded in this system are identified, which we will discuss further below.

5.3.1 China's Quasi-Federalism: Structures and Division of Powers

Federalism has three essential features, though it has various definitions. First, it requires a division of powers between the central (or federal) governments and regional (or constituent) governments. Second, the two levels of government are of equal status (Wheare 1946). Third, the division of powers and equal status are guaranteed by a written constitution (Elazar 1987).

China is a unitary state where the central government has supreme authority over local governments. China thus does not satisfy the 'equal status' rule of federalism. However, there is a constitutionally prescribed division of powers between the central government and local governments. Its governance structure is usually explained by the analytical framework of *tiao-kuai* (Lieberthal and Oksenberg 1988; Schurmann 1968). *Tiao* refers to the vertical–functional relationship between the central and local governments, while *kuai* refers to the horizontal–territorial relationship at different levels of government (Lieberthal 1997). Although the central government maps out strategies and issues directives to local governments, local governments largely control the resources, staff, and information needed to implement the directives. In environmental governance, *tiao* and *kuai* usually underlie the mechanisms for power sharing and bargaining (Alkon and Wong 2018). Therefore, albeit a *de jure* unitary state, China operates under a de facto federal-like system (Zheng 2007).

Different from analysing the environmental federalism in federal states, China is a party-state where the Chinese Communist Party (CCP) controls virtually all the other political organizations and institutions mainly through dominating the selection of cadre leaders.

The structures and division of powers can be illustrated by *tiao* (vertical) and *kuai* (horizontal) lines as shown in Figure 5.4. Vertically, there are five orders of organizations. The highest rank of the party committee, government, and environmental and ecological agencys have constitutional authority over the lower ranks of corresponding agencies separately. In other words, the local organizations are the 'branches' of the central organizations and therefore should be responsible to them. It should be noted that this is different from the federal structure of the United States where the federal government and state governments do not follow a superior–subordinate relationship. Horizontally, the party leads the

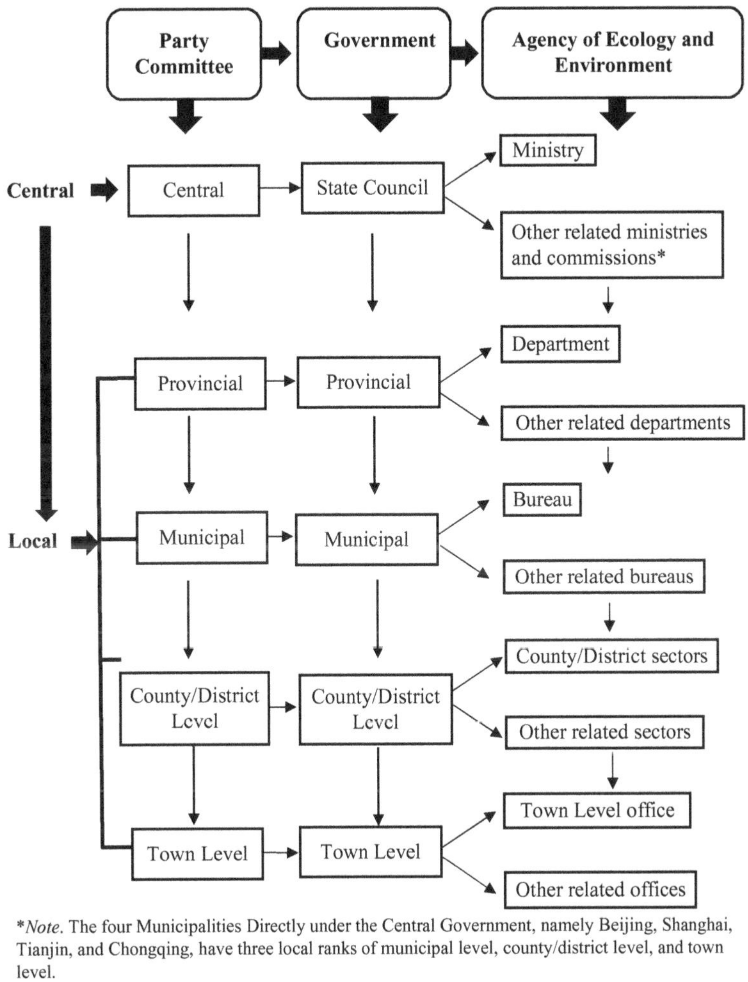

Figure 5.4 Structures of environmental federalism in China.
Note. The four municipalities directly under the Central Government, namely Beijing, Shanghai, Tianjin, and Chongqing, have three local ranks of municipal level, county/district level, and town level.

government which in turn leads its subordinate agencies. For instance, at the highest rank, the Central Party Committee headed by the general secretary leads the State Council steered by the prime minister, which in turn governs the Ministry of Ecology and Environment. It should be noted that the prime minister and the minister of ecology and environment and other ministers are also members of the Central Party Committee. At the nexus of *tiao* and *kuai* exists overlapping authority. For example, the provincial environmental agency is subordinate to the central environmental agency and meanwhile affiliated to the provincial government. Besides, a power shift from the National Development and Reform Commission (NDRC) to the Ministry of Ecology and Environment (MEE) in steering climate governance has strengthened the latter's authority in taking tougher measures to curb climate change. NDRC is an affiliated commission of the State Council. It is often called the 'small State Council' due to its high status as a comprehensive agency for national strategic planning and economic management. It had overseen GHG emissions reduction until MEE took over in 2018. This power shift, decided by the CCP Central Committee and State Council out of consideration for balancing economic growth and climate governance, authorized more power to MEE to implement climate change policies. Over time, the power shift at the highest level has also been transmitted to lower levels, strongly empowering local environmental agencies to tackle urgent environmental problems.

The central government has constitutional authority over, and is responsible for negotiating international commitments and designing national policies, for climate mitigation and adaptation (local governments are not allowed to conduct para-diplomacy), while local governments are mainly responsible for implementing these policies although they enjoy a certain degree of flexibility in formulating locally suited policies.

5.3.2 How the Quasi-Federalist System Works

As is shown in Figure 5.4, what underpins the quasi-federalist system is a multi-level administrative structure that functions through hierarchical central–local governments, party–government power structure, and dual leadership that local-level environmental agencies receive.

In the quasi-federalist system, the central government and local government (including four levels, namely province, municipality, county, and town) do not have equal status. The former has supreme authority and power to formulate national strategies and goals while the latter has little power in national-level decision-making although they have much discretion in the implementation of national policies. This system creates a top-down policy process where the central government defines problems, sets the agenda, formulates national policies, and evaluates and supervises policy implementation while local governments mainly implement national policies. For climate policies, the central government sets targets and assign them to provinces which then break their quotas assigned down to lower levels of local governments. Meanwhile, local governments at each level have significant discretionary power during policy implementation, though the degree of discretion varies in response to different socio-economic conditions of local governments, the urgency of environmental issues, and social norms (Chŏng and Chung 2000; Shin 2017).

Another distinctive feature of the multi-level governance structure is the party–government power structure. At each level, party institutions co-work with government institutions, and the party leader is the most powerful figure, superior to the government leader who is the second most powerful figure. In general, the party committee, at each level, is mainly responsible for making major decisions and the government is mainly in charge of implementation. However, the boundary between party and government is blurred in reality because the party committee has the power to promote or demote officials working in both same-level party institutions and government institutions, and the government leader also serves as the deputy party leader. To understand the climate policies in China, it is necessary to take the party–government power dynamics into consideration.

Local environmental agencies receive dual guidance and supervision. They are not only horizontally responsible for the same-level local governments but also vertically responsible for their immediate higher level environmental agencies (Ma 2017). However, the dual leadership is often at odds with each other in implementing environmental policies because environmental protection and economic development are often at odds with each other (Zhang 2021). The vertical leadership tends to impose strict directives on, and strengthen oversight over, lower levels of environmental agencies while the horizontal leadership is likely to loosen supervision over same-level environmental agencies for the sake of local economic growth. To take tougher measures against climate change, the central party committee and central government proposed strengthening the vertical leadership to counterbalance local interests in 2015.

5.3.3 Mechanisms of China's Quasi-Federalism

How does China's quasi-federalism work? We introduce five main mechanisms here briefly, which will be detailed in the case study section.

5.3.3.1 Target-Responsibility System

The target-responsibility system, which was imported into the Chinese bureaucracy in the 1980s to boost economic development and introduced to environmental management soon afterward, is an adapted, Chinese version of Management by Objectives (Lan and Hu 2008).

As its name implies, the central organization sets targets and assign the sub-targets to corresponding local organizations whose leaders will be rewarded if the targets are achieved, and punished otherwise. In China, the central government is responsible for making a comprehensive work plan for energy conservation and emission reduction, where targets are set and allocated to provincial governments, which will further divide the targets and allocate them to lower levels of government.[4]

5.3.3.2 Inclusion of Environmental Performance in Local Officials' Promotion

Since the reform and opening-up in 1978, economic performance has been an overwhelming factor in local officials' career advancement until the second half of the 2000s when China realized the severity of environmental degradation. Since 2007, environmental performance has been incorporated in the evaluation of local officials' political achievements.

5.3.3.3 Fiscal Incentives

Roughly speaking, China's tax system since 1949 has gone through three phases. The first phase was the period of the planned economy from 1949 to 1978 when taxation was highly centralized. After that much of the power of levying taxes was devolved from the central government to local governments to boost the economy from 1978 to 1994, eventually leading to central government's financial deficiency. Therefore, in 1994, taxation was centralized again, and this system has remained in place to the present day.

To create an incentive for local governments to combat climate change, the *Environmental Protection Tax Law* was enacted 2016, requiring all environmental taxes to be allocated to local governments starting from 2018. In general, this has increased local governments' investment for environmental protection and reduced CO_2 emissions, even though environmental taxes have traditionally played a limited role in environmental regulation (Li et al. 2021).

5.3.3.4 Central Inspection System for Ecological and Environmental Protection

To ensure that the national strategy to combat ecological and environmental deterioration was implemented as required by local governments, the State Council of China created the Central Inspection System for Ecological and Environmental Protection (CISEEP) in 2016 and the Ministry of Ecology and Environment is responsible for the nationwide inspection.

CISEEP prescribes that an inspection task force should be approved by the central party committee and government to supervise and scrutinize what provincial governments have done in environmental and ecological governance.

5.3.3.5 Public Participation

The issues of climate change are so complex that the central government needs to resort to public participation to improve the effectiveness in climate governance.

Measures for Public Participation in Environmental Protection was approved by the then-called Ministry of Environmental Protection.[5] This document stipulates that citizens can report environmental issues by way of writing letters or emails to environmental agencies or calling the environmental hotline. Local environmental agencies are also encouraged to engage social organizations in environmental protection (climate change included) through project grants or purchasing services.

In summary, although these five mechanisms mentioned above were not exclusively created for addressing climate change, they have been playing an essential role in China's quasi-federalist system for climate governance.

5.4 Case Study

In this section, we analyse a case of climate mitigation in China: the pilot low-carbon city initiative.

5.4.1 An Analytical Framework

One distinctive feature of China's environmental quasi-federalism is its selective centralization or decentralization in the policy process between the central government and local governments. It is also worth noting that it is the central government that determines when and where to centralize or decentralize – which to a certain degree demonstrates the 'strategic pragmatism' in China's policy process (Gallagher and Xuan 2019). Centralization here means that the central government has dominant power in making major decisions while decentralization means that the central government devolves much of its power to local governments in the implementation of climate and environmental policies. The five mechanisms are fitted into the different stages of the policy process. The central government has absolute authority over agenda setting. It also has a dominant power in formulating and adopting policies where the target-responsibility system is brought in to establish specific targets that local governments must achieve. However, the implementation is decentralized mainly because local governments are far more familiar with local conditions and are more capable of deploying resources available to accomplish climate mitigation/adaptation tasks. Fiscal incentives and local officials' promotion pressure come in to ease off the GDP-oriented development pattern. Public participation is also promoted. But in the last step, centralization is back again. A central inspection

Table 5.1 *Analytical framework for case studies of Chinese environmental quasi-federalism*

Policy Process	Selective Centralization or Decentralization	Mechanisms
Agenda Setting	Highly Centralized	
Formulation and Adoption	Centralized	target-responsibility system (mainly the 'target' part)
Implementation	Decentralized	1) fiscal incentives 2) incorporation of environmental performance in local officials' promotion 3) public participation
Supervision and Evaluation	Centralized	1) central inspection system for ecological and environmental protection 2) target-responsibility system (mainly the 'responsibility' part)

brigade heads for provinces to supervise and evaluate local environmental performance which will serve as the decisive criterion to judge whether the assigned targets are met (see Table 5.1). In general, major decision-making and evaluation are centralized and implementation is decentralized.

It should be clarified that each mechanism that is fitted to each step of the policy process is not exclusively functioning in that step. Instead, it might straddle two or more stages. For example, although the target-responsibility system is introduced in formulating and adopting the low-carbon policy, it also imposes pressure on local governments' implementation and serves as criteria for policy evaluation.

We will employ this analytical framework to analyse the implementation, evolution, and conflicts/cooperation between the central government and local governments in the following two cases.

5.4.2 Case of Climate Mitigation: Pilot Low-Carbon City Initiative

To curb adverse climate change and promote green development, China launched an incremental and massive pilot low-carbon city policy in 2010.

Up to now, a total number of eighty-seven local governments at different levels throughout China have joined the pilot low-carbon policy experimentation since three rounds of pilot policy were implemented respectively in 2010, 2012, and 2017[6] (see Table 5.2). In the past ten years an increasing number of jurisdictions have adopted low-carbon policy from east to west, scattered to nationwide, and provincial-level governments down to county/district-level governments.

Table 5.2 *Three rounds of pilot low-carbon initiatives*

Starting Time	Leading Central Organization	Number of Local Governments
July 2010	National Development and Reform Commission	13, including 5 provinces and 8 cities
December 2012	National Development and Reform Commission	29, including 2 municipalities directly under the central government (provincial level), 1 province, and 26 cities
January 2017	National Development and Reform Commission	45, including 41 cities and 4 counties/districts

Note. As early as in 2008, Shanghai and Baoding joined a World Wildlife Fund initiative to explore low-carbon urban development.

5.4.2.1 Agenda Setting

The agenda setting of the pilot low-carbon initiative is highly centralized. Although local governments have the right to make suggestions, the central government has the final say on what issues should be prioritized.

In November 2009, the State Council proposed the general target for cutting GHG emissions. Then, NDRC rolled out the pilot low-carbon policy. In the second and third rounds, it was also the NDRC that steered this policy. However, in 2018, MEE was authorized more power and began to dominate the climate change policy. Facing the next five years, MEE is finalizing the guidelines on reaching carbon emissions peak by 2030 in the fourteenth Five-Year Plan, realizing carbon neutrality by 2060, making specialized plans for combating climate change, and accelerating the national carbon trading market.[7]

This kind of centralized agenda setting is reasonable for the urgent battle against climate change in that it can expedite the formulation, adoption, and implementation of climate policies without consuming too much time to negotiate a consensus.

5.4.2.2 Formulation and Adoption of Low-Carbon Policy

The formulation and adoption of the low-carbon policy are also centralized, although there exist much bargaining and negotiation between central and local governments in this process.

Based on suggestions from local governments, the NDRC formulated and adopted the pilot low-carbon city initiative through internal discussion and research, and then issued official written policy documents to local governments.

The central government's centralized power in formulating and adopting the low-carbon policy is mainly reflected in the following four aspects.

First, the NDRC has the final say over which provinces or cities should join the pilot low-carbon initiative although local socio-economic conditions will be considered. Second, the NDRC sets the general target and requires that local governments incorporate tackling climate change into their Five-Year-Plan, set their targets and formulate specific measures for reducing GHG emissions, by considering local natural conditions, resources endowments, and economic foundation. Third, the pilot local governments are required to apply information technology (e.g., establish an integrated system for GHG emissions statistics, monitoring and accounting) to track the pace of curbing CO_2 emissions and to accumulate evidence for local energy conservation and emission reduction policy design. The last aspect lies in NDRC's power to evaluate policy implementation and choose and promote successful cases. All of these are written into government documents.[8]

Supervision of local government has also been reinforced over time. In 2010, when the first batch of local governments was selected to implement the low-carbon policy, the NDRC took a mild and open attitude and did not force local officials to be held accountable if the local GHG emission reduction targets were not met. Yet in 2012, when the second batch of pilot local governments was chosen, the NDRC introduced the target-responsibility system to the national low-carbon pilot policy, which was reformed and intensified in 2017.

The target-responsibility system is generally effective in allocating GHG emission reduction targets to local governments and imposing a certain degree of pressure on local officials to take action to promote low-carbon experimentation. These mandatory targets are all-important because they provide relatively clear goals that local governments are pressured to achieve. Yet they are not always effective due to unreasonable target allocation, the intricate nature of the targets, distraction from parallel programmes, and unreasonable choice of indicators.

Initially, the national target for reducing GHG emissions was broken down and allocated almost equally to each province regardless of the significant differences between them in energy consumption, industrial structure, resource endowments, and technological level. This inevitably twisted the target distribution system. This problem has been gradually alleviated with the constant promotion of a national carbon emission trading system. These targets are also intricate in that some target indicators are unquantifiable, with most of the quantifiable indicators allocated to related sectors by local governments (e.g., industry, energy, building, and transportation) in a clear-cut way. Another issue is that pilot provinces and cities are faced with parallel programmes at the same time and thus confused with and distracted by the overlapping goals of these programmes[9] (Lo 2014). In such situations, the target-responsibility system would fall short of expectations. Indicators of target matter too. Initially, energy intensity was taken as the key indicator. However, energy consumption could still grow even if the energy intensity declined when the economy grew at a faster rate (Lo 2020). To remedy this situation, the indicator of energy consumption was introduced in 2016 at the outset of China's thirteenth Five-Year-Plan.

5.4.2.3 Implementation

The implementation of the low-carbon pilot city policy is decentralized. The NDRC does not provide specific guidance or methods for low-carbon development. Local governments are granted a relatively high degree of discretion over experimenting with locally suited policies to reduce GHG emissions. In policy implementation, local governments are so diverse in population, industrial structure, and energy structure, that they take on different patterns of climate governance (Yi and Liu 2015). The strength of local governments' climate actions is determined by a mix of factors such as governance costs, local leaders' career advancement, collaboration among horizontal sectors, and public opinion. There are pioneers as well as laggards in climate governance innovation. To provide incentives for local governments to address climate change, the central government has established several mechanisms.

Climate governance is undoubtedly a high-cost project. Since the tax-sharing reform in 1994, China's taxation has remained centralized. The central government has more tax revenue and less expenditure responsibility while local governments have less tax revenue and more expenditure responsibility. Additionally, local governments largely control the personnel and finance of their environmental protection agencies. Therefore, local governments tend to budget tightly for climate mitigation and adaptation, and local environmental protection agencies are often constrained by limited finance even though they might have ambitious plans. To ease off this financial predicament, the *Environmental Protection Tax Law* was enacted in 2016 and all the environmental taxes have been distributed to local tax revenue since 1 January 2018, a major decision made by the State Council.

GDP-dominant political achievement for local officials' promotion is another obstacle against pushing climate governance forward. Since 1978, China's provincial leaders' promotion has been highly linked with local economic performance (Li and Zhou 2005). Economic growth has long taken priority over environmental protection. To reverse this path dependence, the central government stipulated that local environmental performance – especially energy conservation and GHG emissions – would be incorporated into the assessment of the political achievements of local officials.

The collaboration among different sectors in local governments has long been a Gordian knot. The building of low-carbon cities is so complex that it entails horizontal cross-sector collaboration, ranging from the economic sector to the environmental sector, from the transportation sector to the energy sector, and from the natural resources sector to the agricultural sector. Horizontal cross-sector collaboration is often difficult because local governments have the same administrative rank, each with different or even conflicting policy objectives. To lubricate and facilitate collaboration, the central government devolved much power to local government leaders, local development and reform commissions, and local environmental agencies.

A low-carbon lifestyle is also an important policy goal. As mentioned above, the complexity of climate governance necessitates strengthening public participation (Liu and Zhang 2012). In official documents, the central government reiterates that local governments should publicize low-carbon development, increase data transparency, and engage citizens in climate governance by establishing diverse channels. Although the overall influence of public participation in combating climate change remains to be seen, empirical research shows that public participation has a positive effect on pushing enterprises to comply with green development policies (Fu and Geng 2019).

These mechanisms have spawned many policy pioneers. Up to now, at least thirty-three pilot provinces and cities have formulated specialized plans for low-carbon development. More than thirteen pilot local governments have formulated specialized plans to address climate change. A total of thirty-seven pilot provinces and cities have announced preliminary targets to peak GHG emissions.[10] The pioneering provinces and cities have taken diverse and innovative approaches to driving low-carbon development, ranging from institutional innovation to the establishment of an information management platform and to market-based measures (see Table 5.3). Their pioneering work captured the attention of the central government, which in turn set up exemplary provinces or cities to promote their experience, a move not only to share and promote successful experience but also to exert pressure on those provinces or cities that did not produce satisfactory climate governance performance.

Yet these mechanisms are not always effective. Policy implementation has also been plagued with laggards. The implementation gap has long been present. The gap is embodied in many forms and can be explained under different aspects, but the root lies in the conflict between local economic interests and national goals for climate governance – a classical intergovernmental dilemma.

Fiscal incentives are not working well in some local governments implementing the low-carbon pilot policy. The belief that economic growth is the top priority has been ingrained in the mental models of some local leaders who do not give due

Table 5.3 *Pioneering low-carbon provinces and cities and their innovative measures*

Pilot Low-Carbon Provinces/ Cities	Innovative Measures
Zhenjiang City	Created dual leadership and accountability (party leader and government leader); established a low-carbon management platform
Guangyuan City	Set up a Bureau of Low-Carbon Development
Yunnan Province	Incorporated low-carbon development into local medium- and long-term socio-economic plan
Shenzhen City	Explored the establishment of a carbon trading market
Shanghai	Promoted low-carbon transportation and designed eco-friendly streets
Beijing	Established cross-district carbon trading system
Chengdu City	Promoted low-carbon transportation
Wuhan City	The first city to have announced the action plan to peak GHG emissions; sought international cooperation
Guiyang City	Promoted circular economy
Suzhou City	Created carbon inventory for local enterprises
Hangzhou City	Established a carbon emission platform for supervision and decision-making
Jinchen City	Reduced CO_2 emissions by gradually replacing boiler combustion with coalbed methane combustion
Guangdong Province	Launched certification of low-carbon products
Chongqing	Launched certification of low-carbon products

weight to ecological and environmental protection. Besides, fiscal incentives are not strong enough to push energy-intensive provinces and cities, (which are predisposed to consume more coal), to decisively cut their CO_2 emissions.

Climate governance performance does not have a substantive effect on local leaders' promotion or demotion, which is mainly determined by local economic growth, the contribution of revenue to the central government, and political considerations (Bo [2002] 2019). Few local leaders have been demoted or prevented from being promoted simply because they did not perform well in achieving climate governance goals. Therefore, local leaders have a relatively large degree of discretion over how many resources will be distributed to reduce CO_2 emissions.

Functional collaboration among different local government departments has remained complex and intricate although local development and reform commissions and environmental departments have been delegated more power (Westman and Broto 2018). For example, some local finance departments were often delayed in appropriating money for inspecting local enterprises' measures to reduce GHG emissions. The openness and sharing of climate governance-related information have not fared well either because some departments have tried to keep their core information to themselves for political or economic interests.

5.4.2.4 Supervision and Evaluation

The fundamental challenge in supervision and evaluation of climate governance is information asymmetry.

Although local governments are required to report work progress to the central government regularly, they usually present achievements but cover up problems. Local governments are also inclined to falsify data to circumvent punishment from the central government (Kostka and Nahm 2017). Therefore, the central government adopts a centralized top-down mechanism of supervision and evaluation.

The Ministry of Ecology and Environment has been supervising and evaluating local low-carbon experiments. 'Soft' and 'hard' measures have both been taken. The former includes creating model low-carbon provinces or cities, and encouraging media, social organizations, and the public to report ecological and environmental issues. The latter is exemplified by the Central Inspection System for Ecological and Environmental Protection (CISEEP) which was inaugurated by the Central Committee and the State Council in January 2016.

CISEEP is an iron-handed top-down inspection system for a comprehensive list of ecological and environmental issues, energy conservation and GHG emissions reduction included. A central inspection workforce would be formed and march into different provinces and state-owned enterprises, and stationed there for a length of time to supervise and evaluate the implementation of national ecological and environmental policies. Up to now, two rounds of inspection have been carried out, respectively in 2016 and 2019. The main task of the central inspection workforce is to find out the problems, penalize the organizations or people involved, and keep track of their remedy measures until the problems are resolved. For example, in the second round of central inspection, serious problems of China Minmetals Group were exposed. One of them was that although the China Minmetals Group included its targets for saving energy and reducing GHG emissions in its development plan for 2019 to 2021, it did not take specific action to achieve the goals. The central inspection team confirmed these problems and urged the China Minmetals Group to make improvement plans which were required to be open to the public for supervision, and whose implementation would receive a follow-up central inspection.

Some empirical research shows that CISEEP is generally effective in pushing local governments to achieve ecological and environmental targets (Jia and Chen 2019; Li et al. 2020). Inequity also occurs due to 'insufficient differentiation based on economic and capacity criteria' (Kostka and Goron 2021).

The low-carbon policy experiments are still ongoing in China. It is arbitrary to take a stand on whether they are successful or not. China's quasi-environmental federalism works out in some respects but fails in others (Cheng et al. 2019; Khan 2013; Lo 2014; Lo, Li, and Chen 2020; Wang et al. 2015). In general, the low-carbon pilot policy has improved the overall GHG emission efficiency of pilot cities though it might take a longer time to achieve the goal, and it might widen the divide between eastern and western regions because the policy has been more effective in eastern pilot cities (Fu, He, and Luo 2021). It seems that its strengths outweigh its weaknesses, but more empirical research should be conducted to reach a more rigorous conclusion.

5.5 Conclusion

China's practice in climate governance partly echoes the 'laboratory of federalism'. In general, a preliminary conclusion can be reached that China's environmental federalism, which features centralized decision-making, evaluation, and supervision, and decentralized implementation, is effective in climate governance even though it falls short in some respects. Since China's practices in climate governance are still ongoing, more empirical research should be done to reach a more rigorous and fine-grained conclusion.

China's quasi-federalist system in climate governance is to a large degree successful in facilitating major decision-making without consuming too much time to reach a consensus, incentivizing local officials to promote low-carbon policies, and exerting centralized and uniform supervision and evaluation of local policies for addressing climate change. Devolved implementation is conducive to experimenting with innovative solutions to climate change.

However, it is also fraught with disadvantages. Inappropriate choice of indicators in the target-responsibility system is likely to skew incentives. Decentralized implementation might also strengthen local governments' tendency to prioritize local economic development over environmental protection. Public participation is generally weak. Local governments are inclined to conceal, manipulate, and falsify data related to environmental quality. Decentralized implementation might fare better when it is combined with centralized policymaking, evaluation, and supervision.

Institutions, ideas, and interests have played a significant role in shaping the relationship between federalism and climate governance in China. The institutional capacity of central and local governments, along with the five mechanisms, constitutes the federal structure in climate governance. The central government's knowledge of science and strategic planning, local governments' expertise, and citizens' awareness of environmental protection help build a growing consensus on fighting against climate change. However, the conflict in policy goals between the national target and local economic growth often hinders the smooth implementation of climate change policies.

Key implications for policymakers can be summarized under four aspects.

First, the specific indicators of the target-responsibility system should be comprehensive, reasonable, and clear. Experts and specialists rather than administrative officers should have the right to determine the whole list of indicators for policy evaluation.

Second, more financial incentives should be given to local governments to cope with climate change. This might include allocating more financial power to local governments and environmental agencies, outsourcing environmental protection projects to a larger degree, and fostering the environmental protection industry.

Third, the public should be included in the policy process to help define problems and supervise implementation. This might, to some degree, reduce the tensions between the central government and local governments.

The last aspect is that the central government should develop and apply more technologies to alleviate the falsification of environment-related data. Examples of this might include setting up real-time monitoring stations and establishing a unified data platform.

Notes

1 The raw data can be accessed from the official website of CAIT Climate Data Explorer www.climatewatchdata.org/ghg-emissions.

2 The rising temperature has led to fewer water resources in northern China and more in southern China.

3 This pledge was made by President Xi Jinping at the 75th Session of the United Nations General Assembly on 22 September 2020.

4 The State Council, Chinese central government, made the first comprehensive plan for energy conservation and emission in 2007, one year after the eleventh Five-Year-Plan of China was announced. Since then, the State Council made two comprehensive plans for energy conservation and emission, respectively in 2011 when the twelfth Five-Year-Plan was launched and 2016 when the thirteenth Five-Year-Plan was unveiled.

5 It was restructured and renamed as the Ministry of Ecology and Environment on 16 April 2018.

6 Local governments which adopted the pilot low-carbon policy include provincial governments, municipalities directly under the central government, city-level governments, and county/district-level governments. More details are offered in Table 5.3.

7 It is from the website of the Ministry of Ecology and Environment of the People's Republic of China. www.mee.gov.cn/ywdt/szyw/202010/t20201013_803022.shtml.

8 The main government document for the pilot low-carbon initiative is the Circular of the National Development and Reform Commission on Carrying out Pilot Projects in Low-carbon Provinces and Cities, respectively in 2010, 2012, and 2017.

9 There are domestic and international parallel programmes. The former includes Eco-City Programme launched by then Ministry of Environmental Protection (now Ministry of Ecology and Environment) and Eco-Garden Programme initiated by the Ministry of Housing and Urban–Rural Development. The latter includes pilot programmes funded by international organizations

and institutions such as the Rockefeller Brothers Fund, Worldwide Fund for Nature, and the United Kingdom Strategic Programme Fund.

10 The statistics are from the Investigation and Summary Report on Pilot Low-Carbon Initiative (2017). Retrieved from www.ncsc.org.cn/yjcg/dybg/201804/P020180920509262040412.pdf.

References

Alkon, Meir, and Audrye Wong. 2018. 'Authoritarian Environmental Federalism'. Available at SSRN 2957002.

Austin, Stephanie E., James D. Ford, Lea Berrang-Ford, Robbert Biesbroek, Jale Tosun, and Nancy A. Ross. 2018. 'Intergovernmental Relations for Public Health Adaptation to Climate Change in the Federalist States of Canada and Germany'. *Global Environmental Change* 52: 226–37. doi: 10.1016/j.gloenvcha.2018.07.010.

Balthasar, Andreas, Miranda A. Schreurs, and Frédéric Varone. 2020. 'Energy Transition in Europe and the United States: Policy entrepreneurs and veto players in federalist systems'. *Journal of Environment & Development* 29 (1): 3–25. doi: 10.1177/1070496519887489.

Bo, Zhiyue. [2002] 2019. *Chinese Provincial Leaders: Economic performance and political mobility since 1949*. Abingdon: Routledge. doi: 10.4324/9781315095721.

Cheng, Jinhua, Jiahui Yi, Sheng Dai, and Yan Xiong. 2019. 'Can Low-Carbon City Construction Facilitate Green Growth? Evidence from China's pilot low-carbon city initiative'. *Journal of Cleaner Production* 231: 1158–70. doi: 10.1016/j.jclepro.2019.05.327.

Chŏng, Chae-ho, and Jae Ho Chung. 2000. *Central Control and Local Discretion in China: Leadership and implementation during post-Mao decollectivization*. New York: Oxford University Press.

Chow, Gregory C. 2015. *China's Economic Transformation*, 3rd ed. Chichester: Wiley.

Dai, Shifeng, and Robert B. Finkelman, eds. 2020. *Coal Geology of China*, 1st ed. Abingdon: Routledge.

Ding, Yihui, Guoyu Ren, and Zongci Zhao, et al., 2007. 'Detection, Causes and Projection of Climate Change over China: An overview of recent progress'. *Advances in Atmospheric Sciences* 24 (6): 954–71. doi: 10.1007/s00376-007-0954-4.

Dorsch, Marcel J., and Christian Flachsland. 2017. 'A Polycentric Approach to Global Climate Governance'. *Global Environmental Politics* 17 (2): 45–64. doi: 10.1162/GLEP_a_00400.

Elazar, Daniel J. 1987. *Exploring Federalism*. Tuscaloosa: University of Alabama Press.

Fang, Jingyun, Yu Guirui, Lingli Liu, Hu Shuijin, and F. Stuart Chapin. 2018. 'Climate Change, Human Impacts, and Carbon Sequestration in China'. *Proceedings of the National Academy of Sciences of the United States of America* 115 (16): 4015–20. doi: 10.1073/pnas.1700304115.

Fu, Jingyan, and Yanyun Geng. 2019. 'Public Participation, Regulatory Compliance and Green Development in China Based on Provincial Panel Data'. *Journal of Cleaner Production* 230: 1344–53. doi: 10.1016/j.jclepro.2019.05.093.

Fu, Yang, Chenyang He, and Ling Luo. 2021. 'Does the Low-Carbon City Policy Make a Difference? Empirical evidence of the pilot scheme in China with DEA and PSM-DID'. *Ecological Indicators* 122: 107238. doi: 10.1016/j.ecolind.2020.107238.

Gallagher, Kelly Sims, Fang Zhang, Robbie Orvis, Jeffrey Rissman, and Qiang Liu. 2019. 'Assessing the Policy Gaps for Achieving China's Climate Targets in the Paris Agreement'. *Nature Communications* 10 (1): 1–10. doi: 10.1038/s41467-019-09159-0.

Gallagher, Kelly Sims, and Xiaowei Xuan. 2019. *Titans of the Climate: Explaining policy process in the United States and China*. Cambridge: MIT.

Hoegh-Guldberg, Ove, Daniela Jacob, Michael Taylor, et al. 2018. 'Impacts of 1.5 C Global Warming on Natural and Human Systems'. In *Global Warming of 1.5°C: An IPCC special report on the impacts of global warming of 1.5°C above pre-industrial levels and related global greenhouse gas emission pathways, in the context of strengthening the global response to the threat of climate change, sustainable development, and efforts to eradicate poverty*, eds. Valérie Masson-Delmotte, Panmao Zhai, Hans-Otto Pörtner, et al., 175–312. IPCC Special Report. www.ipcc.ch/sr15.

IEA (International Energy Agency). 2019. *CO_2 Emissions From Fuel Combustion 2019*. Paris: IEA. doi: 10.1787/2a701673-en.

IPCC (Intergovernmental Panel on Climate Change). 2014. *IPCC 2013: The physical science basis: Contribution of working group I to the fifth assessment report of the Intergovernmental Panel on Climate Change*, eds. Thomas F. Stocker, Dahe Qin, Gian-Kasper Plattner, et al. New York: Cambridge University Press. doi: 10.1017/CBO9781107415324.

Jia, Kai, and Shaowei Chen. 2019. 'Could Campaign-Style Enforcement Improve Environmental Performance? Evidence from China's central environmental protection inspection'. *Journal of Environmental Management* 245: 282–90. doi: 10.1016/j.jenvman.2019.05.114.

Jordaan, Sarah M., Adrienne Davidson, Jamal A. Nazari, and Irene M. Herremans. 2019. 'The Dynamics of Advancing Climate Policy in Federal Political Systems'. *Environmental Policy and Governance* 29 (3): 220–34. doi: 10.1002/eet.1849.

Karapin, Roger. 2020. 'Federalism as a Double-Edged Sword: The slow energy transition in the United States'. *Journal of Environment & Development* 29 (1): 26–50. doi: 10.1177/1070496519886001.

Khan, Jamil. 2013. 'What Role for Network Governance in Urban Low Carbon Transitions?' *Journal of Cleaner Production* 50: 133–39. doi: 10.1016/j.jclepro.2012.11.045.

Kostka, Genia, and Coraline Goron. 2021. 'From Targets to Inspections: The issue of fairness in China's environmental policy implementation'. *Environmental Politics* 30 (4): 513–37. doi: 10.1080/09644016.2020.1802201.

Korsbakken, Jan Ivar, Robbie Andrew, and Glen Peters. 2019. 'Guest Post: China's CO_2 emissions grew slower than expected in 2018'. Carbon Brief, 5 March. www.carbonbrief.org/guest-post-chinas-co2-emissions-grew-slower-than-expected-in-2018.

Kostka, Genia, and Jonas Nahm. 2017. 'Central–Local Relations: Recentralization and environmental governance in China'. *China Quarterly* 231: 567–82. doi: 10.1017/S0305741017001011.

Lan, Zhiyong, and Shuigen Hu. 2008. 'The Performance Evaluation of the Chinese Government: Theories and practice'. *CASS Journal of Political Science* 80 (3): 106–15. https://caod.oriprobe.com/issues/712463/toc.htm?.

Li, Hongbin, and Li-An Zhou. 2005. 'Political Turnover and Economic Performance: The incentive role of personnel control in China'. *Journal of Public Economics* 89 (9): 1743–62. doi: 10.1016/j.jpubeco.2004.06.009.

Li, Panni, Zhongguo Lin, Huibin Du, Tong Feng, and Jian Zuo. 2021. 'Do Environmental Taxes Reduce Air Pollution? Evidence from fossil-fuel power plants in China'. *Journal of Environmental Management* 295: 113112. doi: 10.1016/j.jenvman.2021.113112.

Li, Ruoqi, Yuanchun Zhou, Jun Bi, Miaomiao Liu, and Shanshan Li. 2020. 'Does the Central Environmental Inspection Actually Work?' *Journal of Environmental Management* 253: 109602–. doi: 10.1016/j.jenvman.2019.109602.

Lieberthal, Kenneth. 1997. 'China's Governing System and its Impact on Environmental Policy Implementation'. *China Environment Series* 1: 3–8.

Lieberthal, Kenneth, and Michel Oksenberg. 1988. *Policy Making in China: Leaders, structures, and processes*. Princeton University Press.

Liu, Lingxuan, Bing Zhang, and Jun Bi. 2012. 'Reforming China's Multi-Level Environmental Governance: Lessons from the 11th Five-Year Plan'. *Environmental Science & Policy* 21: 106–11. doi: 10.1016/j.envsci.2012.05.001.

Lo, Kevin. 2014. 'China's Low-Carbon City Initiatives: The implementation gap and the limits of the target-responsibility system'. *Habitat International* 42: 236–44. doi: 10.1016/j.habitatint.2014.01.007.

Lo, Kevin. 2020. 'Governing Energy Consumption in China: A comprehensive assessment of the energy conservation target-responsibility system'. *Energy Transitions* 4 (1): 57–67. doi: 10.1007/s41825-020-00023-0.

Lo, Kevin, He Li, and Kang Chen. 2020. 'Climate Experimentation and the Limits of Top-Down Control: Local variation of climate pilots in China'. *Journal of Environmental Planning and Management* 63 (1): 109–26. doi: 10.1080/09640568.2019.1619539.

Ma, Yun. 2017. 'Vertical Environmental Management: A panacea to the environmental enforcement gap in China?' *Chinese Journal of Environmental Law* 1(1): 37–68. doi: 10.1163/24686042-12340004.

Moore, Scott M. 2017. 'The Dilemma of Autonomy: Decentralization and water politics at the subnational level'. *Water International* 42 (2): 222–39. doi: 10.1080/02508060.2017.1276038.

Nan, Xu, and Liu Jingyang. 2014. 'China Faces a Unique Challenge in Adapting to Climate Change'. *China Dialogue*, 31 March. https://chinadialogue.net/en/climate/6809-china-faces-a-unique-challenge-in-adapting-to-climate-change.

Piao, Shilong, Philippe Ciais, Yao Huang, et al. 2010. 'The Impacts of Climate Change on Water Resources and Agriculture in China'. *Nature* 467 (7311): 43–51. doi: 10.1038/nature09364.

Ross, Robert S. 2019. 'On the Fungibility of Economic Power: China's economic rise and the East Asian security order'. *European Journal of International Relations* 25 (1): 302–27. doi: 10.1177/1354066118757854.

Schurmann, Franz. 1968. *Ideology and Organization in Communist China*, 2nd ed. Berkeley: University of California Press.

Shin, Kyoung. 2017. 'Neither Centre Nor Local: Community-driven experimentalist governance in China'. *China Quarterly* 231: 607–33. doi:10.1017/S0305741017000923.

Spracklen, Dominick V. 2016. 'China's Contribution to Climate Change'. *Nature* 531 (7594): 310–12. doi: 10.1038/531310a.

Steurer, Reinhard, and Christoph Clar. 2015. 'Is Decentralisation Always Good for Climate Change Mitigation? How federalism has complicated the greening of building policies in Austria'. *Policy Sciences: Integrating Knowledge and Practice to Advance Human Dignity* 48 (1): 85–107. doi: 10.1007/s11077-014-9206-5.

Thuiller, Wilfried. 2007. 'Climate Change and the Ecologist'. *Nature* 448 (7153): 550–52. doi: 10.1038/448550a.

Tian, Hongzhen, Taibao Yang, and Qinping Liu. 2014. 'Climate Change and Glacier Area Shrinkage in the Qilian Mountains, China, from 1956 to 2010'. *Annals of Glaciology* 55 (66): 187–97. doi: 10.3189/2014AoG66A045.

Wang, Yufei, Qijiao Song, Jijiang He, and Ye Qi. 2015. 'Developing Low-Carbon Cities through Pilots'. *Climate Policy* 15 (Sup1): S81–S103.

Watts, Nick, W. Neil Adger, Paolo Agnolucci, et al. 2015. 'Health and Climate Change: Policy responses to protect public health'. *The Lancet* 386 (10006): 1861–914. doi: 10.1016/S0140-6736(15)60854-6.

Westman, Linda, and Vanesa Castán Broto. 2018. 'Climate Governance through Partnerships: A study of 150 urban initiatives in China'. *Global Environmental Change* 50: 212–21. doi: 10.1016/j.gloenvcha.2018.04.008.

Wheare, Kenneth Clinton. 1946. *Federal Government*. London: Oxford University Press.

Yi, Hongtao, and Yuan Liu. 2015. 'Green Economy in China: Regional variations and policy drivers'. *Global Environmental Change* 31: 11–19. doi: 10.1016/j.gloenvcha.2014.12.001.

Zhang, Pan. 2021. 'Target Interactions and Target Aspiration Level Adaptation: How do government leaders tackle the "environment–economy" nexus?'. *Public Administration Review* 81(2): 220–30. doi: 10.1111/puar.13184.

Zheng, Yongnian. 2007.*De Facto Federalism in China: Reforms and dynamics of central–local relations*. Series on Contemporary China, vol. 7, eds. Joseph Fewsmith and Zheng Yongnian. Singapore: World Scientific.

6

Climate Governance and Federalism in Ethiopia

YALEMSEW ADELA, ADEFIRES WORKU, AND TILAYE NIGUSSIE

6.1 Introduction

Ethiopia is the only African country that has remained independent for centuries, with its own written script, number system, and calendar (Fiseha and Habib 2010). It is a multi-cultural and multi-religious country with a rural majority that relies on traditional natural resource management. Since Eritrea's independence in the early 1990s, Ethiopia has been a landlocked country with an area of 1.1 million square kilometres and the second-most populous country in Africa, after Nigeria (CSA 2014). Agriculture is the mainstay of the economy, accounting for more than half of the GDP, employing more than 85 per cent of the workforce, and generating over 90 per cent of the country's foreign exchange (Alemu, Oosthuizen, and Van Schalkwyk 2002). Ethiopia's agriculture is dominated by smallholder farmers practising rain-fed mixed crop production, and is typically characterized by low productivity, implying its high vulnerability to various anthropogenic and natural hazards (Devereux 2006; Gebre-Selassie and Bekele 2010; Ketema and Dubale 2020; Worku 2016). Environmental degradation and climate variability and change are two of the major challenges undermining the agriculture sector in particular, and Ethiopia's effort to become a middle income country by 2025, and to make considerable progress in achieving the sustainable development goals by 2030 in general.

Ethiopia has developed and adopted several strategies in response to climate change. The Climate Resilient Green Economy (CRGE) Strategy, which promotes a paradigm shift to low-carbon growth is one of the strategies designed and implemented over the past decade (FDRE 2011). The CRGE Strategy is prepared by drawing on the experiences and achievements of various precursor strategies such as the Growth and Transformation Plan (GTP I) and International Agreements and Protocols to which Ethiopia is a party and signatory. The CRGE is integrated into different development plans such as the GTP-II that has been implemented by sector ministries at federal level and in regional states.

This chapter provides an overview of how the federal structure has been affecting efforts to mitigate and adapt to climate change in Ethiopia. Ethiopia is a federal country comprising eleven regional states and two city administrations. The decentralized governance structure determines the relationship between the federal Government and its component units. The structure and division of powers provides for the federal Government and regional states to formulate and implement policies, plans, and strategies in their respective jurisdictions. The federal, regional, and sub-regional level achievements in the implementation of strategies such as the CRGE strategy, the central goal of which is to promote a green and resilient economy, are influenced by the decentralized administrative structure. The differences in achievements are attributable to constraints on the implementation of ongoing and future climate change mitigation and adaptation interventions, such as inadequate capacity of implementing bodies; scarcity of financial resources; dearth and inaccessibility of information where and when available; variations of priorities and type of interventions; geographical location; weak knowledge management systems, monitoring and evaluation systems.

6.2 Trends, Patterns, and Impacts of Climate Change in Ethiopia: An Overview

6.2.1 Climate Change Trends and Patterns

Despite its growing vulnerability, Ethiopia's contribution to global warming is insignificant – its per capita GHG emissions remain among the lowest both in Africa and in the world (UNFCCC 2005). The assessment made in 2010 indicated that the country emitted 150 Mt CO_2 equivalents (less than 0.3 per cent of global emission). The report also expounded that 50 per cent of the emissions are from agriculture (crop and livestock) and 37 per cent from forestry, followed by industry, power, transport, and buildings contributing on average 3 per cent each. The larger share of emission from agriculture is attributed to the presence of more than 50 million cattle and nearly 100 million different livestock species. These emit substantial amounts of methane and other oxides of nitrogen. Likewise, there is an increasing use of inorganic fertilizers to grow crops, and an expansion of agricultural land which reduces forest cover and grasslands. As described in the CRGE strategy, with the business-as-usual (BAU) scenario, emission will increase to 400 Mt CO_2 equivalent by 2030, and with increasing per capita emission from 1.8 tons to 3 tons. Industrial emissions are projected to increase by more than twelve-fold, and emissions from other sectors will also increase because of continued infrastructure development.

A long-term trend analysis of temperature data shows about 0.2°C rise every decade, where the rise in the minimum temperature is approximately 0.4°C per decade. At the national level, temperature has increased by approximately 1°C since the 1960s. According to the World Bank (2021), the average number of 'hot nights' per annum (the hottest 10 per cent of nights) increased by 37.5 per cent between 1960 and 2003. Similarly, the average number of 'hot days' per year increased by 20 per cent with decreasing number of cold days. Increasing temperatures have been resulting in increased evapo-transpiration and reduced soil moisture and higher rates of warming including the central and highland areas of the country. This report predicts that the mean annual temperature will increase between 0.9 and 1.1°C by 2030, 1.7 and 2.1°C by 2050, and 2.7 and 3.4°C by 2080. In contrast to temperature, the average precipitation has remained reasonably constant. The IPCC mid-range emission scenario shows that compared to the 1961–90 baseline, the mean rainfall variability between years, seasons, and regions ranges between 25 and 50 per cent.

6.2.2 Impact of Climate Change

Ethiopia is one of the countries most vulnerable to climate change. Projections show increasing risk of drought, heavy rains, and flood in various agroecological zones. Extreme climate effects might set back development efforts and accomplishments unless appropriate adaptation measures are put in place. A report by the World Bank discerns that the occurrence, severity, and coverage of droughts has ominously increased over the past few decades, causing significant damage to life and livelihoods (World Bank 2006). In Ethiopia, almost all sectors including agriculture, infrastructure, energy, transport, and health are affected by drought (Adem and Bewket 2011; Mesfin 1984). Recurrent drought has been dramatically decreasing crop production, causing the death of livestock and increases food insecurity and malnutrition, forcing people to be displaced and aggravating environmental degradation making food security a major challenge (NMA 2006).

The Irish Red Cross (2007) report shows that incidences of flood increase eight-fold in Africa where the impact of climate variability and change is high. For instance, the disastrous flash flood in Eastern Ethiopia in 2006 caused causalities and displaced several thousands of people (Irish Red Cross 2007). The frequency of flood and the areas affected have significantly increased over the past few decades (NASA Earth Observatory 2008). Similarly, the Centre for Research on Epidemiology of Diseases (CRED) reports that Ethiopia's vulnerability to climate change has increased. Though droughts and floods affect people from all walks of life, they are especially detrimental to the lives and livelihoods of smallholder farmers and pastoralists (Oxfam 2009). In the presence of multi-faceted challenges including poor socio-economic conditions, fragile ecosystems, and low adaptive capacity, Ethiopia cannot address the challenges on its own.

Because of differences in environmental, institutional, and socio-economic characteristics of regions, sub-regions and communities, the impact of climate change varies from place to place. Pastoral and agropastoral communities have been suffering the most. The growing vulnerability of these communities could also be due to the dryland agroecology they live in. An estimated 75 per cent of Ethiopia's landmass is classified as dryland with high moisture stress. Although drylands exist in all parts of the federal regions, they are pervasive in Afar, Somali, Gambela, Benishangul, Oromia, and Southern Nations and Nationalities. In addition, some of the communities, particularly those designated as emerging regions in the federal system, have comparatively high vulnerability because of weak institutional capacity. Other regions have also experienced severe land degradation because of traditional farming techniques, severe erosion, high soil acidity, and other factors that increase their vulnerability.

6.3 Policy and Institutional Frameworks and Responses to Climate Change in Ethiopia

6.3.1 Policy and Institutional Framework

The Ethiopian Constitution provides articles dealing with environmental management and sustainable development. Article 43, for instance, states that 'The right of Ethiopia to sustainable development shall be secured and ensured by all international agreements and ties concluded, developed or preserved by the State.' Similarly, Article 44(1) says 'all persons have the right to live in a clean and healthy environment'.

In addition to the provisions of the Constitution, the government issued policies, strategies, programs, and legislations that aimed at improving forest management, biodiversity conservation, and reversing the loss of renewable natural resources that otherwise intensify vulnerability to climate change and other hazards. The Environmental Policy (1997), Environmental Protection Organs Establishment Proclamation (295/2002), Environmental Impact Assessment Proclamation (299/2002), Environmental Pollution Control Proclamation(300/2002), Forest Development, Conservation and Utilization Policy and Strategy (2007), Climate Resilient Green Economy Strategy (2011), National Forest Sector Development Programme (NFSDP 2018), National REDD+ Strategy (2018), updated Nationally Determined Contribution (2021), National Adaptation Plan (NAP-ETH 2019), and Forest Development, Conservation and Utilization Proclamation 1065/2018 etc show the efforts the government of Ethiopia has made to protect the environment, sustainably manage renewable natural resources, and reduce vulnerability to climate change and variability.

The African Development Bank states that Ethiopia's government policies, strategies, and institutional frameworks for the management of natural resources and environment are adequate and sound (African Development Bank 2015). Despite differences in implementation capacity, climate-related issues are integrated into sectoral programmes that are implemented at federal, regional, and sub-regional levels. In addition, successive growth and transformational development plans were developed by considering the links between poverty and environment. Recent trends show that the government of Ethiopia remains committed to integrate environmental protection into development planning and implementation processes.

In addition to issuing policies and strategies, the government of Ethiopia has also put in place institutions that are responsible for guiding the implementation of these policies and strategies formulated to achieve environmental goals. In 2013, the government established the Environment, Forest, and Climate Change Ministry, which was restructured in 2018 to function as a Commission. Currently, it is split into two institutions (Environmental Protection Authority and Forest Development) with the former having regulatory responsibilities concerning the environment, and the latter with a mandate to lead the forestry sector development including restoration of degraded landscapes, increase forest cover, reduce deforestation, and thereby contribute to the mitigation of and adaptation to climate change. In fact, not only the environment and forest institutions, but also all sectoral ministries are obliged to include climate change mitigation and adaptation plans in their respective work plan as elaborated in the CRGE strategy. Regional bureaus are also expected to do the same. Ethiopia has established the CRGE Facility to coordinate the mobilization of financial resources needed to implement priority climate and environmental interventions. The facility has enabled Ethiopia to access funds from bilateral and multi-lateral development partners (GCF 2016). For instance, Ethiopia is one of the few countries that secured funds for its large-scale REDD+(**R**educing **E**missions from **D**eforestation & forest **D**egradation) Investment Programme. REDD+ is a global policy framework under the UNFCCC for climate change mitigation in the forest sector.

6.3.2 Responses to Climate Change

As a country vulnerable to climate change, Ethiopia has been trying its best to respond to the problem. The first and earliest response was the decision made to sign the UNFCCC during the 1992 United Nations Conference on Environment and Development held in Rio de Janeiro, Brazil. Ethiopia ratified the Convention on 31 May 1994 by Proclamation No. 97/1994 and issued Proclamation No. 97/1994. In addition, the Kyoto Protocol was ratified on 21 February 2005 and became law through Proclamation No. 439/2005. Both international agreements are considered as an integral part of the law of the country as provided in Article 9 of the Constitution. Furthermore, Ethiopia has signed several environmental agreements including the United Nations Conventions on Biodiversity (UNCBD) and United Nations Conventions to Combate Desertification and Mitigate the Effects of Drought (UNCCD), which are important to advance climate change mitigation and adaptation actions. Ethiopia's commitment to implement the international agreements got traction by conducting the national greenhouse gas emissions and sinks inventory in 1994. This established the basis for the initial national communication submitted to the Secretariat of the UNFCCC in 2001 (FDRE 2001).

Ethiopia has remained a prominent player in international climate change negotiation forums. For instance, based on the decision made by the parties to UNFCCC, it has identified its most urgent and immediate adaptation needs, which led to the preparation of the National Adaptation Plan of Action (NAPA) in 2007 (FDRE 2007), which identified dozens(58 in number) of adaptation options. The NAPA was revised in 2019 to build on ongoing efforts to address the impact of climate change. The goal of NAP–Ethiopia (NAP–ETH) is to reduce vulnerability to the impacts of climate change by building adaptive capacity and resilience. Supported by several institutions, governance structures, and finance, NAP–ETH aims to strengthen holistic integration of climate change adaptation in Ethiopia's long-term development pathway and enhance systems for disaster risk reduction and management in different sectors. In addition to the efforts made to adapt to the impacts of climate change, Ethiopia has also submitted its nationally appropriate mitigation action (NAMAs) to UNFCCC, which focused on appropriate development and management of hydropower, wind power, nuclear, geothermal, electric rail, and urban waste.

Ethiopia's CRGE Strategy issued in 2011, is one of the most important policies to reduce GHG emission by adopting green growth initiatives (FDRE 2011). In contrast to the adaptation plan, the CRGE Strategy focuses on building a climate ressillient and low carbon economy. Ethiopia has also determined to reduce its carbon footprint by achieving 68.8 per cent GHG reduction by 2030. As mentioned earlier, the 2010 emissions assessment showed 150 megatons of carbon dioxide equivalent (Mt CO_2e) which will be reduced to 145 MtCO_2e in 2030. The updated NDC, in addition to the familiar mitigation interventions, has identified forty adaptation options, which eventually will substantially contribute to social-ecological resilience.

Sustainable forest management is one of the pathways adopted to improve the valuation and provision of ecosystem goods and services. In this regard, Ethiopia has committed to restore 22 million ha of degraded land through afforestation and assisted natural regeneration. The Green Legacy Initiative (GLI) launched by Ethiopia's prime minister aims at planting 20 billion seedlings over five years. This is a manifestation of the government's commitment and determination to increase the country's forest cover by reversing deforestation and setting the stage for building a green economy. In this regard, the ongoing landscape restoration efforts are supported by robust forest management policies and strategies.

The various climate change mitigation and adaptation strategies and programmes mentioned above have been led by the federal Government and implemented by regional states. The programmes and actions cascaded and adapted to the context of each region are mandatorily mainstreamed in regional plans. In addition, projects implemented in regions (for example reafforestation undertaken in areas with reduced forest cover because of deforestation and forest degradation) are required to be aligned to regional plans based on the suitability of the area where projects are implemented.

6.4 Climate Change Governance and Federalism in Ethiopia

6.4.1 Ethno-federalism in Ethiopia

Ethiopia's federal system has been in place since 1991, when the Ethiopian People's Revolutionary Democratic Front (EPRDF) overthrew the Socialist regime. The EPRDF established its Constitution, which came into effect in 1994. The preamble states that Ethiopia's federal system is based primarily on ethnicity, with each 'nation, nationality, and people' having the right to have their own region and decentralized administration. According to this provision, Ethiopia's federal system now has eleven regions (up from nine in 1995) and two city administrations. There is a strong tendency to increase the number of regions because the Constitution enshrines the right of the nationalities within a regional state to establish their own state if the proper procedures are followed (Markakis 2006). For instance, the regional states of Sidama and Southwest Ethiopia have been recently formed following a referendum, which ratified the shrinking of the Southern Nations, Nationalities, and People regional state. As in most federal countries, the constituent units in Ethiopia vary in the size of their area and population. The Constitution provides for the management of the variation by recognizing that 'the Member States of the federal Democratic Republic of Ethiopia shall have equal rights and powers' (Institution of the Ombudsman Establishment Proclamation, Year 6 No. 41, Proclamation No. 211/2000).

The Constitution grants ethnically defined regional states the right to self-determination, which devolves political, administrative, and economic power, and unlike in other federal countries, guarantees the right to secede from the federal country. Ethiopia, as an old nation, has gone through political, economic, and social upheavals, territorial expansion and contraction, division and merger of political administration, consolidation, and separation of administrative boundaries. However, the current ethnic-based federalism is the first of its kind in the country's political history.

Chabal and Daloz (1999) and Ottaway (1994) remind us that ethnic federalism is still controversial, with some academics viewing it as a recipe for state disintegration, while others see it as a governance system that allows for different ways of thinking about ethnicity while avoiding conflicts and marginalization, particularly in African politics. Similarly, Ethiopia's ethnic federalism continues to be a hot topic of debate among elites and ordinary citizens alike, attracting both criticism and support. One of the most persistent criticisms of the Constitution concerns its provision of rights to regions to secede from the federation, as well as the fact that it is more primordial in its nature. This is unusual in other federal countries where federalism has been practised as 'union and non-centralization at the same time' (Aalen 2002). Elazar (1987) also argued that federalism is considered to advocate the values of 'unity in diversity', giving the constituent units the right to self-government within the framework of unity, not the right to secede. In contrast, Ethiopia's federal arrangement has been attracting support as it promotes the value of diversity, self-rule, recognition, and wider space for accommodating diverse political views. In fact, the critics also appreciate these values, including the right to self-rule and devolution of power, arguing these values and provisions can be accommodated without stretching the rights of nationalities up to secession.

6.4.2 Division of Power

The federal government is given enumerated and limited powers and responsibilities under the Constitution. It has the mandate to develop and implement national policies, plans, and strategies pertaining to overall economic and social development. Similarly, the Constitution empowers the federal government to develop and implement national policies and strategies in the financial and monetary sectors, as well as in the utilization and conservation of natural resources. Furthermore, the federal government has the authority to establish national standards for the protection of cultural and historical sites, as well as for public health, education, science, and technology. In addition to the more traditional roles of the federal government in the fields of defence, foreign affairs, inter-state and international trade, these powers and obligations are particularly important.

The Constitution assigns exclusive state and concurrent roles to the regions, as well as residual matters. Consequently, by virtue of its mandate to deal with residual matters, powers and responsibilities of the regional governments are likely to increase. Contrary to practice in other federal countries such as India and Nigeria, the constituent units are permitted to have their own constitutions, even though they are subject to the supremacy of the federal constitution. Initially, the federal and regional constitutions were similar. Some years later, however, they have been revised in a manner that represents significant variations (Habib 2010).

6.5 Implication of Ethiopia's Ethnic-Based Federalism on Climate Change Governance

Despite its seemingly linguistic orientation, the ethnic federalism exercised in Ethiopia is based on primordial identity (Abbink 2011). This has aggravated minor differences among various ethnic groups, resulting in conflict (Taye 2017) which caused human casualties and destruction of property in different parts of the country. Though the factors are diverse, one main cause is the establishment of identity (language-based) federalism. This federal system is not limited to political governance, but extends to other sectors such as natural resources, the environment, and climate change management.

As explained above, the Constitution provides for the central and regional governments to have their own legislative, judicial, and executive power and rights related to important political, economic, and social issues (Markakis 2007). However, Fiseha (2018) argues there are no clear distinctions in legislative, judicial, and executive power and rights of the central and regional governments. Such overlaps affect the implementation of climate change mitigation and adaptation interventions at various levels of the federal and regional administration (Fiseha 2018).

Another feature of Ethiopia's federal system related to climate governance is the country's proclivity to create new regions. Following the coming into full force and effect of the Constitution in 1995, two additional regions have been established following a referendum. In view of the aspirations of nationalities to form their own region, one can only imagine the challenges this will pose to the new, old, and central governments in terms of sharing existing scarce resources and guaranteeing rights concerning the utilization and management the same resources. This is because referendums result in reorganizing the use of infrastructure use and determining new modalities for accessing natural resources such as land, forest, water, and human and technological capital, which were previously used as shared resources. For example, the carbon-rich forests that were once administered by the SNNP region are now administered by the new

Southwest Ethiopia region. The transfer of such critical ecosystems to the new regional state will undoubtedly cause temporary loss of momentum in advancing climate change mitigation and adaptation actions and delay the achievement of targets. It also forces the old region to forego significant amounts of funds it hitherto received from national and international treasuries, which used to contribute to the region's effort to achieve sustainable development goals. The old region also loses human capital, land, and infrastructure, all of which would have helped to improve the region's adaptive capacity.

Notwithstanding the above, the same argument can be made of the new region. Even though the referendum may address politically and economically motivated aspirational issues, it also deprives the new region access to existing infrastructure, institutions, technologies, and other assets that have been jointly built over the years. For example, if Hawassa city, which has been developed as a regional centre for the SNNP region, is claimed as the property by Sidama region, it will take the SNNP region and the new regions a long time and a large investment to have such a regional capital. The problem is that the loss and the need for shifting funds to develop new infrastructure will undermine the new region's ability to adapt to climate change, at least until it builds the necessary capacities and capabilities. Moreover, the transition may increase the new region's carbon footprint because the change necessitates more investment in institutions, transportation, and other infrastructure – all of which may increase the rate of deforestation that contributes to GHG emission in the new region. Furthermore, there is no guarantee that the central government will maintain natural forest blocks in the new regions. This is because the new region may establish development priorities that result in land use change, which is expected at the early stages of a region's establishment.

Another important aspect of Ethiopia's federalism in terms of climate change governance is the huge disparity between regions. For example, Afar and Somali regions are more vulnerable to recurring and severe droughts and floods than others because of their geographical location. These two component units of the federal administration possess extensive drylands, which are characterized, among other factors, by moisture stress. In contrast to other regions that have highlands, midlands, and lowlands, which allow for better seasonal rainfall amount and distribution, communities in Afar and Somali regions do not have the conducive climatic endowment. These communities, majorly of pastoral livelihood, do not have the ease move to areas in other regions with favourable climate to avoid seasonal risks and hazards caused by climate change. Ethnic federalism created visible differences between regions, in terms of having access to fertile arable land, water, forest, and other natural capital, which is a source of disproportionate vulnerability to climate change. There is also a huge disparity in human capital between regions. For example, Harari region, with an estimated population of less than 300,000 people, competes with Oromia region, having an estimated population of over 30 million. This has implications in terms of human capital, tax collection, access to resources from the central government, and affects, among other things, responsiveness to climate change. There are also significant differences in institutional capacities and capabilities across regions, which either facilitate or hinder mitigation and adaptation actions aimed addressing the impact of climate change. As a result, one could argue that ethnic federalism has contributed to existing regional differences in terms of preparedness to respond to climate change. Indeed, the Constitution states that members of the federal Government 'shall have equal rights and powers' (Institution of the Ombudsman Establishment Proclamation, Year 6 No. 41, Proclamation No. 211/2000). Such narratives, motivated and driven by identity-based ideology and policy which have been translated to action through the institutionalization of a federal administrative structure, have been increasing tensions between ethnic groups, and increasing risks to vulnerable communities which lack capacity to minimize harm to themselves and their fragile environment. Experiences over the last three decades show an increasing number of communities trapped in vicious circles of vulnerability and uncertainty because of unfounded narrative that spreads hate and division among communities rather than promoting peaceful coexistence and unity. This has also added to the central government's burden, as it must allocate a large amount of budget every year for the provision of safety net/social security services to support these vulnerable communities. If voluntary mobility were easier, the government would have a better chance of assisting vulnerable communities to become self-sufficient, allowing the safety net programme budget to be used for other development interventions.

Another important challenge of the federal system in Ethiopia related to climate governance is associated with the division of power and rights among central and regional governments. The Constitution confers enumerated and limited powers and responsibilities on the central government. Although the central government has the power to formulate national policies, plans, and strategies concerning economic and social development, the right to own and administer land and natural resources that is crucial for climate change adaptation and mitigation is under the jurisdiction of regional governments. The regions have full right to administer land and natural resources in harmony with federal policy, strategic and legal frameworks. That means the federal Government does not have direct control over natural resources (except enacting laws for the utilization and conservation of land and other natural resources) and cannot have the direct right to redistribute benefits derived from natural resources to all citizens to enhance adaptive capacity and ensure sustainable development. The federal Government cannot put a particular critical ecosystem that has national significance under full protection or relocate certain vulnerable groups to a suitable watershed without the agreement of regions. Some regions have listed the ethnic groups in their Constitution as the owners of the region, alienating other groups that have lived in the regions for generations. The state of exclusion and/or under-representation in decision-making processes has implications for climate change mitigation and adaptation and is an important governance issue in terms of inclusiveness and empowerment concerning natural resources management and benefit sharing. It needs to be noted that regional states as part of their power and functions are responsible for administering land and other natural resources in accordance with federal laws. It also needs to be taken into consideration that the Constitution allows the regional states to deal with matters not given expressly to the federal Government alone, or concurrently to the federal Government and the states. This has provided states with unintended expanded powers. In other words, the current ethnic federal system doesn't bind itself to political governance only but often extends to making decisions over the use of natural resources, access to facilities, infrastructure, and markets. Minorities with less access to these facilities and benefits are always the ones who disproportionately carry the risks of climate change.

There are also arguments and critics of Ethiopia's ethnic federal system in relation to the country's rapidly growing population, which has a direct relationship with climate governance. The allocation of budget and other resources from the federal government to states, and the distribution of seats in the federal parliament (House of People's Representatives), is based on the population size of each region. This has fuelled competition among regions to increase the size of their population and has resulted in the country's rapid population growth. There aren't many countries in the world where the population has more than doubled in less than thirty years. Ethiopia's population, which was estimated to be less than 60 million people thirty years ago, is now estimated to be 120 million. Rapid population growth in regions has resulted in overuse of natural resources, which in turn is exacerbating social-ecological vulnerability both in the regions and at national scale.

One can argue that the ethnic-based regions are focusing on maintaining their advantages (for example in natural resources endowment) and their own development. This has resulted in the pursual of fragmented national agenda and poorly coordinated planning concerning cross-cutting issues including responses to climate change. As mentioned earlier, climate change management requires cooperation and concerted action by all government institutions. In contrast, there is growing incoherence among regions and the federal government in terms of setting development priorities. For instance, the recent CRGE strategy progress assessment report revealed a slow implementation of the strategy in some regions, which has a significant impact on achieving the national targets. Another example is the contrasting plan of some regional governments compared to the central government aimed at reducing the number of livestock, which account for the lion's share of GHG emissions. According to the assessment report, the number of livestock is increasing rather than decreasing, as stated in the CRGE Strategy. This is because livestock is a priority economic sector in some regions, primarily in pastoral regions such as Afar and Somalia. Since such and other similar practices do not consider the neighbouring regions, several challenges and problems occur in the regions that increase the vulnerability of different communities to climate change and weaken their climate resilience capacity. For example, the current environmental and social impacts (ESIA) report approval mechanism could describe the challenge. Regions in the federal system are also entitled to approve ESIA reports prepared for a wide range of projects that could have potential impact on people and the environment. Because of the weak capacity of the bureaucracy to manage such complex tasks and the limitations of the ethnic federal system, regions approved thousands of ESIA projects without considering the interregional impacts of the projects. This has caused grievances in communities living in the delivering and receiving ends of the ESIA decisions. For example, different floriculture industries established near Lake Ziway in the Great Rift Valley have faced recurrent damage by youth at different times.

Despite the job opportunities the industries have created for the local people, the pollution in Lake Ziway, which had been once used as a livelihood means through fishing activities, has caused loss of its fish population, leaving many people without income (Teklu et al. 2018). And the fish are no longer preferred by the public due to the belief that toxic chemicals have accumulated in the fish. This means, under the prevailing impact of climate change, the resilience of local people (their capacity to cope with shocks and disasters) is threatened. That leads to violence and distraction of companies, though other politically motivated reasons might have triggered the destructive actions. Similarly, pollution in the Awash River caused by untreated effluent from industries flowing from the central highland through the Great Rift Valley depressions have resulted in severe water quality deterioration, affecting the quality of vegetable production downstream (Tadese, Sonder, and Peden 2003). Not only that, this polluted water source also affects the pastoral communities downstream which are more vulnerable to climate change. The cause of such problems is not only the failure of industries to put in place environmental protection measures, but also how the ESIA and environmental management plans are approved and executed. Such problems are not only limited to the Rift Valley, but also observed in different regions confronted with similar issues.

One key root cause to these problems is the implementation of ethnic federalism, which promotes skewed development. All regions in Ethiopia are under high competition to attract investment in their regions regardless of the projects' nature in terms of the sustainability indicators. Projects require different types of landscapes and climatic conditions to produce their products efficiently. However, what is observed among regions is that they accept every project proposal without thoroughly pre-screening for their suitability in terms of technical, social, and environmental standards. This hastens the unintended environmental and social crises in the respective regions and across the country. Consequently, there are growing trends to take advantage of the loosening project approval mechanisms by local and international investors to install old and second-hand machinery that has been abandoned in other countries. This has been affecting aquatic and soil systems, which have hampered the livelihood activities of communities. Had there been a coordinated approval mechanism among regions, these problems would not have been magnified to such an extent.

Under the present federal administrative arrangement, the Government is responsible for national parks and a few other resources that are found in different parts of the country and is expected to work in close collaboration with regions. However, the federal Government approves ESIA reports for large-scale investment projects implemented at the regional level sometimes without the understanding of the regional respective bureaus. This shows the conflicting application of mandates by the federal and regional governments, which in turn continue to cause undesirable problems at local level. For instance, the massive deforestation of natural forests for large-scale commercial agricultural investment in the Gambella Region (GRAIN 2019), carried out following the approved ESIA by the federal Government, has evicted thousands of farmers from their land, and failed to meet the target. These and other similar unwarranted interventions have affected thousands of people, exacerbating the vulnerability of communities to climate change. Moreover, the mining sector is also confronted with similar bottlenecks which result in multi-faceted environmental, economic, and social problems.

In the recent two decades, Ethiopia has set a green economy policy where the CRGE strategy is an integral component of this policy. As a result of this, Ethiopia has built a couple of small, medium, and large hydropower projects, which are climate-sensitive. However, due to the uncoordinated natural resource management among the regions, most watersheds are highly degraded and significant sediment intrusion into the dams has been reported in different journal articles and media outlets. One barrier to fixing this problem is the nature of federalism, which allows people to use their natural resources without any intervention. In other words, if the hydropower is located in one region, the watershed is stretched in other regions that require watershed management at the upper catchment, such as maintaining the existing vegetation in the basin, implementing soil and water conservation activities, and other requirements. Despite the fact that this is the fundamental agreement in the scientific principles, what is observed on the ground is quite the opposite. Theoretically, one can raise the idea of coordination among regions to manage such kinds of problems. In practical terms, this has been unsuccessful and cannot be achieved without a constitutional amendment. To elaborate, how can one forbid regions from utilizing their resources when they are constitutionally entitled to do so? Is this issue merely approached with coordination as a solution among the regions? This situation needs to be rectified through a constitutional amendment, which provides for mandatory coordination between regions concerning the enhancement of ecosystem services that have implications for important infrastructure that could be affected by the action or inaction of regions either upstream and downstream.

Finally, climate change adaptation and mitigation necessitate massive private sector engagement. The private sector's involvement is critical for mobilizing finance, introducing innovation and technology, and building capacity, which are required to turn challenges into opportunities and thus improve social-ecological resilience. The private sector could play critical roles in creating green jobs, transferring risks to third parties, and thus contributing to the enhancement of the adaptive capacity of communities in regions. However, over the past decades, the progressive division of regions along ethnic lines has slowed and, in some cases, stopped, resulting in casual hit-and-run investments. As propensity to identity politics becomes stronger, manifestations of malpractices and double standards have emerged in decision-making processes. This in turn ensued corrupt practices caused havoc on the governance of actions aimed at reducing the impacts of climate change. Unless and until the above-mentioned issue is resolved, there is serious concern that the problem will become more complex, trigger conflicts, and erode trust among regions and between the federal government, all of which will weaken the state and invite external risks to the country. In general, climate change governance requires a coordinated effort. The key pillars of climate change governance include mitigating and adapting to the risks of climate change. This will ensure that the appropriate management of renewable natural resources and the environment, that is, water, soil, air, and living things will enhance the resilience capacity of communities. This calls for coordinated action among regions.

Despite the challenges and problems vis-à-vis ethnic federalism in Ethiopia's climate change governance, the federal arrangement has resulted in political and economic gains. The opportunities of the federal system to govern climate change issues in the Ethiopian context might be highlighted in the following perspectives. These are: (a) nationalities are recognized and represented in the parliament; (b) power is devolved to regional level, providing space for exercising self-rule including adopting climate change-related policies and practices that match regional contexts; and (c) nations, nationalities, and peoples can promote and develop their culture, religion, social values, and their traditional knowledge system. These opportunities give regions the right to exercise self-rule including practising climate change adaptation and mitigation activities that fit to their contexts. The constitutional right conferred on the nationalities under the federal system to promote and develop their cultures, religion, and other social values including the opportunity to integrate their traditional knowledge system into conventional knowledge system that has profound importance to climate change adaptation and mitigation. Though the right of the regions to formulate their own development plan has caused unprecedented challenges, the federal system provided the regions with the opportunity to focus on their priorities and improve infrastructures and basic social services including access to education and health. At national level, the prospect of multiple political parties competing in a democratic process for parliamentary seats is expected to progressively deliver outcomes that will be important for sustainable development and the fair share of benefits derived from these outcomes, which is critical to ensure social-ecological resilience in Ethiopia.

6.6 Conclusion

Despite its insignificant contribution to global warming, Ethiopia has been suffering due to impacts of climate change. Climate change, coupled with widespread landscape degradation, has had a detrimental impact on agriculture, the primary source of income for most of the population, as well as the rest of the sectors, putting severe strain on the country's hard-won achievements. Ethiopia's government has been attempting to address such issues. Ethiopia introduced its climate-resilient green economy strategy (CRGE strategy) in Durban, South Africa, in 2011. The CRGE was well received by the international community, and the country has been in the forefront in the fight against climate change since then. Ethiopia has also signed the Paris Agreement and submitted its most recent Nationally Determined Contribution (NDC) to the secretariat of the United Nations Framework Convention on Climate Change (UNFCC). Ethiopia's NDC is ambitious and aims to cut emissions by 68.8 per cent by 2030. Institutions have been established to address climate change and environmental issues, and efforts have been made over the past decade to mainstream climate change governance at various levels of the federal and regional structure. Climate change adaptation and mitigation programmes and projects are also underway in various parts of the country.

Despite these developments, effective climate change governance continues to face challenges, undermining achievements and contributing to increased food insecurity and disproportionate vulnerability in some regions. Even though there exist a strand of supplementary and complementary factors, the major problems are linked to the ethnic-based federal arrangement. The demarcation of regional boundaries based on identity has resulted in significant differences in human, institutional, and financial capacity across regions, resulting in a slow process of implementing climate change management activities. The large disparity in population size is linked to scarcity of critical human resources in some regions, which weakens institutions and results in poor law enforcement. Because of their geographical location, the identity-based governance structure has also created regions that are most vulnerable to the effects of climate change. It has also created regions and communities with less natural capital, such as fertile land, adequate water, and forest resources, all of which are critical for increasing social-ecological resilience. These differences play a significant role in the slow progress of some regions and communities in stepping away from the vicious circle of poverty and vulnerability to extreme weather events and environmental degradation.

Furthermore, ethnicity-laden, region-centric development competition among regions to meet their respective plans has resulted in insufficient coordination to manage climate change and environmental issues with cross-regional and national implications. The constitutional right granted to regions to use their natural resources, including critical ecosystems, provided the federal government with less leeway in minimizing the impact of developments in the upperstream region on the downstream regions. Attempts to enact laws and regulations dealing with such issues have so far been unsuccessful, because such initiatives are frequently viewed as a violation of the regional and federal constitutions. The fact that regions have their own development priorities has slowed the achievement of national targets related to building a green economy. For instance, the national plan to reduce the number of livestock is incompatible with the regional plan because the livestock sector is considered the main economic activity in some regions. The same is true for conserving carbon-rich forests or increasing the country's forest cover, as ethnic regions have access to land and the mandate to manage natural resources. Most painfully, cross-regional migration and resettlement programmes, which are critical climate change adaptation strategies, are no longer feasible under Ethiopia's federal arrangement. This is also partly because of the deterioration of trust between nationalities and regions and the central government as a by-product of the narrative of self-rule over the last three decades. There have been numerous instances where the dominant narrative of diversity over unity has resulted in the eviction of dozens of members of other nationalities, making them more vulnerable to climate change.

Despite the challenges and problems, federalism does have benefits vis-à-vis climate change governance in Ethiopia. This federal system encourages nationalities to develop their culture, language, and local institutions, while also encouraging the amalgamation of indigenous knowledge and practices, which have played critical roles in the efforts made to adapt to and mitigate the effects of climate change. The devolution of power to the lower administrative structure facilitates the empowerment of communities at grassroots level and allows marginalized groups' voices to be heard. The take-home message from this analysis is the need to critically review, amend, and remove some articles in the federal and regional constitutions so that responses to climate change effects can be facilitated, coordinated, and improved. This will reinvigorate Ethiopia to realize a matured federal system predicated on multiple factors such as ethnicity, natural resources, agroecology, geography, and others.

References

Aalen, Lovise. 2002. *Ethnic Federalism in a Dominant Party State: The Ethiopian experience 1991–2000.* Bergen: Michelson Institute of Development Studies and Human Rights.

Abbink, John. 2011. 'Ethnic-Based Federalism and Ethnicity in Ethiopia: Reassessing the experiment after 20 years'. *Journal of Eastern African Studies* 5(4): 596–618.

Adem, Alebachew, and Woldeamlak Bewket. 2011. *A Climate Change Country Assessment Report for Ethiopia*. Addis Ababa: Epsilon International R&D.

African Development Bank. 2015. *Transitioning from INDCs to NDCs in Africa*. Abidjan: AFDB.

Alemu, Zerihun Gudeta, Lukas K. Oosthuizen, and H. D. Van Schalkwyk. 2002. 'Agricultural Development Policies of Ethiopia Since 1957'. *South African Journal of Economic History* 17:1–2.

Central Statistics Agency of Ethiopia (CSA), 2014. *Socio Economic Survey*. Addis Ababa.

Chabal, Patrick, and Jean-Pascal Daloz. 1999. 'Africa Works: Disorder as a political instrument'. *Canadian Journal of African Studies* 35(1): 181–3.

Devereux, Stephen. 2006. *Vulnerable Livelihoods in Somali Region, Ethiopia*. Brighton: Institute of Development Studies.

Elazar, Daniel. 1987. *Exploring Federalism*. Tuscaloosa: University of Alabama Press.

FDRE. 2001. *Initial National Communication of Ethiopia to the United Nations Framework Convention on Climate Change*. Addis Ababa: Ministry of Water Resources.

FDRE. 2007. *Ethiopia's National Adaptation Program of Action*. Addis Ababa: Ministry of Water Resources.

FDRE. 2011. 'Ethiopia's Climate-Resilient Green Economy Strategy'.

Fiseha, Assefa. 2018. 'Federalism and the Evolving Conceptions of Unity and Diversity in Ethiopia'. In *Revisiting Unity and Diversity in Federal Countries: Changing concepts, reform proposals and new institutional realities*, eds. Alain-G. Gagnon and Michael Burgess, 27–56. Leiden: Brill.

Fiseha, Assefa, and Mohammed Habib. 2010. 'Ethiopia'. In *Diversity and Unity in Federal Countries*, eds. Luis Moreno and César Colino, 139–66. Montreal & Kingston: McGill–Queen's University Press.

GCF. 2016. 'GCF/B.12/32: Decisions of the GCF Board: Twelfth Meeting of the Board, 8–10 March 2016'. www.greenclimate.fund/document/gcf-b12-32.

Atsbaha, Gebre-Selassie, and Tessema Bekele. 2010. *A Review of Ethiopian Agriculture: Roles, policy, and small-scale farming systems*. Addis Ababa: EDA and KOPIN.

GRAIN. 2019. 'Karuturi Global's New Land Deal in Ethiopia Must be Scrapped'. 13 June. https://grain.org/en/article/6258-karuturi-global-s-new-land-deal-in-ethiopia-must-be-scrapped.

Habib, Muhammad. 2010. *The Ethiopian Federal System: The formative stage*. Addis Ababa: Friedrich-Ebert-Stiftung.

Irish Red Cross. 2007. 'Africa Flood Disasters Increase by 8 Times in Three Years, Red Cross Figures Show'. *Relief Web*, 28 September. https://reliefweb.int/report/ethiopia/africa-flood-disasters-increase-8-times-three-years-red-cross-figures-show.

Ketema, Adamu Mulu, and Kasahun Dubale. 2020. 'Effect of Climate Change on Agricultural Output in Ethiopia'. *Journal of Perspectives on Financing and Regional Development* 8(3): 195–208.

Markakis, John. 2006. *Ethiopia: Anatomy of Traditional Polity*, 2nd ed. Addis Ababa: Shama Books.

Mesfin, W. 1984. *Rural Vulnerability to Famine in Ethiopia 1958–1977*. New Delhi: Vikas.

NASA Earth Observatory. 2008. 'Floods in Ethiopia'. https://earthobservatory.nasa.gov/images/35935/floods-in-ethiopia.

NMA. 2006. *National Adaptation Programme of Action of Ethiopia (NAPA)*. Addis Ababa: National Meteorological Agency.

Ottaway, Marina. 1994. 'The Ethiopian Transition: Democratization or a new authoritarianism?' *Northeast African Studies* 2(3): 67–84.

OXFAM International. 2009. *Band Aids and Beyond: Tackling disasters in Ethiopia 25 years after the famine*. Nairobi: Oxfam International.

Tadese, Girma, Kai Sonder, and Don Peden. 2003. 'The Water of the Awash River Basin a Future Challenge to Ethiopia'. www.iwmi.cgiar.org.

Taye, Bekalu Atnafu. 2017. 'Ethnic Federalism and Conflict in Ethiopia'. www.accord.org.za/ajcr-issues/ethnic-federalism-conflict-ethiopia.

Teklu, Berhan, Amare Hailu, Daniel A. Wiegant, Bernice S. Scholten, and Paul J. Van den Brink. 2018. 'Impacts of Nutrients and Pesticides from Small- and Large-Scale Agriculture on the Water Quality of Lake Ziway, Ethiopia'. *Environmental Science and Pollution Research* 25(14): 13207–16.

UNFCCC. 2005. 'Inventories of Anthropogenic Emissions by Sources and Removals by Sinks of GHG. Sixth Compilation and Synthesis of Initial National Communications from Parties not included in Annex I to the Convention'.

Worku, Adefires. 2016. 'Management of Dry Forests for Social-Ecological Resilience of the Pastoral and Agro-Pastoral Communities in the Dry Zone of Ethiopia'. Diss. Technology University of Dresden, Faculty of Environmental Sciences.

World Bank. 2006. *Ethiopia: Managing water resources to maximize sustainable growth. Country water resources assistance strategy*. Washington DC: World Bank.

World Bank. 2021. 'Climate Risk Profile: Ethiopia'. https://climateknowledgeportal.worldbank.org/sites/default/files/2021-05/15463A-WB_Ethiopia%20Country%20Profile-WEB.pdf.

7

Climate Governance and Federalism in the European Union

MARIACHIARA ALBERTON

7.1 Introduction

Building on its climate change mitigation policies and associated greenhouse gas emissions reductions, the EU aims to become the world's first climate-neutral economy by 2050, with a reduction of 55 per cent in greenhouse gas (GHG) emissions by 2030. Accordingly, in her first speech to the European Parliament, the new president of the European Commission, Ursula von der Leyen, announced the 'European Green Deal' as the EU's new growth strategy involving all economic sectors.[1] The Commission's increased ambition on climate-related activities has been supported by a revised proposal agreed by the European Council for the 2021–7 EU long-term budget to allocate 30 per cent of expenditure to such activities.[2] However, meeting the climate change challenge appears complex in the process of European differentiated integration (De Witte, Ott, and Vos 2017), the (still) problematic 2004 enlargement to include former communist countries, as well as Brexit (Leruth, Gänzle, and Trondal 2019), the current economic downturn and energy crisis.

The EU offers an interesting combination of different federal features while not yet a federation in most senses (*infra*, Section 7.3) manifest in its climate governance. These include a favourable context for decentralized and experimental policymaking, enhanced prospects for triggering dynamic processes of policy diffusion, and availability of multiple levels and venues for policymaking. In this realm, however, a number of knots still need to be disentangled, as Member States' diverging priorities on energy and sometimes conflicting positions on mitigation targets make intergovernmental decision-making in climate policy increasingly difficult. Policy fragmentation and poor or delayed implementation of EU legislation by Member States remain a critical issue.

7.2 EU Climate Change Commitments, GHG Emissions, and Climate Impact

In order to achieve GHG reduction targets, the EU has built a complex climate policy architecture based mainly on three pillars: the Emission Trading System

(ETS), a cap-and-trade system applying to some sectors and aiming to reduce emissions on a European level; the Effort Sharing (ES) instrument, which sets individual Member State's targets in non-ETS sectors; and the Land Use, Land Use Change, and Forestry (LULUCF) regulation, which accounts for emissions and removals stemming from land-use activities.[3] More specifically, the EU has combined binding objectives on emissions with additional binding targets on energy efficiency and renewable use for the year 2020 through its '2020 Climate and Energy Package'.[4] These are embodied in a set of binding acts (including those on ETS, ES, renewable energy, and energy efficiency) mandating a 20 per cent reduction of GHG emissions compared to 1990 levels, along with the achievement of a 20 per cent improvement in energy efficiency and of a 20 per cent share of renewables in the EU final energy consumption.

The first two targets were binding on Member States (MS), while the latter was merely 'indicative'. The following '2030 Climate and Energy Framework' builds on the preceding framework and upgrades and updates the EU's emission reduction and energy targets for the period from 2021 to 2030: at least 55 per cent cuts in GHG compared to 1990 levels (implemented by the EU ETS and ES and the LULUCF Regulation), a 32 per cent share for renewable energy, and a 32.5 per cent improvement in energy efficiency (under the 'Clean Energy for all Europeans Package', consisting of eight legislative acts, among which are those on renewable energy and energy efficiency and the new 'Governance Regulation', *infra*).[5]

Besides increasing targets, major differences between the 2020 and the 2030 frameworks relate to: the shift of renewable energy binding targets from MS to an overall Union target as a sign of a more intergovernmental and 'renationalized' phase (Kulovesi and Oberthür 2020; Rayner and Jordan 2016) compared with the previous phase, which had benefitted from a strong endorsement of those MS leaders and elites (e.g., UK and Germany) convinced of the necessity of an ambitious EU climate change mitigation policy (Rayner and Jordan 2016); the integrated framework for climate and energy planning, reporting, and reviewing under the new Governance Regulation as an important driver of cooperation, coordination, and convergence overseen by the Commission (Ringel and Knodt 2018; Szulecki et al. 2016); the integration of the LULUCF sector into the EU's Climate and Energy policy framework, not previously covered. More recently, the European Climate Law entered into force to turn the political commitment towards having a climate-neutral economy by 2050, included in the 'European Green Deal Communication', into a legal obligation.[6] In addition, the European Commission published the 'Fit for 55' Package to revise key EU policies and legislative acts across various sectors, including energy, transport, and building, and align them with the new 2030 climate target of at least 55 per cent GHG reductions and the 2050 climate-neutrality objective.[7]

Consistent with these commitments, the total GHG emissions in the EU-27 plus the UK have decreased by 1,330 million tonnes of CO_2e since 1990.[8] More recently, the economic downturn in 2020 following the Covid-19 pandemic, though incidental, has sharply reduced emissions and overall energy consumption, with the share of energy consumed from renewable sources likely having increased, and thereby securing achievement of the EU's climate and energy goals for 2020.[9] However, continuing at the rate achieved between 1990 and 2019 would be insufficient to meet the 2030 and 2050 objectives. In addition, current and predicted effects of climate change across the EU reinforce the urgency of mitigation and adaptation actions.[10]

7.3 Climate Governance in the Context of an Evolving EU Integration Process

7.3.1 The Quasi-federal Nature of the EU

The EU has many federal features (Palermo 2019). These include the ideological roots of the EU integration project (Burgess 2000); many EU principles (e.g., the precedence of EU law over national law; the direct effect of EU law in the national legal systems and between citizens; the principles of loyal cooperation, conferral, subsidiarity, and proportionality; the distribution of powers, *infra*); and a consistent part of EU constitutional terminology (e.g., terms such as pre-emption, supremacy, exclusive and concurrent powers, residual clause). In particular, the EU appears to resemble more the model of administrative federalism (Börzel 2005; Burgess 2000; Kincaid 1999; Schütze 2009), as most of legislative powers in the EU are currently shared (e.g., environment, climate, energy), and responsibilities for policy execution mostly rest with the Member States. Significant elements of federalism are still missing in the EU integration process, however. For instance, the hierarchical relationship between the EU and the MS, with the former prevailing over the latter (MacCormick 1999), is far from being settled and requires continual adjustment (*infra*).

Thus, throughout the long-running European integration process, the EU has been described by scholars as: *sui generis* (Mason 1955; Phelan 2012; Wallace and Wallace 2000); a system of multi-level governance (Hooghe and Marks 1996, 2001; Piattoni 2010); an incomplete, supranational constitutional creature (MacCormick 1999; Walker 2012); a supranational federation (von Bogdandy 2012); an asymmetric integration process (Palermo 2019).

The debate has also been addressed by the Court of Justice of the EU (CJEU) and by the constitutional courts of several MS, each asserting a different perspective on the sovereignty issue.[11] These 'judicial dialogues' are not yet concluded and represent a specific feature of the evolving EU process of integration (Krommendijk 2020; De Witte et al. 2016).[12]

7.3.2 Architecture of the EU

The EU's institutional architecture is based mainly on two orders of government: the EU and the Member States. However, the MS's regional and local levels are increasingly gaining institutional representation at EU level (e.g., through the Committee of the Regions). In particular, the European Commission has sought to incorporate regions into the policy process – both to increase policy effectiveness and to enhance its visibility and legitimacy at the regional level (Keating 2017).

In general, the European Council defines the EU's overall political direction and priorities and includes the heads of state or government of the EU MS. MS defend their own national interests in the Council of the European Union, where the relevant national ministries meet and have the authority to commit their governments to the actions agreed on in the meetings, while the interests of the EU are promoted by the European Commission, where politically independent members (from MS' national governments) sit.

The EU Treaties (i.e., the Treaty on the Functioning of the European Union, TFEU, and Treaty on European Union, TEU) codify some typical federal principles, such as loyal ('sincere') cooperation (art. 4.3 TEU), conferral, subsidiarity, and proportionality (art. 5 TEU). Four types of powers are listed (arts. 3–6 TFEU):

(1) exclusive, only the EU can act;
(2) shared between the EU and MS, such as on environment and climate change (art. 191–3 TFEU);
(3) those where the EU sets up arrangements and MS must coordinate;
(4) those where the EU can support, coordinate, or supplement MS's actions.

7.3.3 Climate Change and the EU Division of Powers

Since 2009, the Treaty of Lisbon has introduced a new EU shared power on energy in art. 194 TFEU (Benson and Russel 2015), which was before exerted by the EU on the basis of various provisions scattered throughout the Treaty, paradoxically allowing a wider margin to the EU (Fehling 2021; Jegen 2014).[13] Therefore, climate and energy measures can be based on both environment (art. 191–3) and energy (art. 194) provisions with two main caveats.

First, whereas the ordinary EU legislative procedure applies to both cases of shared competences, MS may adopt more stringent protective measures than those set at the EU level only according to art. 193 TFEU, not under art. 194 TFEU. For instance (art. 191–3), being the legal basis of the Effort Sharing instrument, under this system MS may maintain or introduce more ambitious targets than those set at the EU level.

Second, the sovereignty clause included under art. 192.2 (i.e., environment and climate competence) foresees that some EU measures may be adopted by the Council unanimously, with a special legislative procedure: for instance, provisions primarily of a fiscal nature, measures affecting land use, and those significantly affecting a MS' energy sources choices and the structure of its energy supply (thus derogating to the sovereignty clause of art. 194.2, which prescribes that EU measures 'shall not affect a MS' right to determine the conditions for exploiting its energy resources, its choice between different energy sources and the general structure of its energy supply'). To this regard, art. 194 TFEU, in addition to other political drivers (Bürgin, 2014, Rayner and Jordan 2016), has directly affected the shift from MS's binding renewable energy targets under the 2020 Package to an overall binding target at EU level under the 2030 Framework. In fact, the legal basis used for adopting the new 2018/2001 Directive 'on the promotion of the use of energy from renewable sources' was art. 194, instead of art 191 TFEU (which was the legal basis for previous renewable energy Directive 2009/28). In other terms, the adoption of MS legally binding targets under the new 2018/2001 Dir. would have likely violated the boundaries set under art. 194.2 TFEU.

Considering the interdependencies between environment, climate, and energy policies and the 'grey areas' left by arts. 191–4 TFEU, the relationship between EU and MS powers regarding climate-related acts is subject to varying interpretation and political compromises. An example of this ambiguity is offered by the Governance Regulation, which has a double legal base (i.e., both art. 191 and 194 TFEU), thus leaving room for uncertainty on several points (e.g., the application of sovereignty clauses and of national reinforcements of protection) (Fehling 2021).[14] On the other side, it should be noted that the Governance Regulation aims at linking the EU climate policies and the so-called Energy Union (a framework strategy launched in 2015 to bring about the transition to a low-carbon, secure, and competitive economy) by integrating Member States' planning and reporting obligations with regard to climate and energy (*infra*).[15]

After EU climate and energy binding acts (e.g., directives and regulations) are adopted, MS are responsible for implementing and enforcing them at national level. The European Commission monitors this implementation and has the power to commence infringement procedures (art. 258 TFEU), which can lead to a case being filed before the CJEU.[16] Poor or delayed implementation of EU environmental legislation, including climate and energy related acts, has being a constant feature of the EU legal history.[17] The opposite case of MS contesting the

Commission, also occurs, for example when some MS's appealed against the Commission's powers to review MS's National Allocation Plans of emission allowances (Bogojević 2010, 2013; Damro et al. 2008).[18] Because of the national challenges against the Commission's decisions and the subsequent legal uncertainty, since 2013 this decentralized, bottom-up process has been substituted by an EU-wide cap.[19]

Besides infringement procedures, 'softer' enforcement mechanisms (i.e., 'iterative processes' or 'dialogues' or the 'Open Method of Coordination') exist under EU law in areas where powers remain at the MS level and EU binding measures cannot be adopted. Such mechanisms rely for their success on the cooperation of MS (Smismans 2011). Under this soft approach, objectives are set at EU level through recommendations, standard-setting, benchmarking, peer review, and best practices (Ringel and Knodt 2018), while decentralized implementation responsibilities rely on MS. The European Commission has consecutively applied these mechanisms to build a structured dialogue with the MS especially in the energy efficiency and renewable sectors, where binding national targets could not be adopted (art. 194.2 TFEU). The aim has been to cajole national energy policies towards more ambitious decarbonization targets, and somehow overcome political divisions between different MS on energy and climate priorities within the Council (Knodt, Ringel, and Müller 2020). The 2018 Governance Regulation enhances these soft governance arrangements by incorporating harder elements (*infra*, Section 7.4.1).

7.4 EU Climate Mitigation as a Product of Federal Dynamics and Variables

7.4.1 The EU as an 'Opportunity Structure' for Policy Innovation, Diffusion, and Interactive Learning

As expected, the EU offers an interesting combination of different (federal) features and represents an 'opportunity structure' for policy innovation, rapid policy diffusion, and interactive learning in the field of mitigation (Jänicke and Quitzow 2017; Meyer-Ohlendorf et al. 2014). Over time both competitive and cooperative forms of governance have driven mitigation policies. MS, especially those most economically efficient and the 'frontrunners', have been facilitated in promoting their interests and pioneer policy choices into EU climate action, while the supranational framework has guaranteed a common arena for mutual learning, gradual convergence around common mitigation objectives in light of considerations of solidarity, and 'differentiated responsibilities' (Fehling 2021; Rayner and Jordan 2016), through funding and supporting mechanisms, thus 'leaving no one behind'.[20] The institutional architecture and the mitigation policies of the EU reflect this mix of decentralized, flexible, competitive, bottom-up approaches, in addition to cooperative and supranational coordination elements, and present both opportunities and challenges as illustrated by the following examples.

Over time, climate policy has become an important driver of EU integration, especially after the increasing support for European-level action in this field showed in public opinion polls and by green parties and environmental NGOs (Oberthür and Roche Kelly 2008; Schreurs and Tiberghien 2007).[21] EU leadership has been driven by such a combination of events, and in turn by the (reinforcing) competing role of mutual leadership played by several MS, for instance, Germany, the UK, the Netherlands, and Denmark, but also Finland and Sweden.

Several MS have anticipated and influenced the EU's climate mitigation initiative and consequently that of other MS, pushing European climate mitigation policy forward while at the same time gaining credit for their actions domestically. This has occurred, for instance, by establishing governance frameworks with a long-term outlook through the adoption of national climate laws (including some adaptation measures as well), also referred to as 'flagship laws' (Fankhauser et al. 2015). The UK's pioneering 2008 *Climate Change Act* inspired a range of related national framework laws, although the Paris Agreement may have accelerated this diffusion (Duwe and Evans 2020).[22] Emulation was a major mechanism shaping climate framework laws developed in other MS after the UK example (Evans and Duwe 2021; Meyer-Ohlendorf 2020). Thus, peer behaviour can be confirmed to have encouraged diffusion in this context, revealing that framework legislation entails a 'signalling character' as it provides for further negotiations (Fankhauser et al. 2016) and offers an indicator for further climate change legislation.

Most MS' national mitigation policies and measures have been implemented in response to EU strategies and to binding instruments (e.g., the 2009 Renewable Energy Directive, the 2012 Energy Efficiency Directive, the Effort Sharing Decision).[23] Only 27 per cent of reported national actions are not directly related to a specific EU policy or legislation.[24] However, since MS have adopted climate mitigation policies and legislation of varied ambition, taking advantage of the flexibility of EU policy and of binding instruments on mitigation (directives), as well as of the autonomy of MS in the energy field (i.e., subsidiarity and sovereignty clause), some delays in adopting mitigation measures as well as implementation variances across MS occurred (Fleig et al. 2017).[25] Thus, EU mitigation policies do not convey such a coherent, homogeneous, and ambitious approach as one would have expected, for instance, based on the EU's climate change leadership aspiration (Gupta and Ringius 2001; Massey et al. 2014; Parker and Karlsson 2010; Rayner and Jordan 2016). Especially because of the 2004 problematic enlargement to include formerly communist central and eastern countries, dominated by fossil fuel energy programmes, the EU's mitigation policy ambition has at times been limited (*infra*), and has been only partially reinstated through substantial financial compensation and exemptions (Kulovesi and Oberthür 2020; Peeters and Athanasiadou 2020; Rayner and Jordan 2016).

7.4.2 The 'Competitive Cooperation' between the Council and the Commission

Member States' diverging priorities and sometimes conflicting positions are reflected, in turn, into the European Council and into the Council of the EU, where MS heads and MS ministers respectively reaffirm their sovereign priorities, sometimes in contrast with that of the European Commission. For instance, in the European Council of June 2019, Poland, Hungary, Slovakia, and the Czech Republic opposed the proposed target of zero emissions, thus hampering a 2050 carbon neutrality target for the EU. Ongoing clashes among national sovereignty over energy policies (Herold et al. 2019; Marcinkiewicz and Tosun 2015; Szulecki 2016) hinder to some extent the EU's mitigation aspirations. Moreover, clashes among MS in the Council and the Commission illustrate a peculiarity of the EU's governance system – the 'competitive cooperation' between the Council and the Commission in legislative agenda setting (Bocquillon and Dobbels 2014).

To reconcile these opposing positions and negotiate a pan-EU climate change mitigation goal, the strategy that EU institutions seem to pursue is based on collective actions where all MS participate in the mitigation efforts, while considering national circumstances and concern of fairness and solidarity. This approach has been applied throughout numerous European instruments, such as EU legislation (e.g., EU ETS, Effort Sharing), specific financial mechanisms (e.g., the Modernisation Fund, supporting investments for a just transition in carbon-dependent regions in ten lower-income Member States), financial assistance through existing funding schemes (e.g., structural and investments funds), and increasing instruments combining research, innovation, and funding (e.g., Just Transition Platform, NER 300 programme), which create networks for MS, regions, agencies, and stakeholders to exchange information and knowledge, good practices, and specific assistance to meet collective targets.[26]

Some national sovereignty instances over energy transition continue to hinder not only the integration of European energy policies and the Energy Union, but also the coherence and effectiveness of European mitigation action (Mata Pérez et al. 2019).[27] Nonetheless, the package of flexible instruments described above constitutes a sound attempt to contrast the 2004 problematic enlargement dimension of the EU with regard to energy governance fragmentation and the increasingly difficult intergovernmental decision-making in climate policy (Rayner and Jordan 2016) previously mentioned. Some scholars have also suggested that the Governance Regulation constitutes an attempt by the Commission to overcome the hard confrontation between the two blocs of MS, the *Visegrad* and *Green Growth Groups*, and to improve MS' planning and reporting gaps by introducing an integrated framework and a 'harder form of soft governance', exemplified by the 'blank cheque' (Ringel and Knodt 2018).[28]

In other terms, in the event of insufficient ambitious national plans or progress towards the energy and climate targets on the part of MS, the Commission is entitled to adopt additional measures, legislative acts, and exercising powers at the EU level (Knodt, Ringel, and Müller 2020; Oberthür 2019). So, in case of delivery gaps, EU intervention over MS's energy mix choices would be justified, thus bypassing the sovereignty clause of art. 194.2 TFEU, and within the boundaries of the subsidiarity and proportionality principles (Monti and Martinez Romera 2020). Recalling that the Renewable Energy Directive (2018/2001) and the Energy Efficiency Directive (2018/2002) define Union-wide targets for 2030, instead of national individual ones, the new means of the Commission to advance MS policy implementation under the Governance Regulation balance the additional flexibilities granted in favour of MS (Monti and Martinez Romera 2020; Oberthür 2019).

In this respect, some scholars have also observed a trend towards a gradual increase of European Commission authority over MS external energy policies (a power not explicitly conferred to the EU under art. 194 TFEU), culminating with the adoption of Decision 2017/684.[29] The Commission has started to influence MS' negotiations over energy imports and relations with third parties through soft pre-emptive compliance instruments, such as *ex ante* checks that prevent non-compliance with EU rules, guarantee the integrity of the internal energy market, and allow the Commission to gain supranational governance capacity in the energy realm (Dehousse 2015, Thaler and Pakalkaite 2020). In this context, more recently, the Council has invited the Commission to prepare a new strategy on external action in the field of energy cooperation in light of a rapid shift towards the climate neutrality goal.[30]

Previous examples shed light on the drivers of the EU integration process and on the current challenges for a coherent European mitigation policy. The evolving powers of the European Commission tend to be described in contrast with those of MS, which are mainly shaped by national economic interest (in addition to political cultures and regulatory styles: Hoppe and Wesselink 2014; and constitutional design: Steurer and Clar 2015). On the other side, the Commission's role is confronted with MS interplay and changing equilibrium in the European Council. Thus, some scholars describe these processes as polarised and disconnected instead of being complementary dynamics in a complex EU governance (Schmidt 2016), while others underline that the decentralized and multi-level governance structure of the EU has encouraged a process of mutual reinforcement, where MS and the European Commission are competing (Schreurs and Tiberghien 2007) or mutually supporting (Bürgin 2014) for leadership.

7.4.3 Multilevel Reinforcing Mechanisms: Linking the EU and Regional/Local Governments and Communities

Over time the Commission has enabled and built on multiple dynamics (both vertical and horizontal ones) to exert increasing influence towards ambitious mitigation actions. This multi-level and multi-sectoral approach in targeting industries, mobilizing economic interests, involving sub-national authorities and domestic stakeholders across levels of governments (Szulecki et al. 2016; Wettestad et al. 2012), has empowered EU mitigation policies by directly linking European goals with domestic and local support and with industry interests for climate-friendly technologies. To this extent, the 'Europe 2020' Strategy, the '2020 Climate and Energy Package', and the '2030 Climate and Energy Framework' have put special emphasis on strengthening the interconnections among the industrial sectors, the research community, and financial resources and specific funding programmes (e.g., NER 300 and NER 400; Innovation Fund and Modernization Fund), aiming at EU low-carbon objectives.[31] Some scholars have noted that these multi-level reinforcing mechanisms are equally present in 'frontrunners' and 'laggards' MS (Jänicke and Quitzow 2017) and provide forums for benchmarking, especially in the framework of soft governance mechanisms, as well as an opportunity structure for innovation and experimentation, interactive learning, and best-practices diffusion.

To this same end, the European Commission has pushed for a greater emphasis on regional and local governments of MS in tackling mitigation. In 2008 the European Commission, with support of the Committee of the Regions, launched the *EU Covenant of Mayors* initiative.[32] This initiative has become a well-established network of cities and towns committed to implementation of the EU's GHG-reduction target by 2030 (e.g., by submitting energy and climate action plans and by taking actions in policy areas directly influenced by local administration).[33] Furthermore, the European Commission has recognized the role of MS' provincial and regional government levels as *Covenant Territorial Coordinators* (CTCs) in supporting municipalities with strategic guidance, financial aid (e.g., through ERDF and Cohesion Fund) and technical support. The CTCs, in some cases, even compensate for the void left by the national level, that is, the absence of MS frameworks for local energy planning. Thus, these local-level initiatives, supported by the European Commission, play an important role in reinforcing mitigation policies in pioneer MS (e.g., Germany, Denmark, the UK) and in filling the gaps in laggard MS (e.g., Poland) with weaknesses at the national level (Jänicke and

Quitzow 2017). In addition, a reduction of high administrative fragmentation has been observed in some MS (e.g., in Spain, Italy, and Belgium) (Melica et al. 2018).

Another advantage partially derived by this tighter vertical coordination network is reflected in the improved capability of local level governments to promote and finance sustainable energy projects. Also, the EU Covenant of Mayors initiative and the CTCs system have been observed to catalyse dynamic processes of policy diffusion (Grafakos et al. 2020; Kona et al. 2016; Melica et al. 2018), and to influence local entities persistently, for example, through baselines, guidance documents, regular communications and templates which push towards convergence (Heyvaert 2013). The transparency of the Covenant system also creates opportunities for 'naming and shaming', thus increasing the compliance of local entities (Kona et al. 2016). At the same time, the Covenant promotes some flexibility and differentiation among the local entities aiming at the development and implementation of climate mitigation and sustainable energy actions which are innovative and tailored to local circumstances.

7.5 Adaptation Action in the EU

7.5.1 The EU Initiative and Financial Assistance as Key Factors to Catalyse Adaptation Action in Member States

The EU initiative on adaptation (2013 EU Adaptation Strategy, followed by the new 2021 Strategy) only encourages MS, regional, and local levels to take action, rather than mandating it, as the EU lacks formal authority in a variety of areas related to adaptation (Fleig et al. 2017).[34] Thus, cooperation among MS and coordination with the EU play a prominent role. The European Commission, in particular, provides financial assistance to MS adaptation initiatives, monitors and assesses the national adaptation strategies, and supports the MS and their government levels through the European Climate Adaptation Platform (Climate-ADAPT), which allows the exchange of data, good practices, and information.[35]

It should be noted that in the adaptation field several EU MS adopted strategies and framework legislation including provisions on adaptation earlier than the EU.[36] The increasing costs and damages associated with more frequent extreme weather events such as floods, storms, and heatwaves, recorded in MS and gaining increased public awareness and attention (Lorenzoni and Hulme 2009) have propelled these MS' early adaptation actions.[37] In addition, according to some scholars (Fleig et al. 2017; Russel et al. 2020), diffusion of adaptation policies and laws in other 'laggard' and 'wavering' countries has been observed as a 'Nordic-country effect'; in other terms, the early adoption of such laws in Nordic countries has affected the diffusion of adaptation frameworks in other EU MS. In the EU context, peer behaviour has been quite influential thanks to the spread of ideas, practices, and institutions (Massey et al. 2014).

However, the steady increase over five years (2013–18) of national adaptation strategies and plans in the EU MS is strongly influenced by the EU Adaptation Strategy, adopted by the Commission.[38] In fact, the EU initiative has catalysed action in MS and particularly in those that were in earlier stages of developing an adaptation policy. The EU's facilitative role through providing guidance, funding research and adaptation action under the Strategy has urged and enhanced MS initiatives (Massey and Huitema 2016). In particular, in central and eastern MS, the most important driver for diffusion of adaptation measures was the EU's effort to put adaptation on the MS agenda and the accompanying financial support (Massey et al. 2014).

The EU's efforts to promote adaptation across MS have intensified in recent years, for instance by establishing mechanisms of knowledge sharing and best-practices exchange among public and private stakeholders of MS (i.e., through the Climate-Adapt Platform), by the involvement of MS' local governments to engage in adaptation initiatives on the basis of voluntary commitments, and by providing financial support through existing European funds (e.g., the EU's Solidarity Fund). In this respect, EU funds play an important role as there is a lack of funding, with only half of Member States having budgets attached to their adaptation instruments (i.e., National Adaptation Strategies, NAS, and National Adaptation Plans, NAP).[39] In addition, since the EU only encourages MS to adopt comprehensive adaptation strategies, a recent factor influencing MS' action could be detected in the Commission's intention to adopt a legally binding instrument in the event that the progress of MS is insufficient.[40] Some hints of coercion, in the long run, can be perceived as an additional driver for the spreading of adaptation action in MS, as hard law equips the EU Commission with the power of initiating infringement procedures in case of non-compliance, while soft governance only relies on the active cooperation of MS.

Vertical coordination among European, national, regional, and local authorities is essential, as current financial and knowledge gaps at the local level may hinder local action. However, systematic coordination across all levels of administration has only been observed in some MS, while gaps in the involvement of sub-national governance levels have been detected in other MS.[41] In 2014, the European Commission launched a separate initiative called *Mayors Adapt*, based on the *EU Covenant of Mayors* experience (see above). This had the aim of engaging cities in taking action to adapt to climate change, either by developing comprehensive adaptation strategies or by integrating adaptation to climate change into their relevant existing plans. In 2015, the two initiatives officially merged into the *Covenant of Mayors for Climate & Energy* and now represent successful experiences (Kona et al. 2017) of vertical (local and regional linking to national/EU levels) and horizontal (e.g., national and transnational city networking, learning and best practices sharing) collaboration for mitigation and adaptation actions.[42] At the local level, involvement in the EU Covenant of Mayors for Climate and Energy has proven to be effective in promoting city-level adaptation policymaking and in linking the EU strategy with local action. In some cases, further support is assured by national and regional initiatives (Grafakos et al. 2020). For instance, the Ministry of the Environment of Czech Republic has officially committed to providing strategic guidance, financial and technical support to local authorities that are signatories to the Covenant and has been recognized by the European Commission as a Covenant National Coordinator. These cases further exemplify how, over the years, the EU has established collaborative policy frameworks, networks facilitating mutually supportive schemes, and knowledge sharing and financial support mechanisms across government levels.

7.5.2 Adaptation as a Cross-Cutting Policy Area across Multiple Scales: Progress and Challenges

In this realm, the transnational cooperation among MS (plus third countries) has also increased with the recognition of the importance of adaptation as a cross-cutting policy area. Notably, EU-driven transboundary adaptation action is channelled through four macro-regional strategies, thus involving most MS.[43] For instance, the EU Strategy for the Danube Region emphasizes adaptation to extreme weather events and provides an important platform to foster cooperation on joint monitoring and flood management. At the same time, this cooperation facilitates the collective implementation of existing EU directives and sectoral policy which, in turn, contribute to efforts for adaptation to climate change with regard to water issues. This case is replicated in the other European macro-regions regarding other sectoral policies and related adaptation initiatives. For instance, multiple initiatives addressing adaptation to climate change exist for mountain ranges and for biodiversity (e.g., Alpine space).[44]

These examples illustrate the process of mainstreaming adaptation action into the EU's sectoral policies at different levels, by supporting environmental policy integration practices across multiple scales (Heyvaert 2013; Jordan and Lenschow 2010), and the process of 'multi-level reinforcement of policy action' in the EU climate change adaptation field. Nonetheless, some policy sectors, such as marine and coastal ones, though singled out as priorities in the EU Adaptation Strategy, receive less attention in terms of adaptation mainstreaming and do not fall into these virtuous dynamics mainly because of some MS' conflicting agendas and preferences. Germany, for instance, has strongly opposed any policy action affecting marine and coastal planning issues from the European Commission (Russel et al. 2018). In addition, gaps at national and sub-national levels in introducing adaptation considerations in certain sectors are still relevant. Only a few MS have national policy instruments that promote adaptation at the sectoral level, in line with national priorities and in areas where adaptation is mainstreamed in EU policies.[45] A final aspect concerns knowledge gaps. Investment in the development of knowledge on climate adaptation is more likely in countries that already have a strong research base, and a greater critical mass (e.g., Germany, France, and the UK, as a former MS), while smaller countries and countries with a small research budget make progress by becoming involved in European research projects and by cooperating with other MS that face similar issues (Massey and Huitema 2016; Massey et al. 2014; Russel et al. 2020).

7.6 Conclusion

EU climate governance has been shaped over the years into a very dynamic and progressive process, leading to ambitious policies with ambitious targets. This leadership was developed despite the hindrance of conflicting MS' positions and diverging priorities, a burdensome enlargement process, and still uncountable uncertainties in the wake of Brexit. The institutional structure of the EU has definitely played a major role in the creation, circulation, and development of climate mitigation and adaptation policies, by providing an arena in which leadership can be exerted at multiple levels and multiple times, by fostering experimental and innovative solutions, by triggering numerous horizontal and vertical forums for mutual learning and support, by mobilizing economic interests at all levels and sectors, by providing substantial financial resources and funding programmes that have supported the mitigation and adaptation policies diffusion and implementation at different levels of governance and in most affected MS, and by combining (differentiated) legal obligations for MS and voluntary mechanisms (i.e., soft governance mechanisms).

Over time, this 'multi-impulse' system has endorsed and reinforced a relatively robust EU climate governance, even against the backdrop of past and present challenges and hindrances – notably inherent tensions in EU climate and energy governance; misalignment of policy objectives; tensions between flexible and stable approaches; policy fragmentation and weak implementation.[46] Despite additional current and future pressures due to the economic and energy crises, and besides the warning that more than incremental developments in the EU's climate policy are needed to meet the EU's goal of carbon neutrality by 2050, a number of mechanisms and options exist to maintain progress towards EU's climate ambitions. As explained through the chapter, the EU Governance Regulation offers the opportunity to overcome the EU's dilemma of having ambitious climate policies but only limited authority and capacity in the energy policy field, and the potential to enforce interim targets, as it allows for streamlining and strong coordination under the Commission's oversight. Moreover, the Commission is in the process of reviewing, and where necessary proposing to revise, all relevant policy instruments to deliver the additional emissions reductions for 2030 and achieve the climate-neutrality target by 2050.

Notes

1 Ursula von der Leyen, president-elect of the European Commission, *Speech in the European Parliament Plenary Session*, 16 November 2019, anticipated by her '*Political Guidelines for the next European Commission 2019–2024*', and followed by the Communication from the Commission '*European Green Deal*', COM (2019) 640 final.

2 At the request of the EU MS Heads of State or Government, the European Commission revised the previous *Multiannual Financial Framework for 2021–2027* (COM (2018) 321 final) and presented in May 2020 a package combining the future *Multiannual Financial Framework* and a specific Recovery effort under *Next Generation EU*. The proposal was agreed on by the European Council on 21 July 2020. See: *Special meeting of the European Council (17, 18, 19, 20 and 21 July 2020) – Conclusions*, EUCO 10/20.

3 Respectively: ETS Dir. 2018/410 (amending Dir. 2003/87); ES Reg. 2018/842 (amending Reg. 525/2013 and Decision 406/2009); LULUFC Reg. 2018/941 (amending Reg. 525/2013 and Decision 529/2013).

4 Including the EU ETS Dir., the Effort Sharing (ES) Reg., and the Renewable Energy Dir. (2009/28) and Efficiency Dir. (2012/27). See European Commission. '2020 Climate & Energy Package.' *European Commission*. https://ec.europa.eu/clima/eu-action/climate-trategies-targets/2020-climate-energy-package_en.

5 Energy Performance in Buildings Directive (2018/844); Renewable Energy Dir. (2018/2001); Energy Efficiency Dir. (2018/2002); Governance of the Energy Union and Climate Action Reg. (2018/1999); Electricity Reg. (2019/943); Electricity Dir. (2019/944); Risk Preparedness Reg. (2019/941); ACER Reg. (2019/942). See: European Commission. 2021. 'Clean Energy for all Europeans Package.' *European Commission*. https://ec.europa.eu/energy/topics/energy-strategy/clean-energy-all-europeans_en.

6 'Regulation (EU) 2021/1119 establishing the framework for achieving climate neutrality and amending Regulations (EC) 401/2009 and (EU) 2018/1999'.

7 See: Commission Work Programme 2021 'A Union of Vitality in a World of Fragility', COM/2020/690 final; and the most recent 'Fit for 55' Package, published in July 2021, COM (2021) 550 final.

8 Following the UK's departure from the EU on 31 January 2020, the EU and the UK entered a transition period (1 February–31 December 2020) as agreed on the basis of the 2019 Withdrawal Agreement and the Political Declaration. The EU law applied to the UK until the end of this period. Key provisions of Regulation (EU) No 525/2013 (*Mechanism for Monitoring and Reporting GHG Emissions*) applied to the UK in respect of greenhouse gases emitted during 2019 and 2020. Following the transition period, the EU UK Trade and Cooperation Agreement entered into force on 1 May 2021 and together with the Withdrawal Agreement and its protocols, provides the framework for present EU relations with the UK. For an overview of the different GHG emission estimates published regularly by bodies of the EU, see EEA (European Environment Agency). 2020. 'Complementary Emission Estimates Produced by EU Organisations.' *EEA*. www.eea.europa.eu/themes/climate/different-emission-estimates/emission-estimates-produced-by-eu. See, in particular: EEA. 2020. *Annual European Union greenhouse gas inventory 1990–2018 and inventory report 2020*. EEA. www.eea.europa.eu/publications/european-union-greenhouse-gas-inventory-2020.

9 See: EEA. 2020. *Trends and projections in Europe 2020. Tracking progress towards Europe's climate and energy targets*. EEA. www.eea.europa.eu/themes/climate/trends-and-projections-in-europe.

10 See: EEA. 2020. *Impacts, vulnerability and risks*. EEA. www.eea.europa.eu/themes/climate-change-adaptation/impacts-vulnerability-and-risks.

11 The CJEU has stated several federal principles, such as the precedence (primacy) of EU law over any national source of law (*Costa v ENEL*, 1964, Case 6/64) and the direct effect of European law in the national frameworks of MS and between citizens (*Van Gend en Loos v Netherlands*, 1963, Case 26/62). On the other side, some national constitutional courts have adopted a dualistic approach to the relationship between the domestic and the EU's legal order that guarantees the predominance of fundamental principles included in the national constitution (i.e., theory of counter-limits). See the following cases of the German (Solange I and II, 1974), Italian (Granital, 1984, Frontini, 1973), French (Decision No. 2004–505 DC) and Spanish (DTC 1/2004) Constitutional Courts.

12 See more recently: BVerfG, Judgment of the Second Senate of 5 May 2020. 2 BvR 859/15, paras. 1–237. www.bverfg.de/e/rs20200505_2bvr085915en.html. following the CJEU judgment C-493/17 (http://curia.europa.eu/juris/liste.jsf?language=en&num=C-493/17); in addition, see the so-called Taricco saga (Taricco I C-105/14, and Taricco II C-42/17).

13 Treaty of Lisbon in force since 1 December 2009. On the novelties introduced by the Lisbon Treaty with regard to energy see: ClientEarth legal briefing, January 2010 'The Impact of the Lisbon Treaty on Climate and Energy Policy: An environmental perspective', available at: www.clientearth.org/latest/documents

14 The Court of Justice of the EU has clarified some doubts related to art. 192.2 TFEU in the case C-5/16, *Republic of Poland v European Parliament and Council of the EU.*

15 European Commission (2015) 'Framework Strategy for a Resilient Energy Union with a Forward-Looking Climate Policy', COM (2015) 80 final.

16 Directives and regulations are legal acts adopted by the EU institutions and addressed to the EU Member States (art. 288 of the TFEU). While directives are binding as to the result to be achieved and need to be transposed into national law of EU Member States, regulations are binding in their entirety and directly applicable in all EU Member States.
Two examples concerning mitigation are the cases brought against some MS for not transposing the EU ETS within the prescribed time, respectively Italy (C-122/05) and Finland (C-107/05.

17 At the end of 2019, there were a total of 327 open infringement cases relating to the environment, the highest number of any area, in addition to 109 cases relating to energy. The 2019 annual report 'Monitoring the Application of EU Law' of the European Commission is available at: www.impel.eu/wp-content/uploads/2020/08/report-2019-annual-report-monitoring-application-eu-law_en.pdf.

18 The Visegrad Group (Poland, Hungary, Slovakia, and Czech), Estonia, and Latvia. (Case T-183/07). For a comment see: Honor Mahony, 2007. 'Latvia Becomes Sixth Country to Fight EU Emissions Caps'. *EU Observer*. http://euobserver.com/?aid=24557.

19 See EU webpage: EEA. 'National allocation plans.' *EEA*. https://ec.europa.eu/clima/eu-action/eu-emissions-trading-system-eu-ets/development-eu-ets-2005-2020/national-allocation-plans_en.

20 For example, the Just Transition Mechanism (JTM), a key tool to ensure that the transition towards a climate-neutral economy is promoted in a fair way. See: EEA. 'The Just Transition Mechanism: Making sure no one is left behind.' *EEA*. https://ec.europa.eu/info/strategy/priorities-2019-2024/european-green-deal/finance-and-green-deal/just-transition-mechanism_en. See also the Communication from the Commission '*European Green Deal*', COM (2019) 640 final.

21 Eurobarometer (https://ec.europa.eu/commfrontoffice/publicopinion/index.cfm). Rounds of Eurobarometer survey on climate change were carried out in 2008, 2009, 2011, 2013, 2015, and 2017. In the 2019 survey, 93 per cent of EU citizens see climate change as a serious problem.

22 Revised in 2019.

23 (2019) Reports by Member States to the European Environment Agency (EEA) under the EU Monitoring Mechanism Regulation.

24 European Topic Centre on climate change mitigation and energy. 2019. *Overview of reported national policies and measures on climate change mitigation in Europe in 2019.* Available at: www.eionet.europa.eu/etcs/etc-cme/products/etc-cme-reports/etc-cme-report-5-2019-overview-of-reported-national-policies-and-measures-on-climate-change-mitigation-in-europe-in-2019.

25 See: European Court of Auditors. 2017. 'EU Action on Energy and Climate Change.' Review 1/2017, https://op.europa.eu/webpub/eca/lr-energy-and-climate/en.

26 The beneficiary MS are Bulgaria, Croatia, Czechia, Estonia, Hungary, Latvia, Lithuania, Poland, Romania, and Slovakia. See: EEA. 'Modernisation Fund.' *EEA.* https://ec.europa.eu/clima/eu-action/funding-climate-action/modernisation-fund_en. With regard to structural and investment funds See: EEA. 'European Structural and Investment Funds.' *EEA.* https://ec.europa.eu/info/funding-tenders/funding-opportunities/funding-programmes/overview-funding-programmes/european-structural-and-investment-funds_en.

27 Conflicts among EU MS governments' different energy positions increased when the idea of an 'Energy Union' started to emerge in EU diplomacy. A more recent example of diverging MS's interests and positions is provided by the Council of October 2020, which was forced to postpone the agreement on a new emission target for 2030 (on the contrary, strongly supported by the Commission, the EU Parliament, and the Green Growth Group) and the submission of the EU's updated NDC to the UNFCCC to December 2020 (European Council Conclusions on Covid-19 and Climate Change, 15 October 2020).

28 The Visegrad group plus Bulgaria and Romania. The *Green Growth Group* consists of the following EU MS: Belgium, Denmark, Estonia, Finland, France, Germany, Italy, Luxembourg, the Netherlands, Portugal, Slovenia, Spain, Sweden, and the UK, plus Norway that have been collaborating since 2014 to make EU climate policy more ambitious and sustainable.

29 Decision (EU) 2017/684 'Establishing an information exchange mechanism with regard to intergovernmental agreements and non-binding instruments between Member States and third countries in the field of energy', repealing Decision 994/2012/EU.

30 See points 9–12, 14 and in particular point 18 of the *Council Conclusions on Climate and Energy Diplomacy: Delivering on the external dimension of the European Green Deal* (doc. 5263/21): 'Council invites the Commission and the High Representative to prepare, by the end of 2021, a new strategy on international energy engagement, in accordance with the goals set out above and taking into account the specificities of particular regions and countries while fostering energy partnerships, and developing regional energy cooperation, particularly in the EU's Neighbourhood.'

31 See respectively: European Commission. 2010. *Europe 2020: a Strategy for smart, sustainable and inclusive growth*, COM (2010) 2020; and the following funding programmes: EEA. 'NER 300 programme.' (https://ec.europa.eu/clima/eu-action/innovation-fund/ner-300-programme_en); EEA. 'Innovation Fund.' (https://ec.europa.eu/clima/eu-action/innovation-fund_en; EEA. 'Modernisation Fund.' (https://ec.europa.eu/clima/eu-action/funding-climate-action/modernisation-fund_en.)

32 See: Covenant of Mayors for Climate & Energy. 'Covenant initiative.' www.covenantofmayors.eu/about/covenant-initiative/origins-and-development.html.

33 Sustainable Energy action plans (SEAP) and Sustainable Energy and Climate Action Plans (SECAP), including climate risk assessments (and adaptation measures, *infra* Section 7.4.2).

34 Respectively, COM (2013) 0216 final and COM (2021) 82 final.

35 See: Climate-ADAPT. 'About Climate-ADAPT.' *EEA.* https://climate-adapt.eea.europa.eu/about.

36 The following Adaptation Strategies were adopted earlier than the EU Strategy: Finland (2005); Spain and France (2006); Denmark, Hungary, Netherlands, UK and Germany (2008); Sweden (2009); Belgium (2010); Lithuania and Ireland (2012). Climate Change Acts including adaptation measures were adopted, for instance, by Finland (2015) and Denmark (2019). See EEA. 2019. 'Number of Countries That Have Adopted a Climate Change Adaptation Strategy/Plan.' *EEA.* www.eea.europa.eu/airs/2016/environment-and-health/climate-change-adaptation-strategies.

37 See EEA: www.eea.europa.eu/highlights/why-does-europe-need-to/climatechangeimpactineurope.pdf/view.

38 European Commission. 2018. *Evaluation of the EU Strategy on Adaptation to Climate Change.* Commission staff working document. https://eur-lex.europa.eu/legal-content/EN/TXT/PDF/?uri=SWD:2018:461:FIN&from=EN.

39 *National Adaptation Strategies* (NASs) address overarching issues, recognize the importance of expected climate change impacts and the need to adapt, and facilitate the process of coordinating the adaptation response, increasing awareness of adaptation and stakeholder involvement, assessing risks and vulnerabilities, and identifying knowledge gaps. *National Adaptation Plans* (NAPs) implement NASs and organize activities for achieving their objectives, typically through sectoral implementation.

40 European Commission. 2018. *Evaluation of the EU Strategy on Adaptation to Climate Change.* Commission staff working document. https://eur-lex.europa.eu/legal-content/EN/TXT/PDF/?uri=SWD:2018:461:FIN&from=EN.

41 'Commission Staff Working Document: Adaptation preparedness scoreboard country fiches', accompanying the document Report from the European Commission to the European Parliament and the Council on the implementation of the EU Strategy on adaptation to climate change (SWD/2018/460 final).

42 www.covenantofmayors.eu/about/covenant-initiative/covenant-in-figures.html.

43 A 'Macroregional strategy' is an integrated framework to address common challenges faced by a defined geographical area relating to MS and third countries located in the same geographical area which thereby benefit from strengthened cooperation contributing to achievement of economic, social, and territorial cohesion. See: http://ec.europa.eu/regional_policy/en/policy/cooperation/macro-regional-strategies.

44 See: Climate-ADAPT. 'Alpine Space.' *EEA*. https://climate-adapt.eea.europa.eu/countries-regions/transnational-regions/alpine-space.

45 SWD/2018/460 final, *cit.*

46 More recently, the EU's energy ministers have decided to extend subsidies for fossil gas in June 2021 on approval of the Energy Council's position for the revision of the energy infrastructure legislation (TEN-E), a move highly criticized by climate groups and which is not in line with EU's climate targets (Climate Action Network Europe 2021, https://caneurope.org/eu-energy-ministers-decide-extend-subsidies-fossil-fuels-revised-energy-infrastructure-legislation-stranding-eus-climate-objectives-european-green-deal).

References

Benson, David, and Daniel Russel. 2015. 'Patterns of EU Energy Policy Outputs: Incrementalism or punctuated equilibrium?' *West European Politics* 38 (1): 185–205.

Bocquillon, Pierre, and Mathias Dobbels. 2014. 'An Elephant on the 13th Floor of the Berlaymont? European Council and Commission relations in legislative agenda setting'. *Journal of European Public Policy* 21(1): 20–38.

von Bogdandy, Armin. 2012. 'Neither an International Organization Nor a Nation State: The EU as a supranational federation.' In *The Oxford Handbook of the European Union,* eds. Erik Jones, Arnand Menon, and Stephen Weatherill: 761–76. Oxford: Oxford University Press.

Bogojević, Sanja. 2010. 'Litigating the NAP: Legal challenges for the Emissions Trading Scheme of the European Union'. *Carbon and Climate Law Review* 3: 217–25.

Bogojević, Sanja. 2013. 'EU Climate Change Litigation, the Role of the European Courts, and the Importance of Legal Culture'. *Law and Policy* 53(3): 184–207.

Börzel, Tanja A. 2005. 'What Can Federalism Teach Us About the European Union? The German experience'. *Regional and Federal Studies* 15(2): 245–57.

Burgess, Michael. 2000. *The Building of Europe, 1950–2000.* London: Routledge.

Bürgin, Alexander. 2014. 'National Binding Renewable Energy Targets for 2020, But Not for 2030 Anymore: Why the European Commission developed from a supporter to a brakeman'. *Journal of European Public Policy* 22(5): 690–707.

Damro, Chad, Iain Hardie, and Donald MacKenzie. 2008. 'The EU and Climate Change Policy: Law, politics and prominence at different levels'. *Journal of Contemporary European Research* 4(3): 179–92.

Dehousse, Renaud. 2015. 'The New Supranationalism'. Paper prepared for the ECPR General Conference, Montreal, 26–9 August 2015. https://ecpr.eu/Filestore/PaperProposal/281383a5–0285-4417-a613-eed8cd5d36bd.pdf>.

De Witte Bruno, Juan Mayoral, Urzsula Jaremba, Marlene Wind, and Karolina Podstawa. eds. 2016. *National Courts and EU Law*. Cheltenham: Edward Elgar.

De Witte Bruno, Andrea Ott, and Ellen Vos, eds. 2017. *Between Flexibility and Disintegration: The trajectory of differentiation in EU law*. Cheltenham and Northampton: Edward Elgar.

Duwe, Matthias, and Nicholas Evans. 2020. *Climate Laws in Europe: Good Practices in Net-Zero Management.* Berlin and Den Haag: Ecologic Institute. www.ecologic.eu/17233.

Evans, Nick, and Matthias Duwe. 2021. Climate Governance Systems in Europe: The role of national advisory bodies. Ecologic Institute. www.ecologic.eu/18093.

Evans, Simon. 2019. 'Analysis: UK renewables generate more electricity than fossil fuels for the first time'. *Carbon Brief.* www.carbonbrief.org/analysis-uk-renewables-generate-more-electricity-than-fossil-fuels-for-first-time.

Fankhauser, Samuel, Caterina Gennaioli, and Murray Collins. 2015. 'The Political Economy of Passing Climate Change Legislation: Evidence from a survey'. *Global Environmental Change* 35: 52–61.

Fankhauser, Samuel, Caterina Gennaioli, and Murray Collins. 2016. 'Do International Factors Influence the Passage of Climate Change Legislation?' *Climate Policy* 16 (3): 318–31.

Fehling, Michael. 2021. 'Energy Transition in the European Union and its Member States: Interpreting federal competence allocation in the light of the Paris Agreement'. *Transnational Environmental Law* 10(2): 1–25.

Fleig, Andreas, Nicole M. Schmidt, and Jale Tosun. 2017. 'Legislative Dynamics of Mitigation and Adaptation Framework Policies in the EU'. *European Policy Analysis* 3(1): 101–24.

Grafakos, S., G. Viero, D. Reckien, et al. 2020. 'Integration of Mitigation and Adaptation in Urban Climate Change Action Plans in Europe: A systematic assessment'. *Renewable and Sustainable Energy Reviews* 121: 1–20.

Gupta, Joyeeta, and Lasse Ringius. 2001. 'The EU's Climate Leadership: Reconciling ambition and reality'. *International Environmental Agreements: Politics, Law and Economics* 1(2): 281–99.

Herold, Anke, Vanessa Cook, and Yifaat Baron, et al., 2019. 'EU Environment and Climate Change Policies State of Play: Current and future challenges'.www.europarl.europa.eu/thinktank/en/document/IPOL_STU(2019)63842.

Heyvaert, Veerle. 2013. 'What's in a Name? The Covenant of Mayors as transnational environmental regulation'. *Review of European, Comparative & International Environmental Law* 22(1): 78–90.

Hoppe, Robert, and Anna Wesselink. 2014. 'Comparing the Role of Boundary Organizations in the Governance of Climate Change in Three EU Member States'. *Environmental Science & Policy* 44: 73–85.

Hooghe, Liesbet, and Gary Marks. 1996. 'European Integration from the 1980s: State-centric v. multi-level governance'. *Journal of Common Market Studies* 34(3): 341–78.

Hooghe, Liesbet, and Gary Marks. 2001. *Multi-Level Governance and European Integration*. Lanham MD: Rowman & Littlefield.

Jänicke, Martin, and Rainer Quitzow. 2017. 'Multi-level Reinforcement in European Climate and Energy Governance: Mobilizing economic interests at the sub-national levels'. *Environmental Policy and Governance* 27(2): 122–36.

Jegen, Maya. 2014. 'Energy Policy in the European Union: The power and limits of discourse'. *Les cahiers européens de Sciences Po* 2: 2–22.

Jordan, Andrew, and Andrea Lenschow. 2010. 'Policy Paper Environmental Policy Integration: A state of the art review'. *Environmental Policy and Governance* 20(3): 147–58.

Keating, Michael. 2017. 'Europe as a Multi-level Federation'. *Journal of European Public Policy* 24(4): 615–32.

Kelemen, Daniel R. 2007. 'Built to Last? The durability of EU federalism'. *The Past and Future of European Institutional Integration* 14(25): 51–66.

Kincaid, John. 1999. 'Confederal Federalism and Citizen Representation in the European Union'. *West European Politics* 22(2): 34–58.

Knodt, Michèle, Mark Ringel, and Rainer Müller. 2020. '"Harder" Soft Governance in the European energy union'. *Journal of Environmental Policy & Planning* 22(29):1–14.

Kona, Albana, Giulia Melica, and Paolo Bertoldi, et al., 2017. 'Covenant of Mayors in figures: 8-year assessment'. Joint Research Centre. https://publications.jrc.ec.europa.eu/repository/handle/JRC106725.

Kona, Albana, Giulia Melica, and Brigitte Koffi, et al., 2016. 'Covenant of Mayors: Greenhouse gas emissions achievement and projections'. Luxembourg: Publications Office of the European Union. doi: 10.2790/351463.

Krommendijk, Jasper. 2020. 'It Takes Two to Tango: An introduction'. *European Papers* 5 (2): 745–54.

Kulovesi, Kati, and Marise Cremona. 2013. 'The Evolution of EU Competences in the Field of External Relations and Its Impact on Environmental Governance Policies'. Transworld Working Paper 17.

Kulovesi, Kati, and Sebastian Oberthür. 2020. 'Assessing the EU's 2030 Climate and Energy Policy Framework: Incremental change toward radical transformation?' *Review of European, Comparative & International Environmental Law* 29(2): 151–66.

Leruth, Benjamin, Stefan Gänzle, and Jarle Trondal, eds. 2019. *Differentiated Integration and Disintegration in a Post-Brexit Era*. London: Routledge.

Lorenzoni, Irene, and Mike Hulme. 2009. 'Believing is Seeing: Laypeople's views of future socio-economic and climate change in England and in Italy'. *Public Understanding of Science* 18(4): 383–400.

MacCormick, Neil. 1999. *Questioning Sovereignty. Law, state and nation in the European Commonwealth*. Oxford: Oxford University Press.

Marcinkiewicz, Kamil, and Jale Tosun. 2015. 'Contesting Climate Change: Mapping the political debate in Poland'. *East European Politics* 31(2): 187–207.

Mason, Henry. 1955. *The European Coal and Steel Community: Experiment in supranationalism*. Leiden: Nijhoff.

Massey, Eric, Robbert Biesbroek, Dave Huitema, and Andy Jordan. 2014. 'Climate Policy Innovation: The adoption and diffusion of adaptation policies across Europe'. *Global Environmental Change* 29: 434–43.

Massey, Eric, and Dave Huitema. 2016. 'The Emergence of Climate Change Adaptation as a New Field of Public Policy in Europe'. *Regional Environmental Change* 16: 553–64.

Melica, Giulia, Paolo Bertoldi, Albana Kona, et al. 2018. 'Multi-level Governance of Sustainable Energy Policies: The role of regions and provinces to support the participation of small local authorities in the Covenant of Mayors'. *Sustainable Cities and Society* 39: 729–39.

Meyer-Ohlendorf, Nils. 2020. *A European Climate Law: Analysis of the European Commission proposal.* Berlin: Ecologic Institute.

Meyer-Ohlendorf, Nils, Matthias Duwe, Katharina Umpfenbach, and Keighley McFarlane. 2014. *The Next EU Climate and Energy Package: EU climate policies after 2020.* Ecologic Institute. www.ecologic.eu/11045.

Monti, Alessandro, and Beatriz Martinez Romera. 2020. 'Fifty Shades of Binding: Appraising the enforcement toolkit for the EU's 2030 renewable energy targets'. *Review of European, Comparative & International Environmental Law* 29(2): 221–31.

Oberthür, Sebastian. 2019. 'Hard or Soft Governance? The EU's climate and energy policy framework for 2030'. *Politics and Governance* 7(1): 17–27.

Oberthür, Sebastian and Claire Roche Kelly. 2008. 'EU leadership in international climate policy: achievements and challenges'. *The International Spectator* 43(3): 35–50.

Palermo, Francesco. 2019. 'Federalism and the European Union: Asymmetry, policies and some recurring federal dilemmas'. In *A Research Agenda for Federalism Studies*, ed. John Kincaid: 198–208. Cheltenham: Edward Elgar.

Parker, Charles, and Christer Karlsson. 2010. 'Climate Change and the European Union's Leadership Moment: An inconvenient truth?' *Journal of Common Market Studies* 48 (4): 923–43.

Peeters, Marjan, and Natassa Athanasiadou. 2020. 'The Continued Effort Sharing Approach in EU Climate law: Binding targets, challenging enforcement?' *Review of European, Comparative & International Environmental Law* 29(2): 201–11.

Pérez, Mata, Daniel Scholten, and Karen Smith Stegen. 2019. 'The Multi-speed Energy Transition in Europe: Opportunities and challenges for EU energy security'. *Energy Strategy Reviews* 26. doi: 10.1016/j.esr.2019.100415.

Phelan, William. September 2012. 'What is *Sui Generis* about the European Union? Costly international cooperation in a self-contained regime'. *International Studies Review* 15(3): 367–85.

Piattoni, Simona. 2010. *The Theory of Multi-Level Governance. Conceptual, empirical, and normative challenges.* Oxford: Oxford University Press.

Rayner, Tim, and Andrew Jordan. 2016. 'Climate Change Policy in the European Union'. *Oxford Research Encyclopedia, Climate Science.*

Ringel, Mark, and Michèle Knodt. 2018. 'The Governance of the European Energy Union: Efficiency, effectiveness and acceptance of the Winter Package 2016'. *Energy Policy* 112:(112) 209–20.

Russel, Duncan, Roos M. den Uyla, and Laura de Vito. 2018. 'Understanding Policy Integration in the EU: Insights from a multi-level lens on climate adaptation and the EU's coastal and marine policy'. *Environmental Science and Policy* 82: 44–51.

Russel Duncan, Sergio Castellari, and Alessio Capriolo, et al., 2020. 'Policy Coordination for National Climate Change Adaptation in Europe: All process, but little power'. *Sustainability* 12(13): 1–18.

Schmidt, Vivien. 2016. 'The New EU Governance: New intergovernmentalism, new supranationalism, and new parliamentarism'. *IAI Working paper* 16(11): 1–16.

Schreurs, Miranda A., and Yves Tiberghien. 2007. 'Multi-level Reinforcement: Explaining European Union leadership in climate change mitigation'. *Global Environmental Politics* 7(4): 19–46.

Smismans, Stijn. 2011. 'From Harmonization to Co-ordination? EU law in the Lisbon governance architecture'. *Journal of European Public Policy* 18(4): 504–24.

Steurer, Reinhard, and Christoph Clar. 2015. *'Is Decentralisation Always Good for Climate Change Mitigation? How federalisation has complicated the greening of building policies in Austria'*. Institute of Forest, Environmental and Natural Resource Policy. Discussion Paper 4-2015.

Szulecki, Kacper, Severin Fischer, Anne Therese Gullberg, and Oliver Sartor. 2016. 'Shaping the Energy Union: Between national positions and governance innovation in EU energy and climate policy'. *Climate Policy* 16(5): 548–67.

Schütze, Robert. 2009. *From Dual to Cooperative Federalism: The changing structure of European law*. Oxford: Oxford University Press.

Thaler, Philipp, and Vija Pakalkaite. 2020. 'Governance through Real-Time Compliance: The supranationalisation of European external energy policy'. *Journal of European Public Policy* 28(2): 208–28.

Walker, Neil. 2012. 'The European Union's Unresolved Constitution'. In *The Oxford Handbook of Comparative Constitutional Law*, eds. Michel Rosenfeld and András Sajó: 1185–204. Oxford: Oxford University Press.

Wallace, Helen, and William Wallace, eds. 2000. *Policy-making in the European Union*, 4th ed. Oxford: Oxford University Press.

Wettestad, Jorgen, Per Ove Eikeland, and Måns Nilsson 2012. 'EU Climate and Energy Policy: A hesitant supranational turn'. *Global Environmental Politics* 12(2): 67–86.

8

Climate Governance and Federalism in Germany

PETER ECKERSLEY, KRISTINE KERN, WOLFGANG HAUPT, AND HANNAH MÜLLER

8.1 Introduction

As a highly industrialised country with a large manufacturing sector, many parts of Germany are still reliant on coal and other traditional industries, and in 2019 the country's per capita greenhouse gas emissions were above the EU average. However, Germany was one of the first countries in the world to take environmental and climate policy seriously; it has had a strong Green Party for several decades and has established numerous government departments and agencies to tackle environmental problems across different levels (Jänicke 2011). Its high-profile energy transition (*Energiewende*) strategy – an explicit policy to shift away from nuclear and fossil-based energy and towards renewable sources – won recognition around the world and was initially very popular domestically (von Hirschhausen 2014).

A key factor underpinning the energy transition's initial success was the country's federal structure, which facilitated cooperation and buy-in from different public bodies across tiers of government (Eckersley 2018b; Weidner and Mez 2008). This structure and the close relationships between different levels of government are the products of the country's unique historical development and contrast markedly with other federal countries such as the USA. As this chapter will show, however, the bonds that supported this collaboration may be weakening, as the policies necessitated by the threat of climate change become more controversial, particularly in regions that are still heavily reliant on traditional industries.

The chapter sets out how Germany's federal system and industrial interests play a key role in shaping climate and energy politics within the country, and the concomitant impact of these evolving politics on the policies of individual states (*Länder*) and the federation. It draws on an extensive literature search of academic studies, federal and *Land* government websites, grey literature and ten expert interviews with officials in the state administrations of Mecklenburg–Western

Pomerania, North Rhine–Westphalia, Hamburg, Brandenburg and Baden–Wurttemberg. This research revealed notable contrasts between different parts of Germany, which led us to group the *Länder* according to their reliance on different energy sources (see also Eckersley et al. 2021). We set out this categorisation in Section 8.3 of the chapter and use it to highlight how different contexts at the state level shape policy processes and outputs across the federation.

Section 8.2 provides a brief overview of how climate policy in Germany has evolved over recent decades, and sets out the main challenges that the country faces in both mitigation and adaptation. We then sketch out how the formal and informal institutions associated with German federalism shape climate policymaking processes, with a specific focus on the activities of several states (*Bundesländer,* or just *Länder*) in both the west and the east of the country. This discussion will show how an increasing reluctance in some *Länder* to adopt a more ambitious climate and energy strategy is likely to make it difficult for the country to introduce initiatives that are sufficiently ambitious to meet its climate objectives (see also Heering and Gustavson 2021; Ohlhorst et al. 2014; Scheiner 2017). As such, climate policy in Germany is a microcosm of the global approach to tackling these issues; the federation provides a high-level framework within which the constituent *Länder* operate, but the actions of individual states and municipalities reflect their own economic and political interests.

Therefore, to return to the questions set out in Chapter 1, this chapter highlights how multiple forums for policymaking within federal systems present both opportunities and challenges for ambitious climate policy. This is because federal structures enable governing units at different levels to seize the initiative and fill the void created by inaction elsewhere but may also reduce pressures on more reluctant actors to respond and thereby impede policy coordination. Relatedly, although decentralised structures empower the *Länder* to pursue their own policies and may help innovative ideas to diffuse horizontally between states and municipalities, these initiatives have not always complemented each other or contributed towards a coherent and effective response to climate change.

8.2 Climate Change in Germany

Germany is a highly advanced industrial economy and the third largest exporter in the world after China and the USA (World Bank 2019). Manufacturing accounts for 23 per cent of national output, and much of this sector relies on energy-intensive processes; carmakers such as Volkswagen, BMW, Audi, Mercedes-Benz, Opel and Porsche are all based in the country. In 2019, Germany's per capita greenhouse gas (GHG) emissions of 10.4 tonnes per annum were significantly above the EU average of 8.23 (OECD 2019, 2020). In 2016–17 renewable sources met 13.4 per cent of energy demand across the country, although some areas are far more dependent on fossil-based fuels than others. Much of Germany's climate strategy, including its GHG emissions reduction targets, is shaped by its membership in the European Union; however, as the bloc's largest and most powerful member state, it also exerts significant influence over the direction of EU policy (Jänicke and Wurzel 2019; Szuleki et al. 2016).

Despite producing above-average GHG emissions, several factors led Germany to be portrayed as an energy and climate leader in the 1990s and 2000s (Kern et al. 2004; Scheiner 2017; Steuwer and Hertin 2021). These included its early development of institutions such as climate and energy agencies; a strong Green Party (particularly in the west); and the *Energiewende* strategy that facilitated a rapid shift towards solar and wind power in many parts of the country (Weidner and Mez 2008). A key part of this strategy involved subsidising small-scale renewable electricity generation through feed-in-tariffs (FiTs), which led to a major increase in solar PV, wind and biomass installations (Mendonca et al. 2010). The *Energiewende* exemplified the concept of 'ecological modernisation', which Germany's federal government adopted from 1998 onwards, in order to replace higher-polluting sectors with low-carbon industries ahead of its international competitors and therefore gain a first-mover advantage (Jänicke 2011).

Underpinning this approach was the idea that economic growth and environmental protection were mutually reinforcing. Although there was some opposition from those southern *Länder* that relied heavily on nuclear power, the *Energiewende* was initially very popular with the German public (von Hirschhausen 2014). Indeed, the country's energy transition served as a model that other developed countries sought to emulate (Hennicke and Welfens 2012), to the extent that the German term began to be used in English-language debates (Beveridge and Kern 2013). Germany was also one of the few developed economies to meet its commitments under the Kyoto Protocol (in its case a reduction in GHG emissions of 21 per cent between 1990 and 2012), which also suggested that its approach was successful. However, most of the country's initial progress in reducing carbon emissions was due to the closure of outdated heavy industry and fossil-fuel power facilities in the former GDR; the country's progress in climate mitigation slowed markedly after these 'wall-fall' benefits were exhausted from the late 1990s onwards (Schleich et al. 2001). The high-level figures also mask regional variations within Germany: as Section 8.4 will show, some *Länder* are still highly dependent on fossil fuels.

Nonetheless, Germany has pledged to cut its GHG emissions by at least 65 per cent by 2030 and 88 per cent by 2040 (compared to the 1990 baseline), and achieve climate neutrality by 2045. Following the adoption of climate protection legislation in eight of the sixteen *Länder*, the federal government passed its own *Climate Act* in late 2019 and amended this law by introducing more stringent targets in June 2021 (Bundesregierung 2021).[1] As Table 8.1 shows, the updated federal targets now exceed those set out in most *Land*-level legislation.

Table 8.1 *GHG emissions enshrined in* Land *and federal climate protection acts*

State	Enactment	GHG emission reduction goals (base year 1990)					
		2020	**2025**	**2030**	**2040**	**2045**	**2050**
North Rhine–Westphalia	23.01.2013/ 01.07.2021	25%		65%	88%	Greenhouse gas neutrality	
Baden–Württemberg	17.07.2013/ 14.10.2020	25%		42%			90%
Rhineland–Palatinate	23.07.2014	40%					climate neutrality minimum 90%
Bremen	24.03.2015	40%					80–95%
Berlin	22.03.2016	40%		60%			85%; climate neutrality
Schleswig–Holstein	07.03.2017	40%		55%	70%		80–95%
Thuringia	18.12.2018			60–70%	70–80%		80–95%
Hamburg	20.02.2020	(40%)		55%;			95%; climate neutrality
Bavaria	13.11.2020		5.5 t. / cap.	below 5 t./ cap.			climate neutrality
Lower Saxony	09.12.2020			55%			80–95%
Germany	**15.11.2019/ 25.06.2021**			**65%**	**88%**	**climate neutrality**	

Reflecting geographical and meteorological factors, climate change will affect different parts of Germany in different ways. In a study conducted for the Federal Environmental Agency, Buth et al. (2015) found that it would probably have a major impact on the following:

- soil quality;
- biodiversity (particularly related to the spread of invasive species);
- agricultural growth periods;
- forests;
- fish stocks;
- river and flash flooding (with concomitant risks on transport, buildings, commerce and other critical infrastructures), particularly in urban centres and districts along the Elbe, Weser, Ems and the Lower Rhine rivers;
- coastal erosion in northern areas caused by storm surges and sea level rise, although all of Germany's main population centres are inland and therefore the direct risk to human habitation is lower than in many other federal countries;
- glacial melt from the Alps and landslides in the far south of the country;
- heatwaves and heat stress, particularly in the southwest of the country.

Overall, therefore, we can see how issues associated with climate mitigation and adaptation affect regions differently across Germany. In terms of mitigation, those *Länder* that are more reliant on traditional industries and fossil fuel extraction and combustion face major economic challenges in the next phase of the *Energiewende*, whereas the effects of climate change will be distributed asymmetrically across the country.

8.3 Climate Change and Federalism in Germany

Along with other members of the European Union (EU), climate policy in Germany is shaped to a high degree by decisions taken in Brussels, including initiatives such as the EU's 2030 climate and energy framework (which includes binding targets for GHG emissions reductions and renewable energy generation[2]), its emissions trading scheme, procurement regulations and the Green Deal. For example, the European Commission's *European Climate Law* included an EU-wide GHG emissions reduction target of at least 55 per cent by 2030 (compared to 1990 levels). The EU also contributes some of the resources that support policy development and implementation through its funding programmes (such as for research or regional development), institutions (such as the European Environment Agency) and initiatives (such as the Covenant of Mayors). Within this context, Member States develop their own strategies and – in federal countries like Germany at least – the constituent units work with municipalities to implement policy objectives.

8.3.1 The Division of Powers in German Federalism

Germany has a long tradition of decentralised governance with origins in the Middle Ages and did not become a unified nation-state until 1871. The country's decentralised approach continued until the Nazis took power in 1933 and was reinstituted in the west of the country after the Second World War (Eckersley 2017) – although the GDR did have a centralised and hierarchical administrative structure until unification in 1990 (Wollmann 2021). Following the end of the Cold War, structures within the 'new' eastern states were modelled on those in the 'old' western part of the country, with the result that each *Land* possesses the same legislative powers (Wollmann 2021), including the right to pass climate legislation.

Reflecting the period in which it was written, the post-war constitution (*Grundgesetz*) did not allocate legal responsibility for climate change or renewable energy to specific tiers of government. Indeed, protecting the environment was only recognised as a public function in 1994 (Erbguth and Schlacke 2014). Furthermore, the integrated nature of German federalism means that the legal boundaries that demarcate specific powers to tiers of government are somewhat blurred compared to many other federal countries (Scheiner 2017). It is also worth noting that municipalities do not have to take action on climate change unless the respective *Land* government has passed primary legislation stipulating which specific tasks they must carry out and provides them with the necessary resources.

In contrast to policy areas where the *Länder* have exclusive legislative powers (such as culture and education), climate policy falls under the so-called concurrent legislation principle, which prevents individual states from introducing new regulations where the federal government has already passed a law. Some environmental sectors, including nature protection, are exempt from these restrictions, but in climate policy most decisions are made in Brussels or Berlin. In many areas of climate policy (for example efficiency standards for buildings, or regulations on renewable energy), EU directives are transposed into national law and implemented by the *Länder* (see Table 8.2).

8.3.2 The Role of the Länder *in Implementing Federal Law*

The *Länder* governments and their administrations are also responsible for implementing most federal laws (Behnke 2020). Thus, although the *Länder* have limited formal decision-making powers, they can decide on organisational issues, procedures and control. This is particularly important for the environmental sector (Gallata and Newig 2017; Newig et al. 2014), where initiatives require coordinated input from a whole host of societal actors to be effective (Wurzel et al. 2013).

Table 8.2 *Climate responsibilities across tiers of government in Germany*

<table>
<tr><th>Role of the states (Länder) in different forms of decision-making</th><th>European Union</th><th>Federal government (Bund)</th><th>States (Länder)</th></tr>
<tr><td>Joint decision-making and responsibilities – mandatory
EU directives are transposed into national law and implemented by the states</td><td colspan="3">Renewable energy:
EU Renewable Energy Directive; Federal Renewable Energy Act (Erneuerbare–Energien–Gesetz), State Ordinances on Wind Energy (Windenergie–Erlasse)
Efficiency standards for buildings:
EU Energy Performance of Buildings Directive; Federal Energy Act for Buildings (Gebäudeenergiegesetz); State administrative ordinances</td></tr>
<tr><td>Joint decision-making and responsibilities – optional
EU directives are transposed into national law and implemented by the states
Federal government may allow but not mandate states to enact own legislation</td><td colspan="2">GHG emissions reduction goals and climate change acts
European Climate Act
Federal Climate Protection Act (Bundes–Klimaschutzgesetz)</td><td>GHG emissions reduction goals
State climate change acts (in 10 out of 16 states)
State energy and climate plans and strategies</td></tr>
<tr><td>Exclusive decision-making and responsibilities
Decision-making restricted to one (or two) levels</td><td>CO_2 emissions performance standards for new cars and vans
(EU regulation)</td><td>Phase-out of nuclear energy
Revision of the Federal Nuclear Energy Act (Gesetz zur Änderung des Atomgesetzes)</td><td>Implementation of climate policies (organisation, procedures, control)</td></tr>
<tr><td>May exclude the states</td><td colspan="2">EU Emission Trading System
EU ETS Directive; Federal GHG Emission Trading Act (THG–Emissionshandelsgesetz)</td><td>Today, states have almost no responsibilities</td></tr>
</table>

Moreover, where federal legislation explicitly clarifies that the *Länder* can pass legislation on specific issues, they are also free to do so. Examples include the federal *Climate Change Act*, which allows the *Länder* to introduce their own climate change acts, or the federal *Renewable Energy Heat Act*, which requires that heating and cooling for *new* buildings must come partly from renewable sources

(*Bundesministerium für Wirtschaft und Energie* 2009). *Länder* governments are not able to set higher standards in this area, but they are free to introduce their own regulations for *older* buildings – and indeed Baden–Württemberg has done so – because the federation has not (yet) introduced such legislation. Therefore, the *Länder* are able to be more ambitious in areas where no EU or federal legislation currently exists, or where federal legislation allows them to introduce more stringent regulations. However, the 'concurrent legislation' principle nonetheless restricts the ability of the *Länder* to legislate in many areas related to climate policy, including strategic energy and transportation systems.

Alongside spatial planning and regional development, the *Länder* also have jurisdiction over local government – an area which can play a key role in climate policy, due to the importance of policy coordination across tiers of governance for successful implementation (Eckersley 2018b). Nonetheless, many common principles apply across Germany, and a particularly important rule ensures that *Land* governments have to provide municipalities with sufficient resources to engage in new policy sectors or undertake new functions. Since many *Länder* are reluctant to do this, action on climate change remains voluntary for local government in most of the states, which has led to some municipalities introducing more ambitious policies than others.

Recognising the often-decisive role that the availability of resources can play in shaping whether municipalities can implement and enforce climate policies effectively, the federal government has sometimes bypassed the *Länder* to provide significant funds directly to local authorities to help with their climate strategies. For example, since 2008 the *Kommunalrichtlinie* initiative has financed the development of local climate protection plans and strategies, and the *Masterplan* scheme provides additional funding to leading municipalities (Kern 2019). Such schemes have ensured that municipalities in those parts of Germany with less ambitious *Land* or local governments have still been able to make some progress (Göpfert 2014). However, the principle of 'concurrent legislation' means that once the federation acts to tackle a particular issue, the *Länder* are often unable to develop more ambitious legislation of their own.

Given that the *Länder* vary significantly in terms of population size and density, as well as their political and economic interests, it is perhaps unsurprising that the federation has felt the need to bypass the state level on occasions.

8.3.3 Integrated and Cooperative Federalism

Despite contrasts between different *Länder*, Germany is often portrayed as a classic example of 'cooperative federalism' (Benz 2007; Börzel 2005; Hegele and

Behnke 2017): studies have found that tiers of government tend to operate in a more complementary and collaborative way than in many other federal countries, particularly the USA (Müller 1998). This collaboration is underpinned and reinforced by various constitutional and institutionalised structures that date from the end of the Second World War (Scharpf et al. 1976). For example, the *Länder* governments are represented in the *Bundesrat* (the second parliamentary chamber at the federal level), which gives them significant veto power over federal legislation (see Scheiner 2017 for an analysis of how this has shaped the federal response to climate change). Other provisions also serve to facilitate cooperation, such as the system of fiscal equalisation that ensures revenues are redistributed from wealthier to poorer states, and a system of shared taxes (Auel 2014; Scherf 2010). *Bund–Länder* associations (*Bund–Länder Arbeitsgemeinschaften*) in various policy sectors reinforce this collaborative approach, including in the climate, energy, mobility and sustainability sectors (Bundesregierung 2019; Flaskühler 2018).

Critics argued that these 'interlocking' arrangements were opaque and undemocratic (because individual actors could not be held accountable for specific policy decisions), and that they resulted in a suboptimal and bureaucratic decision-making 'trap', because individual *Länder* could exert significant influence over policy and therefore decisions were taken based on the 'lowest common denominator' (Adelberger 2001; Monstadt and Scheiner 2014; Scharpf 1988; Schultze 1999). In an attempt to address these problems, the federal government introduced reforms in 2006 and 2009, which aimed to clarify the roles of different tiers of government and limit the influence of the *Bundesrat* in federal law-making. However, although these changes did demarcate clearer responsibilities in some areas, some argued that they had little impact on the overall system (Scharpf 2009; Zohlnhöfer 2009).

Nonetheless, despite its collaborative policymaking culture, the German federal system does allow different *Länder* to adopt contrasting strategies and initiatives to try and achieve similar objectives within a common overall framework. In addition, in cases where different actors agree on policy goals, the system can help to coordinate activity and therefore result in more effective implementation (Wollmann 2004a) – particularly when problems span tiers of government and policy sectors. This occurred with the Covid-19 pandemic (Bouckaert et al. 2020; Kropp and Schnabel 2021; Kuhlmann and Franzke 2021) and also applies to climate change, where support for the *Energiewende* helped to mobilise resources and activity so that initiatives could be implemented and enforced across different levels (Eckersley 2018a).

8.3.4 Energy Mixes and Climate Governance in the German Länder

Notwithstanding the similarities in each state's legal and constitutional position, the sixteen *Länder* vary significantly in terms of their geographic size, population, socioeconomic make-up, energy mix and levels of greenhouse gas emissions (see

Table 8.3). For example, the largest state, North Rhine–Westphalia (NRW), is almost the same geographical size as the Netherlands and has a similar population (17.9 m), whereas the smallest, Bremen, has fewer than 700,000 inhabitants. The two southernmost *Länder* (Bavaria and Baden–Württemberg) are notably wealthier than most other parts of Germany, particularly the East.

Despite Germany's initial progress in the *Energiewende*, and the federal government embracing the concept of 'ecological modernisation' from 1998 onwards (Jänicke 2011; Kern et al. 2008), some parts of Germany still rely heavily on fossil fuels. This has contributed to varying levels of ambition at the *Land* level, in terms of climate and energy legislation and institutions, and the resources they make available to help municipalities develop strategies and implement policy initiatives. For example, eight of the sixteen *Länder* had adopted climate acts that enshrined GHG reductions targets into primary legislation before the federal government took this step in late 2019. Despite this similarity, however, these laws differ in ambition (see Table 8.1).

Those *Länder* still relying on coal-generated electricity have been less likely to introduce such initiatives. In contrast, states that were more reliant on nuclear power prior to the *Energiewende* have made a more rapid and effective transition to cleaner energy and generally provide more support to municipalities to help with this shift. As of late 2021, the new SPD–Green–FDP coalition government announced that it would 'ideally' like to bring forward the previous coal phase-out date of 2038 to 2030 (SPD 2021), but it appeared unlikely that this target would feature in federal legislation. Notably, the *Länder* in which power stations that burn the hard 'black' and the more polluting 'brown' lignite coal constitute a powerful coalition within the German federal system (particularly via their representation in the *Bundesrat*).

Various factors are likely to have shaped these different *Länder* approaches to climate policy. For example, numerous studies have found that larger political–administrative units (such as big cities or big nation-states) are usually able to call upon more resources to develop and implement policy (Hoff and Strobel 2013; Kern 2019; Reckien et al. 2018; Salvia et al. 2021). Therefore, we might expect the biggest *Länder* to be more active in this area. In addition, political factors probably play a role: in jurisdictions where the Green Party has significant representation or forms part of the governing coalition, governments at all levels are more likely to act on the climate (Abel 2021; Wurster and Köhler 2016).

In addition, analysis of the energy base of each *Land* suggests that this is a key factor shaping how different *Länder* have tried to address the issue of climate change. Based on their relative reliance on different sources of energy, and considering the three city-states separately, we group the sixteen *Länder* into five categories that help to explain these contrasting approaches. Table 8.4 summarises these groupings and they are represented geographically in Figure 8.1 (see also

Table 8.3 *Population, GHG emissions and renewable energy in the German* Länder. *Adapted from* Agentur für Erneuerbare Energien *(<www.foederal-erneuerbar.de/landesinfo/bundesland>)*

State	Population (2018)	GHG emissions (2015) in mill. tons	GHG emissions/ cap. (2015) in tons	GHG emissions/ cap. (1990–2015) in %	Share of renewables in primary energy consumption (2016–17, %)
Baden–Württemberg	11,069,533	76.73	6.9	– 22.3	13.0
Bavaria	13,076,721	90.86	6.9	– 24.1	18.2
Berlin	3,644,826	16.02	4.4	– 43.0	4.0
Brandenburg	2,511,917	62.30	24.8	– 26.9	6.1
Bremen[3]	682.986	13.55	19.8	– 2.7	18.4
Hamburg[4]	1,841,179	15.46	8.4	+10.9	4.1
Hesse	6,265,809	40.08	6.4	– 26.4	9.8
Lower Saxony	7,982,448	83.99	10.5	– 19.7	19.0
Mecklenburg–Western Pomerania	1,609,675	15.19	9.4	– 18.9	37.0
North Rhine–Westphalia	17,932,651	278.85	15.5	– 22.3	4.8
Rhineland–Palatinate	4,084,844	31.49	7.7	– 42.4	12.7
Schleswig–Holstein	2,896,712	25.82	8.9	– 30.7	33.1
Saarland	990,509	22.12	22.3	– 14.4	4.6
Saxony	4,077,937	51.79	12.7	– 40.0	9.1
Saxony–Anhalt	2,208,321	34.35	15.5	– 25.0	18.7
Thuringia	2,143,145	16.64	7.8	– 38.9	24.5
Germany	***83,019,200***	***858.66***	***10.4***	***– 28.4***	***13.4***

Note. Bremen[3] and Hamburg[4]: see Endnotes 3 and 4.

Table 8.4 *Energy-based typology of the German* Länder

Coal states (coal regions in western and central Germany)	
Brandenburg; North Rhine–Westphalia; Saarland; Saxony; Saxony–Anhalt	Traditional coal states; high CO_2 emissions/cap.; exporters; small renewable energy sector
Nuclear/solar energy states (southern states)	
Bavaria; Baden–Württemberg	Traditional nuclear states; relatively low CO_2 emissions/cap.; growing renewable sector (particularly solar)
Wind energy states (northern, coastal states)	
Lower Saxony; Schleswig–Holstein; Mecklenburg–Western Pomerania	Traditional nuclear states; relatively low to medium CO_2 emissions/cap.; growing renewable sector (particularly wind); becoming energy exporters;
Energy importing states (central states)	
Rhineland–Palatinate; Hesse; Thuringia	Dependent on energy imports; medium CO_2 emissions/cap.; growing renewable sector
City-states	
Berlin; Hamburg; Bremen	Low potential for renewable energy generation; relatively low CO_2 emissions/cap. due to population density; dependent on fossil fuel energy production (coal and gas) and imports

Source: Taken from Eckersley et al. (2021).

Eckersley et al. 2021 for a more detailed breakdown of each state). In most cases, these groupings also reflect geographic and climatic factors, which themselves shape the energy resources that are available locally and the strength of certain industries in lobbying and policymaking processes.

8.3.5 Coal States

These five *Bundesländer* rely heavily on either hard 'black' coal or 'brown' lignite for energy production. Although they have reduced GHGs significantly since 1990, this was from a very high base. In addition, three of the coal states (Saxony, Saxony–Anhalt and Brandenburg) experienced significant industrial decline following unification in 1990, which accounts for a major proportion of their drop in emissions. North Rhine–Westphalia is the only coal state to pass a climate change act that commits it to reductions in GHG emissions. Green Party representatives in the state legislatures of Brandenburg, Saxony–Anhalt, Saarland and Saxony introduced similar legislative initiatives in their respective *Land* parliaments, but they were rejected by the governing majority on each occasion (Eckersley et al. 2021).

Figure 8.1 Energy-based typology of the German *Länder*.

Two coal states – Brandenburg and Saxony–Anhalt – have invested heavily in wind power. Brandenburg aims to increase the share of renewables in energy generation to 32 per cent by 2030 (MWE B 2012) and Saxony–Anhalt provides a range of funding sources to support municipalities in climate policy development and implementation (MWW SA 2014). The other three coal states have been less ambitious, particularly in recent years. For example, after NRW's Social Democratic–Green government was replaced by a centre–right coalition of Christian and Free Democrats in 2017, its progress on mitigation stalled, redirecting its investment from mitigation to adaptation (Eckersley et al. 2021; interviews with *Energieagentur NRW* and the NRW Environment Ministry). For its part, Saarland produces the lowest share of energy from renewable sources among the non-city-states (Statistisches Amt Saarland 2018). Saxony has introduced funding schemes for climate protection and adaptation initiatives (S SMWAV 2013), but it had still not set any specific targets for further GHG reductions by spring 2021, and its parliament rejected a Green Party proposal to introduce a climate protection act in 2018 (Abgeordnetenwatch.de 2018). As such, we can see how these *Länder* lag behind most of the rest of Germany in their climate policies.

8.3.6 Nuclear/Solar Energy States

Traditionally, Bavaria and Baden–Württemberg in southern Germany relied heavily on nuclear energy, which meant that they had relatively low GHG emissions per capita. However, since the *Energiewende* initially prioritised the phasing out of nuclear power, both faced significant challenges to bridge the gap between energy supply and demand through renewable sources, primarily solar photovoltaics. In 2018, for example, over 40 per cent of Germany's installed solar PV capacity was in Bavaria and Baden–Württemberg (AEE 2019). Both *Länder* also promoted themselves as climate leaders among subnational units through 'paradiplomacy' (see Ralston 2013 re Bavaria), and membership of the Under2-Coalition of states and regions driving climate action. Indeed, Baden–Württemberg was a founding member of this network, and its Green Party premier is one of four European co-chairs.

Bavaria claims to have established the world's first environmental ministry in 1970 (BSUV n.d.) and has helped to fund municipal climate initiatives for many years (Bayerische Staatsregierung 2009; Kern 2008). Although the *Land* government did not propose a Climate Protection Act until 2019, this committed Bavaria to climate neutrality by 2050 (BSUV 2020), thereby increasing competition between leading *Länder* in terms of environmental ambition. Bavaria has made significant progress in terms of solar PV installations, but this is largely due to private investments incentivised by feed-in tariffs introduced by the federal government and financed by energy customers up until 2017, rather than a specific *Land* initiative.[5]

Baden–Württemberg was also a forerunner in environmental protection; it established an environment ministry in 1975 and a climate protection and energy agency in 1994, and in 2013 it became the second *Land* to pass a Climate Protection Act, committing to a 90 per cent reduction in GHG emissions by 2050.[6] Like Bavaria, it has overseen a significant expansion in renewable (particularly solar) energy generation in recent years (Diekmann et al. 2019). However, total GHG emissions in the state have fallen more slowly than in other parts of the country (partly due to a considerable population increase), and the *Land* only met its initial target of a 25 per cent reduction by 2020 due to the impact of Covid-19 on transport emissions (Statistisches Landesamt Baden–Württemberg 2021).

8.3.7 Wind States

Traditionally, Germany's three northernmost states (Lower Saxony, Schleswig–Holstein and Mecklenburg–Western Pomerania) have relied on nuclear power and imported electricity. However, their coastal location and climatic conditions have enabled them to shift towards wind power (both on- and offshore) more easily than their inland counterparts.

In contrast to three of the other eastern *Länder*, Mecklenburg–Western Pomerania did not have substantial carbon-intensive industry during the GDR period and was therefore not as badly affected by the 1990s deindustrialisation or the later energy transition. Indeed, its geographical location has facilitated the creation of large green manufacturing and services sectors (Diekmann et al. 2019). However, parliamentary attempts by opposition representatives from the Green Party to introduce a Climate Protection Act were rejected by the ruling SPD–CDU coalition government.

More people are employed in low-carbon industries in Lower Saxony than in any other German state, and over 40 per cent of electricity generated in the *Land* comes from renewable sources (N MUEBK 2020). The SPD–CDU state government passed a climate protection act in 2020, which commits the *Land* to legally binding targets of 80–95 per cent reductions in GHG emissions and a complete transition to renewable energy by 2050 (Landtag Niedersachsen 2019). At the same time, however, Lower Saxony still provides significant subsidies for oil and natural gas and provides fewer funding opportunities for municipal climate action than some other states (Eckersley et al. 2021).

With long coastlines on both the North and Baltic Seas, Schleswig–Holstein is very well located to benefit from wind power and has been able to exploit this advantage by overseeing a major expansion in installations. Like Mecklenburg–Western Pomerania, the *Land* is largely rural and has little heavy industry, meaning per capita GHG emissions are relatively low. The state's 2017 *Climate Protection Act* committed it to GHG reductions of 40 per cent by 2020, 55 per cent by 2030, 70 per cent by 2040 and 80–95 per cent by 2050.[7] In 2019, Schleswig–Holstein and Baden–Württemberg were judged to be the two leading states in Germany for renewable energy (Diekmann et al. 2019).

8.3.8 Energy Importer States

Three states in southern-central Germany (Rhineland–Palatinate, Hesse and Thuringia) have small (albeit growing) renewable sectors but are largely reliant on energy imports. None of them have had large fossil fuel or nuclear sectors, and therefore the *Energiewende* presents them with a smaller economic and political challenge than some other states.

Hesse was one of the first *Länder* to adopt an active environmental policy, with a comprehensive sustainability strategy in 2008 that also covered issues of climate protection. The state also introduced GHG reduction targets of 30 per cent by 2020, 40 per cent by 2025 and 90 per cent by 2050, against the baseline year 1990 (HMUKLV 2017). In addition, Hesse has introduced a major funding scheme through which municipalities can acquire grants to finance climate-related initiatives (HMUKLV 2019). However, its government has not adopted a Climate Protection Act, and therefore its climate targets are not anchored in legislation.

Rhineland–Palatinate adopted a Climate Protection Act in 2014, which established legally binding targets to reduce GHG emissions by 40 per cent by 2020 and 90–100 per cent by 2050 against the baseline year of 1990.[8] As the *Land* has over 2,300 municipalities, nearly all of which have fewer than 2,000 inhabitants, the *Land* government encourages and facilitates significant horizontal collaboration between local authorities, in order to help them access necessary resources and increase their capacity to develop and implement effective policy.

Thuringia reduced its GHG emissions by over 60 per cent between 1990 and 2020 – a larger percentage drop than in any other *Land* – and renewable sources (mostly wind and solar) now account for 59 per cent of electricity production within the state (TLS 2019). In 2018 it became the only *Land* in the former GDR to have adopted a climate protection act outside Berlin: this sets out a series of legally binding staged targets for GHG emission reductions, culminating in 80–95 per cent by 2050.[9] The state also provides funding schemes for municipalities to invest in climate protection and adaptation initiatives, including applications for European Energy Award accreditation.

Despite their progress, however, these three *Länder* do not have the wind resources of the northern *Länder*, and private actors have invested less in solar power than in the southern *Länder*. As a result, they will probably continue to rely on energy imports for the foreseeable future and therefore have less control over the way in which this energy they consume is generated.

8.3.9 City-States

Due to being densely populated 'city-states', Berlin, Hamburg and Bremen face different climate challenges than the other *Länder* and have sought to respond to them in different ways. Berlin made substantial progress in reducing GHG emissions during the early 1990s, but still relied on fossil fuels for 90 per cent of its energy by 2016, when hard coal generated over 40 per cent of the city's electricity (BSDETCP 2019). Given this situation, it remains unclear whether Berlin can meet its target of climate neutrality by 2050, as set out in its 2016 *Energy and Climate Act*. However, it has been more active in the area of adaptation than many other states, proposing a range of initiatives that aim to reduce the impact of extreme weather events – particularly heatwaves and storms – on the city's infrastructure (SUVK 2016).

Bremen also still relies heavily on coal-fired power stations for its electricity – although its substantial steelmaking sector, which accounts for around half of the state's GHG emissions, skews its climate-related statistics to a large extent. The state parliament did pass an energy act as early as 1991, focusing on energy conservation and efficiency, and then adopted climate protection legislation in

2015 that included GHG reduction targets of 80–95 per cent by 2050 (Eckersley et al. 2021). The *Land* also has an adaptation strategy (published in 2018) that seeks to address concerns about water management, heatwaves and flood risks (Freie Hansestadt Bremen 2018).

Like the other city-states, Hamburg relies heavily on fossil fuels, particularly coal, for its energy; renewables accounted for a mere 4 per cent of consumption in 2017. Indeed, the new Moorburg coal-fired power station, which began generating electricity in 2015, has meant that the city's GHG emissions have increased by 20 per cent in the last five years (Eckersley et al. 2021). However, Hamburg has set ambitious targets in its climate protection act: a 55 per cent reduction in CO_2 emissions by 2030 (compared to 1990) and climate neutrality by 2050,[10] and has also adopted a climate adaptation strategy (BFHH 2013).

8.4 Discussion

Despite the existence of an elite consensus about the serious threat that climate change poses, the above illustrations show how approaches to the issue are becoming more fragmented and diverse across Germany, because some *Länder* want to make faster progress than others. This is driven by political, economic and geographic factors within each *Land*. In political terms, those *Länder* where the Green Party has formed part of the coalition government have normally adopted a Climate Change Act and more ambitious mitigation policies. In contrast, in those areas where the Greens have generally been in opposition, their attempts to introduce climate legislation have been stymied (Eckersley et al. 2021). Economically, a greater dependence on fossil fuels in the energy mixes of some *Länder* has made them more reluctant to engage in far-reaching mitigation activities (such as phasing out coal). Ultimately, these economic factors (and the powerful lobby groups that represent political and industry interests) are themselves shaped by geography: the governments of those *Länder* where renewable energy resources are more plentiful (such as the windy north or sunnier south) have been more enthusiastic about the energy transition than their counterparts in central Germany.

Another related issue here is the extent to which *Land* and municipal governments can access the resources necessary to develop and implement ambitious policies. Public institutions in wealthier cities and states often have more money to spend on policy initiatives, and their residents may also be more likely to view climate change as a priority issue (Moser and Kleinhückelkotten 2018). Leaving aside the three city-states, those *Länder* that have performed better on climate mitigation tend to be wealthier (see Table 8.3). Politically speaking, such factors should make it easier for governments to introduce ambitious climate policies. Access to other resources, particularly relevant knowledge and expertise, can also shape a subnational government's climate strategy (Lerman et al. 2021).

As with most other federal systems, Germany's institutional architecture enables the *Länder* to develop their own strategies and legislation in response to these specific circumstances, but they are restricted by the fact that policies in key sectors such as the emissions trading system, vehicle emissions standards and energy infrastructure are made at the EU or federal levels (see Table 8.2). The flexibility that they can exercise within this framework has led to examples of collaboration, policy diffusion and copying, such as with the proposing and passing of Climate Acts in many *Länder*, or the development of climate and energy agencies to support implementation at the regional and local levels. However, the consensual nature of German federalism means that federal policy often moves at the speed of the slowest participant. Some *Länder* will try to prevent federal laws from being sufficiently ambitious to bring about significant emissions reduction. For example, *Land* opposition to phasing out coal-fired electricity generation, as well as federal regulations on the siting of wind turbines in rural areas or the construction of north–south power lines to transmit renewable electricity, could prove significant (Eichenauer 2018; Neukirch 2020).

Some scholars suggest that the inflexibilities and number of veto points within the federal system are the main obstacle to German climate change action (Scheiner 2017), in line with the 'joint decision-making trap' argument. However, others argue that powerful lobby groups such as car manufacturing and the coal industry have managed to shape decision-making by influencing the occupants of key federal economy and transport ministries in the SPD and centre–right parties (Töller 2019). Both factors would seem to be important: the federal system provides industry groups with multiple venues through which they can pursue their interests and persuade policymakers to slow down progress, particularly if politicians perceive that action to protect the climate could entail electoral costs. Interestingly, however, the Green Party entered the federal government after the September 2021 elections for the first time since 2005. Since the Greens have far fewer links with the fossil fuel and automobile industries than the SPD, CDU and CSU that formed the previous federal coalition, and their presence in *Land* governments has resulted in more ambitious climate policies, such an eventuality could herald a major shift in Germany's overall strategy.

Although Germany's climate policy hitherto has been largely based on the principles of collaboration, the federal government may need to take a more balanced approach to take it to the next level. Its major climate actions have resulted from coordinated action across tiers of government, combined with support from the federal or *Länder* level for municipalities that may otherwise have been unable or unwilling to act. The legal framework within which the

Länder operate – particularly the requirement to provide municipal governments with the necessary resources to undertake new functions – means that they do influence how climate policy is implemented within Germany. However, stronger direction from the federal government might be necessary if the country is to achieve its objective of climate neutrality by 2045 (Heering and Gustavson 2021; Steuwer and Hertin 2021).

8.5 Conclusion

Thus far, Germany's federal system has facilitated a coordinated approach to climate policymaking across and between tiers of governance, whilst also enabling the *Länder* to pursue their own strategies for climate mitigation and adaptation. Since there has been broad agreement on the need to take action in many key areas (such as the closure of polluting industries in the former GDR and the initial scaling up of renewable energy capacity), joint decision-making has yet not proven to be a trap in the climate and energy sectors. Indeed, there has been a degree of competition between some *Länder* for the unofficial title of Germany's climate leader – although ambitions largely reflect the political and economic conditions within each *Land*. As a result, policy can move forward within the constituent parts of the federation in the absence of consensus, but this has meant that the *Länder* are travelling at different speeds. Indeed, closer analysis reveals the extent to which local context has shaped the strategies, polices and approaches to implementation in the different *Länder*. The availability of renewable energy resources and a *Land*'s existing dependence on fossil fuels has a significant effect on mitigation policy, and the extent to which a *Land* is vulnerable to the impact of climate change influences its adaptation approach.

In this sense, the governance of climate change in Germany is something of a microcosm of global efforts to tackle the issue: individual states adopt their own strategies within a wider institutional framework, and these approaches reflect their economic situation and political priorities. In both cases this results in a fascinating mix of different policies that reflect local and regional contexts, but which may be insufficient to achieve their objectives: Germany's federal targets for reducing GHG emissions on the one hand, and the Paris Agreement's pledge to keep global temperature increases below 2°C on the other. Indeed, having implemented most of the 'low-hanging' climate policies, Germany now needs to take more far-reaching decisions around transport and energy infrastructure to continue making progress. Given the distribution of sectoral interests within the federal government and between the *Länder*, it will be much more difficult to reach a political consensus on these issues. As such, we can see how the traditional 'joint decision trap' critique might well apply to the case of climate change once decisions become more controversial: the need to seek consensus in a bureaucratic decision-making system that involves many veto players could ultimately result in suboptimal policy. Although strong intergovernmental structures did help to coordinate and mobilise activity throughout the policy chain in the initial phases of the *Energiewende*, and this strengthened the hand of the state in implementing policy objectives, it has become increasingly difficult to maintain momentum. Germany's federal system has contributed to progress slowing down, because it provides different interests with multiple venues to push their cause and veto more ambitious initiatives (Töller 2019). Individual states remain free to develop more far-reaching policies within the constitutional framework, but the new federal government may need to adopt a more coercive approach in order to ensure that other parts of the country do not fall too far behind. Following the approach adopted after Germany took the decision to phase out nuclear energy, both the federal government and the EU are also likely to provide substantial financial support to those *Länder* that will be most affected by the shift away from coal, in order to reduce the societal impact of this transition.

Overall, therefore, Germany's experience suggests that federal systems can have beneficial effects for policy experimentation, coordination and implementation when there is a political consensus, but they may be less effective where it is difficult to reach agreement between key actors.

Notes

1 Bundes-Klimaschutzgesetz [KSG] [Federal Climate Protection Act], 17 December 2019, Bundesgesetzblatt [BGBl] I at no 48, p. 2513. www.buzer.de/s1.htm?g=Bundes-Klimaschutzgesetz.

2 These targets are 'binding' in the sense that the EU can impose fines on any country that fails to achieve them.

3 Bremen's economy relies heavily on carbon-intensive steel industries. As the smallest of the sixteen *Länder* in terms of population, the GHG emissions produced by this sector distort its overall per capita figures significantly.

4 Hamburg's GHG emissions decreased steadily between 1990 and 2015, after which its Moorburg coal power station was put into operation and reversed this decline dramatically. With around 8.5 million tons of CO_2 per year, this single plant accounts for around half of the state's total annual emissions.

5 The 2017 federal Renewable Energy Act replaced these subsidies with a tendering system that makes renewable installations much less financially attractive (Fell 2017).

6 Klimaschutzgesetz Baden-Württemberg [KSG BW] [Climate Protection Act Baden-Württemberg], 17 July 2013, Gesetzesbeschluss des Landtags Baden-Württemberg at 15/3842. https://vm.baden-wuerttemberg.de/fileadmin/redaktion/m-mvi/intern/Dateien/PDF/Klimaschutz_Gesetzesbeschluss_Klimaschutzgesetz-1.pdf.

7 Energiewende- und Klimaschutzgesetz Schleswig-Holstein [EWKG] [Schleswig-Holstein Energy Transition and Climate Protection Act], 7 March 2017, Gesetz- und Verordnungsblatt für Schleswig-Holstein [GVOBl] at no. 4, p. 124. www.gesetze-rechtsprechung.sh.juris.de/jportal/?quelle=jlink&query=EWKSG+SH+%C2%A7+1&psml=bsshoprod.psml&max=true.

8 Landesklimaschutzgesetz [LKSG] [State Climate Protection Act], 19 August 2014, Gesetz- und Verordnungsblatt für das Land Rheinland-Pfalz [GVBl] 2014 at p. 188. http://landesrecht.rlp.de/jportal/portal/t/onc/page/bsrlpprod.psml?pid=Dokumentanzeige&showdoccase=1&js_peid=Trefferliste&documentnumber=1&numberofresults=22&fromdoctodoc=yes&doc.id=jlr-KlimaSchGRPrahmen&doc.part=X&doc.price=0.0&doc.hl=1#focuspoint.

9 Thüringer Klimagesetz [ThürKlimaG] [Thuringia Climate Act], 18 December 2018, Gesetz- und Verordnungsblatt für den Freistaat Thüringen [GVBl] 2018 at p. 818. http://landesrecht.thueringen.de/jportal/?quelle=jlink&query=KlimaSchG+TH&psml=bsthueprod.psml&max=true.

10 HmbKliSchG, op. cit. at note 5.

References

Abel, Dennis. 2021. 'The Diffusion of Climate Policies among German Municipalities'. *Journal of Public Policy* 41 (1): 111–36. doi: 10.1017/S0143814X19000199.

Abgeordnetenwatch.de. 2018. 'Gesetz zur Förderung des Klimaschutzes in Sachsen'. *Abgeordnetenwatch.de*, 5 September. www.abgeordnetenwatch.de/sachsen/6/abstimmungen/gesetz-zur-foerderung-des-klimaschutzes-sachsen.

Adelberger, Karen. 2001. 'Federalism and its Discontents: Fiscal and legislative power-sharing in Germany, 1948–1999'. *Regional & Federal Studies* 11 (2): 43–68. doi: 10.1080/714004687.

AEE (Agentur für Erneuerbare Energien). 2019. 'Bundesländer-Übersicht zu Erneuerbaren Energien: Installierte Leistung Photovoltaik'. *AEE.* https://perma.cc/3JHS-DGM9.

Auel, Katrin. 2014. 'Intergovernmental Relations in German Federalism: Cooperative federalism, party politics and territorial conflicts'. *Comparative European Politics* 12 (4–5): 422–43. doi: 10.1057/cep.2014.13.

Bayerische Staatsregierung. 2009. *Klimaprogramm Bayern 2020.*

Behnke, Nathalie. 2020. 'Administrative Autonomy of the German *Länder* over Seven Decades of the Basic Law'. In *Reformbaustelle Bundesstaat,* eds. Felix Knüpling, Mario Kölling, Sabine Kropp, and Henrik Scheller. 187–202. Wiesbaden: Springer.

Benz, Arthur. 2007. 'Inter-Regional Competition in Cooperative Federalism: New modes of multi-level governance in Germany'. *Regional and Federal Studies* 17 (4): 421–36. doi: 10.1080/13597560701691797.

Beveridge, Ross, and Kristine Kern. 2013. 'The "Energiewende" in Germany: Background, development and future challenges'. *Renewable Energy, Law and Policy* 4 (1): 3–12. www.jstor.org/stable/24324649.

BFHH (Bürgerschaft der Freien und Hansestadt Hamburg). 2013. 'Aktionsplan Anpassung an den Klimawandel'. *Hamburg.de.* www.hamburg.de/anpassungsstrategie.

BMUB (Federal Ministry for the Environment, Nature Conservation, Building, and Nuclear Safety). 2016. *Climate Action Plan 2050: Principles and goals of the German Government's climate policy.* Berlin: BMUB. www.bmu.de/fileadmin/Daten_BMU/Pools/Broschueren/klimaschutzplan_2050_en_bf.pdf.

Börzel, Tanja A. 2005. 'What Can Federalism Teach Us about the European Union? The German experience'. *Regional and Federal Studies* 15 (2): 245–57. doi: 10.1080/13597560500115626.

Bouckaert, Geert, Davide Galli, Sabine Kuhlmann, Renate Reiter, and Steven Van Hecke. 2020. 'European Coronationalism? A hot spot governing a pandemic crisis'. *Public Administration Review* 80 (5): 765–73. doi: 10.1111/puar.13242.

Bundesministerium für Wirtschaft und Energie. 2009. *Gesetz zur Förderung Erneuerbarer Energien im Wärmebereich.* www.erneuerbare-energien.de/EE/Navigation/DE/Recht-Politik/Das_EEWaermeG/das_eewaermeg.html.

Bundesregierung. 2019. *Klimaschutzprogramm 2030 der Bundesregierung zur Umsetzung des Klimaschutzplans 2050.* Bundesregierung. www.bundesregierung.de/resource/blob/975226/1679914/e01d6bd855f09bf05cf7498e06d0a3ff/2019-10-09-klima-massnahmen-data.pdf.

Bundesregierung. 2021. *Climate Change Act 2021: Intergenerational contract for the climate*. Bundesregierung.
www.bundesregierung.de/breg-de/themen/klimaschutz/climate-change-act-2021-1936846.

Buth, Mareike, Walter Kahlenborn, Jonas Savelsberg, et al. 2015. *Vulnerabilität Deutschlands gegenüber dem Klimawandel.* Umwelt Bundesamt. www.umweltbundesamt.de/en/publikationen/vulnerabilitaet-deutschlands-gegenueber-dem.

BSDETCP (Berlin Senate Department for the Environment, Transport, and Climate Protection). 2019. *BEK 2030: Berlin energy and climate protection programme 2030.* Berlin: BSDETCP. www.berlin.de/sen/uvk/_assets/klimaschutz/publikationen/bek2030_broschuere_en.pdf.

BSUV (Bayrisches Staatsministerium für Umwelt und Verbraucherschutz). 2020. ‘Bayerische Klimaschutzoffensive’. *BSUV.* www.stmuv.bayern.de/themen/klimaschutz/klimaschutzgesetz/index.htm.

BSUV. n.d. ‘Ein halbes Jahrhundert Bewahrung unserer Lebensgrundlagen: das Bayerische Umweltministerium’. *BSUV.* www.stmuv.bayern.de/ministerium/aufgaben/geschichte_umweltministerium.htm.

Conradt, David P. 2001. *The German Polity*, 7th ed. New York: Longman.

Deutscher Bundestag. 2016. *Zur Kompetenz- und Lastenverteilung zwischen Bund, Ländern und Kommunen hinsichtlich der europäischen Klima- und Energieziele.* Deutscher Bundestag, Wissenschaftliche Dienste. www.bundestag.de/resource/blob/480032/b7046e177327b2294f37f07d196eb239/wd-8-062-16-pdf-data.pdf.

Diekmann, Jochen, Wolf-Peter Schill, Andreas Püttner et al. 2019. *Vergleich der Bundesländer: Analyse der Erfolgsfaktoren für den Ausbau der Erneuerbaren Energien 2019: Indikatoren und Ranking*. Berlin: Deutsches Institut für Wirtschaftsforschung. www.diw.de/documents/publikationen/73/diw_01.c.698754.de/diwkompakt_2019-145.pdf.

Eckersley, Peter. 2017. ‘Cities and Climate Change: How historical legacies shape policy-making in English and German municipalities’. *Politics* 37 (2): 151–66. doi: 10.1177/0263395716670412.

Eckersley, Peter. 2018a. ‘Who Shapes Local Climate Policy? Unpicking governance arrangements in German and English cities’. *Environmental Politics* 27 (1): 139–60. doi: 10.1080/09644016.2017.1380963.

Eckersley, Peter. 2018b. *Power and Capacity in Urban Climate Governance: Germany and England compared.* Oxford: Peter Lang.

Eckersley, Peter, Kristine Kern, Wolfgang Haupt, and Hannah Müller. 2021. *The Multi-level Context for Local Climate Governance in German Municipalities: The role of the federal states.* Erkner: Leibniz Institute for Research on Society and Space. https://leibniz-irs.de/fileadmin/user_upload/Transferpublikationen/2021/IRS_Dialog_multi-level_governance.pdf.

Eichenauer, Eva. 2018. ‘Energiekonflikte: Proteste gegen Windkraftanlagen als Spiegel demokratischer Defizite’. In *Energiewende: Politikwissenschaftliche Perspektiven*, eds. Jörg Radtke and Norbert Kersting, 315–42. Wiesbaden: Springer. doi: 10.1007/978-3-658-21561-3.

Erbguth, Wilfried, and Sabine Schlacke. 2014. *Umweltrecht*, 5th ed. Baden-Baden: Nomos.

Fell, Hans-Josef. 2017. *The Shift from Feed-in-Tariffs to Tenders is Hindering the Transformation of the Global Energy Supply to Renewable Energies.* Berlin: Energy Watch Group Policy Paper. www.energywatchgroup.org/wp-content/uploads/FIT-Tender_Fell_PolicyPaper_EN_final.pdf.

Flaskühler, Christina Agnetha. 2018. *Föderale Klimaschutzgesetzgebung in Deutschland im Lichte des wohlgeordneten Rechts. Forum Umwelt-, Agrar- und Klimaschutzrecht*, vol. 14, ed. Ines Härtel. Baden-Baden: Nomos.

Freie Hansestadt Bremen. 2018. 'Anpassungsstrategie an die Folgen des Klimawandels für das Land und die Städte Bremen und Bremerhaven'. *Freie Hansestadt Bremen.* www.bauumwelt.bremen.de/klimaschutz/klima-energie/klimaanpassung/anpassungsstrategie-an-die-folgen-des-klimawandels-fuer-das-land-und-die-staedte-bremen-und-bremerhaven-60162.

Gallata, Judith A. M., and Jens Newig. 2017. Policy Implementation through Multi-level Governance: Analysing practical implementation of EU air quality directives in Germany. *Journal of European Public Policy* 24 (9): 1308–27. doi: 10.1080/13501763.2017.1314539.

Göpfert, Christian. 2014. 'Der Beitrag von Klimaschutzkonzepten zum kommunalen Klimaschutz: Eine empirische Analyse zusätzlicher Effekte von im Rahmen der Kommunalrichtlinie geförderten Klimaschutzkonzepten auf die lokalen Klimaschutzaktivitäten'. *Wuppertaler Studienarbeiten zur nachhaltigen Entwicklung* 6. http://hdl.handle.net/10419/95870.

Heering, Jonas, and Thane Gustavson. 2021. Germany's *Energiewende* at a Crossroads. *German Politics and Society* 39 (2): 47–69. doi: 10.3167/gps.2021.390203.

Hegele, Yvonne, and Behnke, Nathalie. 2017. 'Horizontal Coordination in Cooperative Federalism: The purpose of ministerial conferences in Germany'. *Regional & Federal Studies* 27 (5): 529–48. doi: 10.1080/13597566.2017.1315716.

Hennicke, Peter, and Paul J. Welfens. 2012. *Energiewende nach Fukushima: Deutscher Sonderweg oder weltweites Vorbild?* Munich: Oekom Verlag.

HMUKLV. 2017. *Integrierter Klimaschutzplan Hessen 2025.* Wiesbaden: HMUKLV. https://umwelt.hessen.de/sites/default/files/media/hmuelv/integrierter_klimaschutzplan_web_barrierefrei.pdf.

HMUKLV (Hessisches Ministerium für Umwelt, Klimaschutz, Landwirtschaft und Verbraucherschutz). 2019. *Richtlinie des Landes Hessen zur Förderung von kommunalen Klimaschutz- und Klimaanpassungsprojekten sowie von kommunalen Informationsinitiativen.* Wiesbaden: HMUKLV. https://umwelt.hessen.de/sites/default/files/media/hmuelv/klima-richtlinie_barrierefrei_0.pdf.

Hessisches Statistisches Landesamt. 2016. *Nachhaltigkeitsstrategie Hessen Ziele und Indikatoren: Fortschrittsbericht 2016.* Wiesbaden: Hessisches Statistisches Landesamt. www.hessen-nachhaltig.de/files/content/downloads/ziele_und_indikatoren/ziele_und_indikatoren_fortschrittsbericht_2016.pdf.

Hoeft, Christoph, Sören Messinger-Zimmer, and Julia Zilles, eds. 2017. *Bürgerproteste in Zeiten der Energiewende: Lokale Konflikte um Windkraft, Stromtrassen und Fracking.* Studien des Göttinger Instituts für Demokratieforschung zur Geschichte politischer und gesellschaftlicher Kontroversen, vol. 12, ed. Stine Marg. Biekefeld: Transcript.

Hoff, Jens V., and Bjarne W. Strobel. 2013. 'A Municipal "Climate Revolution"? The shaping of municipal climate change policies'. *Journal of Transdisciplinary Environmental Studies* 12 (1): 4–16.

Hörter, Annerose, Julia Schirrmacher, Martin Beer, Britta Sommer, and Judith Utz. 2018. *Schlüsselakteure bewegen kommunalen Klimaschutz: Erfolgreicher Klimaschutz dank*

Schlüsselakteuren: Projektergebnisse und Handlungsempfehlungen für lokale Klimaschutzakteure. Europa-Universität Flensburg and Zentrum für Nachhaltige Energiesysteme. http://schluesselakteure.de/wp-content/uploads/2018/05/Leitfaden_Erfolgreicher-kommunaler-Klimaschutz-dank-Schl%C3%BCsselakteuren.pdf.

Jänicke, Martin. 2011. 'German Climate Change Policy: Political and economic leadership'. In *The European Union as a Leader in International Climate Change Politics*, eds. Rüdiger Wurzel and James Connelly, 129–46. Abingdon: Routledge.

Jänicke, Martin, and Rüdiger Wurzel. 2019. 'Leadership and Lesson-Drawing in the European Union's Multilevel Climate Governance System'. *Environmental Politics* 28 (1): 22–42. doi: 10.1080/09644016.2019.1522019.

Kern, Kristine. 2008. 'Sustainable Development Initiatives in Federal States in Germany'. In *In Pursuit of Sustainable Development: New governance practices at the sub-national level in Europe*, eds. Susan Baker and Katarina Eckerberg, 122–44. London/New York: Routledge.

Kern, Kristine. 2019. 'Cities as Leaders in EU Multilevel Climate Governance: Embedded upscaling of local experiments in Europe'. *Environmental Politics* 28 (1): 125–45. doi: 10.1080/09644016.2019.1521979.

Kern, Kristine, Wolfgang Haupt, Peter Eckersley, and Hannah Müller. 'Climate Change Acts on the Move: Diffusion and dynamics in the German multi-level System'. *Unpublished manuscript*.

Kern, Kristine, Stephanie Koenen, and Tina Löffelsend. 2004. 'Red–Green Environmental Policy in Germany: Strategies and performance patterns'. In *Germany on the Road to 'Normalcy': Policies and politics of the first Red–Green federal government (1998–2002)*, 1st ed., ed. Werner Reutter, 183–206. New York: Palgrave Macmillan.

Kropp, Sabine, and Schnabel, Johanna. 2021. 'Germany's Response to COVID-19: Federal coordination and executive politics'. In *Federalism and the Response to COVID-19*, eds. Rupak Chattopadhyay, Felix Knüpling, Diana Chebenova, Liam Whittington, and Phillip Gonzalez, 84–94. London: Routledge.

Kuhlmann, Sabine, and Jochen Franzke. 2022. 'Multi-level Responses to COVID-19: Crisis coordination in Germany from an intergovernmental perspective'. Local Government Studies 48 (2): 312–334.

Landtag Niedersachsen. 2019. *Gesetzentwurf: Niedersächsisches Gesetz zur Förderung des Klimaschutzes und zur Anpassung an die Folgen des Klimawandels*. Available at: www.landtag-niedersachsen.de/Drucksachen/Drucksachen_18_05000/04501-05000/18-04839.pdf.

Lerman, Laura Visintainer, Wolfgang Gerstlberger, Mateus Ferreira Lima, and Alejandro G. Frank. 2021. 'How do Governments, Universities, and Companies Contribute to Renewable Energy Development? A municipal innovation policy perspective of the triple helix'. *Energy Research & Social Science* 71: 101854. doi: 10.1016/j.erss.2020.101854.

Mendonça, Miguel, David Jacobs, and Benjamin Sovacool. 2010. *Powering the Green Economy: The feed-in tariff handbook*. London: Earthscan.

Monstadt, Jochen, and Stefan Scheiner. 2014. 'Allocating Greenhouse Gas Emissions in the German Federal System: Regional interests and federal climate governance'. *Energy Policy* 74: 383–94.

Moser, Stephanie, and Silke Kleinhückelkotten. 2018. 'Good Intents, but Low Impacts: Diverging importance of motivational and socioeconomic determinants explaining pro-environmental behavior, energy use, and carbon footprint'. *Environment and Behaviour* 50 (6): 626–56. doi: 10.1177/0013916517710685.

Müller, Edda. 1998. 'Differences in Climate Change Policy in Germany and the United States from a Political Science Perspective'. *Energy & Environment* 9 (4): 463–77.

MWE B (Ministerium für Wirtschaft und Europaangelegenheiten des Landes Brandenburg). 2012. *Energiestrategie 2030 des Landes Brandenburg*. Potsdam: MWE B. https://mwae.brandenburg.de/media/bb1.a.3814.de/Energiestrategie2030_2012.pdf.

MWW SA (Ministerium für Wissenschaft und Wirtschaft des Landes Sachsen-Anhalt). 2014. *Energiekonzept 2030 der Landesregierung von Sachsen-Anhalt*. Magdeburg: MWW SA. https://mule.sachsen-anhalt.de/fileadmin/Bibliothek/Politik_und_Verwaltung/MLU/MLU/04_Energie/Energiepolitik/00_Startseite_Energiepolitik/Energiekonzept_2030.pdf.

Neukirch, Mario. 2020. 'Grinding the Grid: Contextualizing protest networks against energy transmission projects in Southern Germany'. *Energy Research & Social Science* 69: 101585. doi: 10.1016/j.erss.2020.101585.

Newig, Jens, Edward Challies, Nicolas Jager, and Elisa Kochskämper. 2014. 'What Role for Public Participation in Implementing the EU Floods Directive? A comparison with the water framework directive, early evidence from Germany and a research agenda'. *Environmental Policy and Governance* 24 (4): 275–88. doi: doi.org/10.1002/eet.1650.

N MUEBK (Niedersächsisches Ministerium für Umwelt, Energie, Bauen und Klimaschutz). 2020. *Energiewendebericht 2019*. Hannover: N MUEBK. www.umwelt.niedersachsen.de/download/155141/Energiewendebericht_2019.pdf.

OECD. 2019. *Historical Population*. https://stats.oecd.org/Index.aspx?DataSetCode=HISTPOP.

OECD (Organisation for Economic Co-Operation and Development). 2020. *Environment Database: Greenhouse Gas Emissions*. https://stats.oecd.org/Index.aspx?DataSetCode=AIR_GHG.

Ohlhorst, Dörte, Kerstin Tews, and Miranda Schreurs. 2014. 'Energiewende als Herausforderung der Koordination im Mehrebenensystem'. In *Im Hürdenlauf zur Energiewende*, eds. Achim Brunnengräber and Maria Rosaria Di Nucci, 93–104. Wiesbaden: Springer.

Ralston, H. A. 2013. *Subnational Partnerships for Sustainable Development: Transatlantic cooperation between the United States and Germany*. Cheltenham: Edward Elgar.

Reckien, Diana, Monica Salvia, Oliver Heidrich, et al. 2018. 'How are Cities Planning to Respond to Climate Change? Assessment of local climate plans from 885 cities in the EU-28'. *Journal of Cleaner Production* 191: 207–19.

Salvia, Monica, Diana Reckien, Filomena Pietrapertosa, et al. 2021. 'Will Climate Mitigation Ambitions Lead to Carbon Neutrality? An analysis of the local-level plans of 327 cities in the EU'. *Renewable and Sustainable Energy Reviews* 135: 110253. doi: 10.1016/j.rser.2020.110253.

Scharpf, Fritz W. 1988. 'The Joint-Decision Trap: Lessons from German federalism and European integration'. *Public Administration* 66 (3): 239–78. doi: 10.1111/j.1467-9299.1988.tb00694.x.

Scharpf, Fritz W. 2009. *Föderalismusreform: kein Ausweg aus der Politikverflechtungsfalle?* Frankfurt: Campus Verlag.

Scharpf, Fritz W., Bernd Reissert, and Fritz Schnabel. 1976. *Politikverflechtung: Theorie und Empirie des kooperativen Föderalismus in der Bundesrepublik*. Kronberg: Scriptor.

Scheiner, Stefan. 2017. *Interessen der Bundesländer in der deutschen Klimapolitik: Föderale Konfliktverarbeitung in drei Handlungsfeldern. Schriften zum Föderalismus*, vol. 5, eds. Roland Lhotta and Frank Decker. Baden-Baden: Nomos.

Scherf, Wolfgang. 2010. 'Die kommunalen Finanzen in Deutschland'. In *Kommunalpolitik in den Deutschen Ländern: Eine Einführung*, eds. Hans-Georg Wehling and Andreas Kost, 367–88. Wiesbaden: Springer. doi: 10.1007/978-3-531-92034-4.

Schleich, Joachim, Wolfgang Eichhammer, Ulla Boede, et al. 2001. 'Greenhouse Gas Reductions in Germany: Lucky strike or hard work?' *Climate Policy* 1 (3): 363–80. doi: 10.1016/S1469-3062(01)00022-5.

Schultze, R. O. 1999. 'Föderalismusreform in Deutschland: Widersprüche – Ansätze – Hoffnungen'. *Zeitschrift für Politik* 46(2): 173–94

S SMWAV (Sächisches Staatsministerium für Wirtschaft, Arbeit und Verkehr). 2013. *Energie- und Klimaprogramm Sachsen 2012*. Dresden: S SMWAV. www.klima.sachsen.de/download/EKP2012.pdf.

SPD. 2021. *Mehr Fortschritt wagen: Bündnis für Freiheit, Gerechtigkeit und Nachhaltigkeit.* Sozialdemokratische Partei Deutschlands. www.spd.de/koalitionsvertrag2021.

Statistisches Landesamt Baden-Württemberg. 2021. *Treibhausgas-Emissionen um 8.7% gesunken*. Statistisches Landesamt Baden-Württemberg. www.statistik-bw.de/Presse/Pressemitteilungen/2021191.

Statistisches Amt Saarland. 2018. *Energiebilanz und CO_2-Bilanz des Saarlandes 2016.* Saarbrücken: Statistisches Amt Saarland. www.statistischebibliothek.de/mir/servlets/MCRFileNodeServlet/SLHeft_derivate_00005581/SL_Bericht_EIV4_2016-j.pdf.

Steuwer, Sibyl, and Julia Hertin. 2021. 'Climate Policy in Germany: Pioneering a complex transformation process'. In *Climate Governance across the Globe: Pioneers, leaders and followers*, eds. Rüdiger K. W. Wurzel, Mikael Skou Andersen, and Paul Tobin, 160–81. Abingdon: Routledge.

SUVK (Senatsverwaltung für Umwelt, Verkehr und Klimaschutz). 2016. *Programm zur Anpassung an die Folgen des Klimawandels in Berlin*. Berlin.de. www.berlin.de/sen/uvk/klimaschutz/anpassung-an-den-klimawandel/programm-zur-anpassung-an-die-folgen-des-klimawandels.

Szuleki, Kacper, Severin Fischer, Anne Therese Gullberg, and Oliver Sartor. 2016. 'Shaping the "Energy Union": Between national positions and governance innovation in EU energy and climate policy'. *Climate Policy* 16 (5): 548–67. doi: 10.1080/14693062.2015.1135100.

TLS (Thüringer Landesamt für Statistik). 2019. *Statistischer Bericht: Energiebilanz und CO_2: Bilanz Thüringens 2016*. Erfurt: TLS. https://statistik.thueringen.de/webshop/pdf/2016/05402_2016_00.pdf.

Töller, Annette Elisabeth. 2019. 'Kein Grund zum Feiern! Die Umwelt- und Energiepolitik der dritten Regierung Merkel'. In *Zwischen Stillstand, Politikwandel und Krisenmanagement: Eine Bilanz der Regierung Merkel 2013–2017*, eds. Reimut Zohlnhöfer and Thomas Saalfeld, 569–90. Wiesbaden: Springer.

Von Hirschhausen, Christian. 2014. 'The German "Energiewende": An introduction'. *Economics of Energy & Environmental Policy* 3 (2): 1–12.

Weidner, Helmut, and Lutz Mez. 2008. 'German Climate Change Policy: A success story with some flaws'. *Journal of Environment & Development* 17 (4): 356–78.

Wollmann, Hellmut. 2004a. 'Local Government Reforms in Great Britain, Sweden, Germany and France: Between multi-function and single-purpose organisations'. *Local Government Studies* 30 (4): 639–65.

Wollmann, Hellmut. 2004b. 'The Two Waves of Territorial Reforms of Local Government in Germany'. In *Redrawing Local Government Boundaries. An international study of politics, procedures and decisions*, ed. John Meligrana, 106–29. Vancouver: UBC Press.

Wollmann, Hellmut. 2021. 'Transformation of Public Administration in East Germany Following Unification'. In *Public Administration in Germany*, eds. Sabine Kuhlmann, Isabella Proeller, Dieter Schimanke, and Jan Ziekow, 253–69. Cham: Palgrave Macmillan. doi: 10.1007/978-3-030-53697-8.

World Bank. 2019. 'Exports of Goods and Services (BoP, current US$)'. *World Bank Data*. https://data.worldbank.org/indicator/BX.GSR.GNFS.CD.

Wurster, Stefan, and Christina Köhler. 2016. 'Die Energiepolitik der Bundesländer: Scheitert die Energiewende am deutschen Föderalismus?' In *Die Politik der Bundesländer: Zwischen Föderalismusreform und Schuldenbremse*, 2nd ed., eds. Achim Hildebrandt and Frieder Wolf, 283–314. Wiesbaden: Springer. doi: 10.1007/978-3-658-08303-8.

Wurzel, Rüdiger K. W., Anthony R. Zito, and Andrew J. Jordan. 2013. *Environmental Governance in Europe: A comparative analysis of new environmental policy instruments*. Cheltenham: Edward Elgar.

Zohlnhöfer, Reimut. 2009. 'Der Politikverflechtungsfalle entwischt? Die Effekte der Föderalismusreform I auf die Gesetzgebung'. *Zeitschrift für Politikwissenschaft* 19 (1): 39–76. doi: 10.5771/1430-6387-2009-1-39.

9

Climate Governance and Federalism in India

ADITYA VALIATHAN PILLAI AND NAVROZ K. DUBASH

9.1 Introduction

The art of governing India is in no small part the navigation of tensions in Indian federalism. The country is large – both in geographic size and population – has an extraordinary cultural diversity, and is composed of constituent units with radically different economic histories. Indian federalism has moved back and forth between periods of central dominance to coordinate divergent interests and state autonomy to satisfy local demands.

These oscillations, over the seven decades since independence, influence India's prospects for climate action today. The federation began life with a strong national government ('the Centre') sustained by ruling party alignment in the Centre and states, and animated by a focus on centralized industrial growth (Tillin 2019). Centrifugal forces occasioned by the assertion of regional identities then led to a new era, in which upstart regional parties focused on populist agriculture and electricity policy (Dubash and Rajan 2001), among other areas, to bolster their electoral prospects. At the end of the twentieth century, economic liberalization heralded the birth of competition and growing disparities between states (Sáez 2002; Subramaniam and Kumar 2012). We have now returned to a period of central dominance (Aiyar and Tillin 2020), which, as we shall see, is beginning to shape India's energy transition.

It is on this ever-shifting federal terrain that a modern edifice of climate governance must be built. Where, then, is the firm ground? Are there enduring characteristics of Centre–state relations that let us arrive at a relatively stable description of Indian climate governance? In this chapter, we put forward a synthetic account of the forces shaping climate governance in India's federal architecture, building on descriptions of environmental federalism (Arora and Srivastava 2019; Chakrabarti and Srivastava 2015; Huang and Gupta 2014); state actions in climate policy (Dubash and Jogesh 2014; Jörgensen et al. 2015; Kumar 2018); and several recent policy moves by both the Centre and states.

India's highly centralized form of federalism, once famously categorized as *'sui generis*, (Sarkaria et al. 1988), offers a context for climate governance that differs from most other chapters in this volume. Though much more centralized than the classical US model, it is many steps short of Germany's top-down administrative federalism. India's states enjoy substantial autonomy in many areas of mitigation and adaptation, but the federal government holds the reins of state finances, constitutes the bulk of bureaucratic capacity, and exercises jurisdictional authority in several areas of climate policy. This creates an awkward climate governance gap because many climate actions require sustained attention and policymaking from the constituent units. This asymmetry makes close and fairly nuanced forms of cooperation between the Centre and states a structural necessity in climate matters.

The federal system has begun to adapt to this structural weakness. The Centre has been involved in developing new financial mechanisms and supplementing state capacity for climate action. The states have, in response, occasionally taken fragments of the national agenda and adapted them to local political contexts that were hitherto innocent of 'climate' politics – at least phrased as such – thereby fulfilling the potential for localized experimentation latent in federations. Some of these experiments have shaped the national policy landscape through vertical diffusion, having been picked up by the central government and injected into the idiom of governance across India's states.

What emerges is a federal system that episodically attempts to rebalance itself through new institutions, experimentation, and diffusion. But this righting reflex is the function of a series of uncoordinated variables – spurts of policy activity driven by foreign policy objectives, independent developments in fiscal federalism, and opportunistic states – and falls short of a new compact in Indian federalism. Climate policies thus episodically appear, and fade away, across the federal landscape.

After providing an overview of climate change in India and its policy responses, this chapter describes India's federal architecture and environmental governance processes before showing how the federal system is adapting to the climate challenge. In the conclusion, it reflects on the inherent vulnerabilities of this form of climate governance.

9.2 Climate Change and India

India's importance in the climate crisis is rooted in the country's size, growth rate, and low economic starting point. It is a country of 1.2 billion people with weak development indicators (Conceição 2019) but has grown rapidly in recent decades as it pushes to join the ranks of middle-income countries. Its per capita income, adjusted for purchasing parity, grew 162 per cent in the two decades since

2000 (World Bank 2020).[1] Emissions have grown concomitantly, slightly more than doubling between 1994 and 2016. Three-fourths of India's emissions are from the energy sector (40 per cent of all emissions are from electricity generation and 9 per cent from road transport); agriculture is around 15 per cent; and industrial processes and product use nearly 8 per cent (Government of India 2021). In 2018, it was the third largest emitter (Global Carbon Project 2019). At the same time, its per capita emissions are a third of the global average, placing equity concerns at the centre of its assessment of mitigation responsibilities (Dubash et al. 2018).

India has long held a diplomatic stance that seeks to avoid constraints on its development. The use of per capita metrics to underscore equity concerns around decarbonization are, consequently, an early and consistently prominent feature of Indian climate politics (Dubash et al. 2018). Its position has been central to the evolution and practice of the principle of 'common but differentiated responsibility' (Sengupta 2013), which places the onus of emissions reductions on developed countries that, India argues, caused global warming through excessive per capita emissions as they grew (C. Dasgupta 2019).

Despite a stable diplomatic scaffolding for its interests, India has been responsive to shifts in global climate politics. A spurt of climate policymaking from 2007 to 2009, triggered by mitigation pressures on developing countries during the Copenhagen Conference of Parties (COP), resulted in a National Action Plan on Climate Change (NAPCC) that laid the institutional and programmatic foundations for action across a set of eight 'missions' that covered both mitigation and adaptation themes. The central government used the NAPCC to draw the contours of state climate governance in this period by requiring the creation of State Action Plans on Climate Change (SAPCCs), thus placing climate change on the agenda of central and state governments for the first time (Pillai and Dubash 2021).

India made its first numerical emissions pledge at Copenhagen, followed by expanded and relatively more ambitious pledges at the Paris COP in 2015 (Dubash et al. 2018). India's 2022 NDC centred on pledges to reduce its emissions intensity by 45 per cent between 2005 and 2030; increase non-fossil fuel sources to about 50 per cent of installed electricity capacity by 2030 (with technology transfers and international finance); and create an additional carbon sink of 2.5–3 bn tons of CO_2 equivalent. It has also increased engagement in the broader regime complex for global climate governance, with participation in multi-lateral forums for hydrofluorocarbons and aviation regulation (Ghosh 2019) among others.[2]

Domestically, India does not have formal climate legislation at either the federal or state levels. Major political moments, such as the ratification of the Paris Agreement, have not stimulated legislative activity either through consideration of climate in new laws or amendments to existing laws. An important reason is the centrality of 'co-benefits' – the simultaneous achievement of development and climate gains – as a motivating narrative (Dubash 2013), which has thus far been accommodated within the scope of existing legislation. Consequently, the institutional structure is composed of a thicket of sectoral climate plans and relevant legislation that antedate the rise of climate change in the governance discourse of the late 2000s. Such laws include the *Electricity Act* of 2003,[3] which creates a legal basis for the nation-wide promotion of renewables; an energy conservation law (2001); and legislation on forests, water, air, biodiversity, and the like. The current governance approach is thus contingent on the creative interpretation and deployment of a variety of sectoral institutions and frameworks.

This architecture must adapt to the challenges of a country particularly vulnerable to climate change. The IPCC (Intergovernmental Panel on Climate Change) projects that countries in South and Southeast Asia will be most at risk from coastal flooding at all levels of warming, with India ranking among the most vulnerable globally (Hoegh-Guldberg et al. 2018, 231). Its long coastline will witness an increase in severe cyclonic storms while the densely populated Indo-Gangetic Plain will face the retreat of the Himalayan glaciers that nourish it (Krishnan et al. 2020). An economy dependent on agriculture will have to manage an increased propensity for droughts and a decrease in summer monsoonal rainfall (Krishnan et al. 2020). Cities will suffer from heat stress, with Kolkata potentially suffering from 'deadly' heat waves annually with 2 degrees of warming (Hoegh-Guldberg et al. 2018, 242). Poorer populations are more likely to be exposed to compounding climate effects, from droughts to water stress, habitat degradation, and lower crop yields (Hoegh-Guldberg et al. 2018, 245). India's federal structure – particularly the states' ability to respond and the Centre's capacity to even out capacity differences – assumes particular salience in this context.

9.3 India's Top-Heavy Federalism

India was forged at a moment of political upheaval and partition, forcing the Constituent Assembly to adopt a centralized federal model to ensure stability. This was also in keeping with emerging trends in federalism in the wake of the Second World War, when large federations began building empowered central governments capable of delivering welfare in pensions, insurance, and healthcare (Tillin 2019). In post-independent India, the inclination was for a powerful central government that sought to shape provincial policy through central planning (Tillin 2019).[4]

The Sarkaria Commission Report, a prominent reform effort to address irritants in federal relations, described what resulted as 'a *sui generis* system of two-tier polity in which the predominant strength of the Union is blended with the essence of co-operative federalism'. Several features of the Constitution, they thought, 'appear to have been deliberately designed to institutionalize the concept of co-operation' (Sarkaria et al. 1988, 1.3.28) rather than full-fledged state autonomy. Other commentators have described the structure as 'quasi-federal' (Wheare 1964, 28). As we show below, the states enjoy legislative powers in several areas, but these domains are often influenced by the Centre. This places India somewhere between the classical model of devolution in the USA and Germany's highly integrated 'administrative federalism' (see Hueglin and Fenna 2015 and Mueller and Fenna 2022 for more on the varieties of federalism).

The original Constitution did not recognize the environment as a distinct area of governance, but related sections, subsequent amendments, and patterns in central legislation have since lent the federal government a dominant role. Article 253 of the Constitution importantly allows the Centre to legislate on the subject matter of international treaties, regardless of whether they concern matters exclusively under state jurisdiction (Huang and Gupta 2014). This provision has been particularly important in the development of Indian environmental law and has paved the way for landmark central legislation including the *Environment Protection Act 1986* (Chakrabarti 2015). The Act gives sweeping powers to the central government, allowing the Centre to 'take all such measures as it deems necessary' to protect the environment and gives it power to lay down 'standards for emission or discharge of environmental pollutants from various sources whatsoever' – though 'environmental pollutants' has not yet been interpreted to include carbon by any court.[5]

In climate governance, the division of powers in the Constitution (laid out in Schedule VII) leaves both the Centre and states with important roles, calling for a mix of top-down direction-setting and resource flows, and bottom-up state-driven policy. The Constitution gives the Centre a hand in several realms of climate governance such as mines and petroleum, industry, and interstate waters. The 42nd amendment to the Constitution, passed during a period of unprecedented centralization and a suspension of democratic rights, placed forests and wildlife in the concurrent list of the Constitution (Chakrabarti 2015) – allowing both levels of government to legislate, with the Centre prevailing in the event of a conflict. The legacy of Indira Gandhi, recently interpreted as an environmentalist prime minister (Ramesh 2017), underpins the Centre's ability to dictate the use of forest lands, and consequently influence related areas of agriculture and water governance. The Centre also enjoys residual powers that allow it to legislate in areas not explicitly listed in the Constitution.

On the other hand, several important areas of climate governance – such as local government, agriculture, and water governance – are the constitutional preserve of state governments. But state policy in these areas is constantly shaped by numerous national schemes (Ministry of Finance 2020) and centrally designed 'model legislations' offered to states. Electricity, which produces over two-fifths of India's emissions (Government of India 2018) and is a crucible of state politics, is in the concurrent list, with the Centre historically defining the framework within which states operate[6] while the states determine policies and the extent to which central guidance is implemented. The states thus play the primary role in adaptation and a substantial role in mitigation, thereby underscoring the need for new federal institutions to rebalance governance capability to the states in these areas.

The fiscal power of the Centre accentuates its structural dominance. States are responsible for implementation, but major sources of tax revenue are allocated to the federal government (Finance Commission of India n.d.). This imbalance has led to corrective institutional channels such as the Finance Commission, a Constitutional body that recommends tax devolution, its distribution between states and various conditional grants, and the Planning Commission, which until 2014 prescribed funds for state development plans (Aiyar and Kapur 2019). The prominence of conditional transfers in this mechanism further restricts state autonomy because they prescribe policy in nearly every governance area (Parikh and Weingast 1997; Rao and Singh 2004). Tillin captures the extent of central dominance (Figure 9.1) by showing that, on average, states raise only 45 per cent of their revenue from sources under their jurisdiction. An important feature is the wide disparity in states' fiscal autonomy.

This asymmetrical federal layout in legal and fiscal realms emphasizes the importance of institutional forums in reaching an agreement on the pace, depth, and cost of climate governance. Climate-specific interactions are channelled through the National Steering Committee on Climate Change (NSCCC), a body composed of several senior central bureaucrats and some chief bureaucrats from the states. The body is less a platform for deliberation than one designed to monitor state actions by ensuring 'uniformity and coherence' in the SAPCCs, provide guidance on individual projects, and approve financial allocations for them (MEFCC 2017). Alongside the NSCCC sits an assemblage of non-climate forums that could affect climate outcomes, such as an annual meeting of energy ministers (Ministry of Power 2018); a forum of electricity regulators (Forum of Regulators 2005); and a currently inactive Inter-state Council that could play a role in several areas including the vexing question of governing interstate river disputes (Chokkakula 2019).

9.4 Emergent Climate Corrections to Top-Heavy Federalism

Recent developments in the federal system show that the Centre and states are redefining their roles to advance climate policy within the constraints of India's

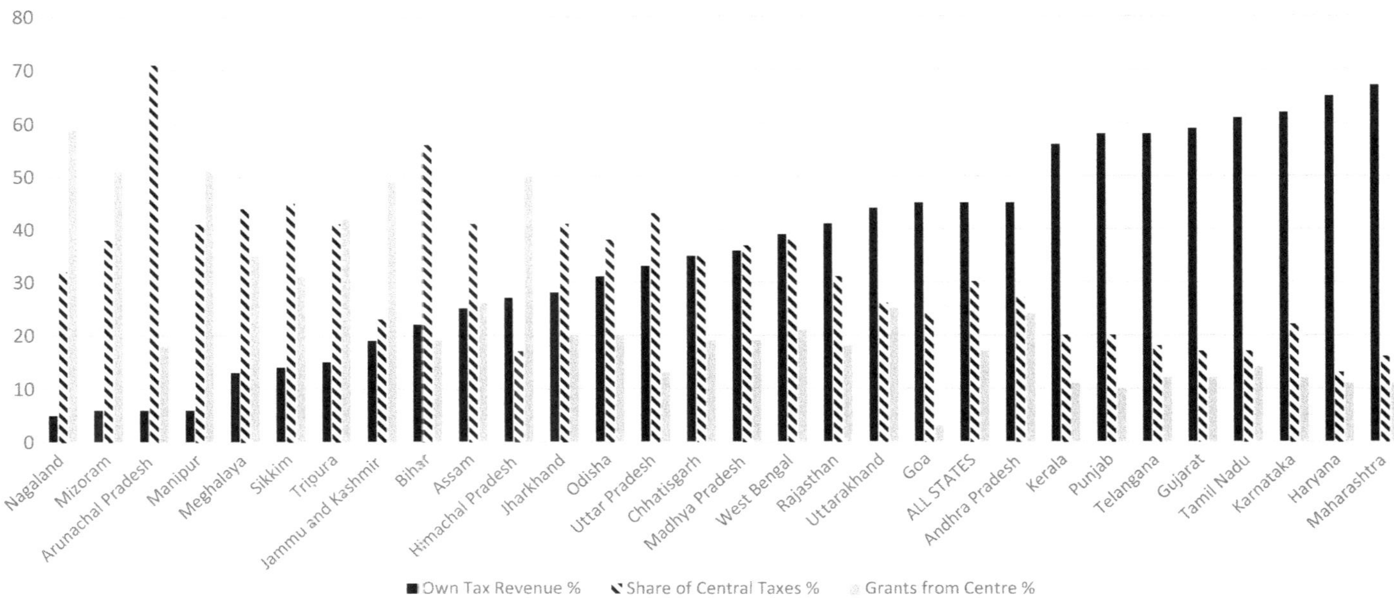

Figure 9.1 State dependence on central transfers in 2016–17 (adapted from Tillin 2019, p. 72).

federal inheritance. This includes the emergence of new priorities in fiscal federalism; national frameworks and capacity to stimulate state action; and political work in the states to make climate priorities locally relevant, policies that are occasionally picked up at the national level.

9.4.1 Federal Finance Flows

The structure of fiscal federalism described in the previous section leaves climate actions partially contingent on new and repurposed financial channels. Over the last decade, fiscal mechanisms have begun to evolve shades of climate responsiveness though these are undeniably subtle tones on a broader canvas. As we show below, they include the incorporation of vulnerabilities into tax devolution by the Finance Commission, the adaptation of some large central government schemes to climate goals, instruments for project-specific central support, and the deployment of central state-owned enterprises to underwrite the renewable transition.

9.4.1.1 Environmental Focus in Finance Commissions

Moves from recent Finance Commissions to include forestry and disaster management variables in their decision framework are among the more important developments in India's nascent climate federalism. The 14th Commission (2015–20) took a bold step in incorporating the state's forest cover as a variable in deciding the quantum of devolution, giving it a 7.5 per cent weightage among four other criteria. The 15th Commission (2021–6) then increased the weight to 10 per cent (XIV Finance Commission 2013; XV Finance Commission 2020). These Commissions argued that states must be compensated for the large opportunity cost in maintaining forests, thus potentially creating political space for pro-forest policies. The 15th Commission, interestingly, uses India's international commitment to increasing forest cover, among other things, to justify its actions – representing a minor link between international processes and India's fiscal structure.

The 15th Commission also reframes the task of disaster management by forcefully arguing for a move away from a disaster-response paradigm to one based on preparedness. They establish substantial fiscal transfers based on metrics related to state capacity, risk exposure, and vulnerability. Given the consultative nature of the Finance Commission, these developments might indicate growing political recognition of climate risks. Indeed, the interim report of the 15th Commission notes that both state and central governments 'argued that issues relating to environment and climate change need to be given greater impetus' during consultations (XV Finance Commission 2019, 4).

9.4.1.2 Central Schemes in Climate-Relevant Sectors

The Finance Commission's priorities sit alongside the Centre's role in shaping state priorities through its national programmes. For example, of the thirty-five mitigation actions listed in India's 2018 Biennial Update Report to the UNFCCC, over one-quarter (ten) have the central government intervening in areas under state control such as agriculture, local government, and industry.[7] Seventeen interventions fall in the concurrent list that are the joint responsibility of the Centre and states while the rest fall under exclusive central or state control (Government of India 2018, 142–50).

This is in keeping with a long-standing tradition of central involvement in state responsibilities through Centrally Sponsored Schemes (programmes designed and largely funded by the Centre) and state plans (Parikh and Weingast 1997). For the period of India's 11th Year Plan (2007–11), CSSs accounted for 40 per cent of central transfers to the states (Aiyar and Tillin 2020), putting it in a league similar to devolution and grants from the Finance Commission. In 2014–15, there were '66 CSSs ... financing all the major social policy programmes of the time' (Aiyar and Kapur 2019, 192). CSSs will thus almost inevitably play a role in stimulating future climate action in the states but suffers from unidirectionality; there are no institutional mechanisms that allow states to contribute to design decisions (Tillin 2019).

9.4.1.3 Federal Financial Experiments

In the wider landscape of state climate finance, these large financial conduits combine with smaller, but still notable, institutional innovations. The NSCCC, the designated body for federal interactions on climate projects, gives the states more discretion in defining their climate priorities than central schemes or Finance Commission flows, but for far smaller sums.[8] States conceive climate programmes and present them to the senior bureaucrats of the NSCCC, which might then offer assistance, approval, and funding (Parliamentary Committee on Estimates 2018). It must be pointed out, however, that minutes of NSCCC meetings obtained through Right to Information requests reveal displeasure within the NSCCC from senior environment ministry officials, who criticized project proposal quality and lamented the slow utilization of funds (MEFCC 2017).

A second institutional innovation comes in the subtle backroom role the Centre plays in adding renewable capacity across the states. The Centre's power trading companies are, for example, the listed buyers for about half of all solar capacity auctioned in India and act as a buffer between private generators and financially precarious state utilities (India RE Navigator 2020). The Centre also drives the hectic pace of India's renewable capacity expansion and indirectly shapes the states' electricity mixes; our analysis of solar capacity auctions from 2010 to 2020 shows that central enterprises were responsible for auctions of almost twice as much solar capacity as state agencies.[9] At the more punitive end, it also tries to reassure investors and developers by imposing penalties on states for defaults on renewable contracts (by withholding fund transfers), thereby leveraging its dominant financial position (Atal et al. 2018).

9.4.2 Central Frameworks and Capacity

The federal government has also played a role in establishing normative frameworks for climate action in the states, indeed pushing them to think about climate change when it was not a priority, while also occasionally directing specialized capacity to state capitals. This notionally balances the uneven levels of interest in climate change in state capitals and their nearly universal bureaucratic constraints.

9.4.2.1 Central Frameworks for State Climate Policies

Unlike several instances in this volume, subnational policymaking in India did not develop organically. It was instead mandated by the central government during a period of heightened climate activity. The creation of the NAPCC forced the environment ministry to consider ways of seeding climate policy in the states, resulting in SAPCCs in thirty-two states and federal territories by 2018 (Parliamentary Committee on Estimates 2018).[10] Though the process has had limited effects, with some observers criticizing them for 'falling woefully short of dealing with the climate-related challenges India is facing' (Kumar 2018, 36), the process put climate policy on the agenda of state governments.

This top-down process has had its drawbacks, notably in the smothering effect of the NAPCC. The SAPCCs were found to replicate the NAPCC, likely because few states embarked on rigorous investigations of their vulnerabilities (Dubash and Jogesh 2014; Kumar 2018). Additionally, central influence constrained the planning exercise by forcing states to prioritize adaptation over mitigation actions to prevent them from undercutting India's international negotiation position (Dubash and Jogesh 2014).

The SAPCC effort was further weakened because they did not lead to a dedicated flow of central financing. States were instead expected to meet expenses through their approved 12th Plan outlays for discretionary expenditure and several smaller pools of central finance (MEFCC 2014). The absence of a large capital infusion seems to have diminished the states' enthusiasm (Kumar 2018). While SAPCC projects could receive funding on a project-by-project basis through the

NSCCC mechanism, this is of small quantum, has high transaction costs, and is subject to central approval.

The Centre reportedly also advised states to dovetail their actions with central schemes like the massive national rural employment programme (Kumar 2018). There were early indications that some states actively experimented with this approach and considered combinations with external donor funding (Dubash and Jogesh 2014). This fiscal tension is compounded by an alleged perception in Delhi that states were hoping to execute a money-grab to finance other developmental initiatives, 'out of greed and not specific need', through the SAPCCs (Kumar 2018, 24). The emergence of a separate channel of climate funds seems unlikely in the fiscal precarity induced by Covid-19.

The federal government has also tried to play a catalytic role by setting policy frameworks in the mitigation arena. It has established clear expectations of a speedy transition by setting ambitious national renewable targets and urging state regulators to force a rapid shift in distribution utility purchase decisions. This approach has, however, revealed institutional tensions. Most state regulators have notified purchase obligations well below suggested trajectories, and financially distressed distribution utilities have remained largely uncompliant (Vembadi et al. 2018). The Centre, in response, suggested an amendment to the framework *Electricity Act 2003* that allowed the Centre to mandate rather than suggest purchase obligation trajectories in the states while increasing penalties for non-compliance.

The Centre's agenda-setting role extends to other important areas as well. In the electric vehicle (EV) domain, it established a subsidy scheme for the manufacture of EVs and has signalled ambitious national targets (Arora 2018), thus stimulating recent policy activity in several states. The Centre was also the first mover with regard to energy efficiency, establishing national institutions and paving the way for the creation of a decentralized network of Energy Service Companies – though the efficacy of this model has been questioned (Harrison and Kostka 2018).

9.4.2.2 The Centre's Influence on Capacity

The Centre also occasionally attempts to address deficiencies in state-level bureaucratic capacity. The challenge is particularly acute in the complex and evolving area of climate policymaking. A longstanding technical advisor to state governments on climate matters notes that state governments have failed to spend monies channelled through the NSCCC mainly because they are unable to conceptualize and execute large climate projects.[11]

Signals from the central government play a role in mobilizing state bureaucrats. Bhardwaj and Khosla (2021) show that performance in delivering high-profile climate-related CSSs come with perks in the form of promotions, monetary rewards, and prestige. The structure of Indian state bureaucracies, led by Indian Administrative Service (IAS) officers eager to make the leap to Delhi, encourages 'allegiance to the bureaucratic hierarchy at the Centre' and pushes the IAS cream to prioritize central schemes in their interaction with subordinate state bureaucrats (Aiyar and Kapur 2019, 210). The Centre's agenda-setting role cannot, however, compensate for a glaring lack of capacity at lower levels. Dasgupta and Kapur (2017) surveyed India's Block Development Offices, an important village-level unit of governance, to find 42 per cent of posts vacant. The localized nature of the climate challenge will likely amplify capacity deficits faced by precariously poised local bureaucrats.

To address some of these challenges, the central government occasionally funnels expertise to the states. For example, the central government convened technical advisors such as UNDP, the UK's Department for International Development, and Germany's GIZ after the SAPCCs were announced, asking these organizations to assist states in plan development.[12] These organizations employed consultants and civil society organizations in what amounted to a short-term fix to the capacity constraint (Dubash and Jogesh 2014). This is not a one-off, with state governments receiving assistance for ongoing SAPCC revisions as well.[13]

Taken together, the federal contribution to climate federalism is notionally catalytic. It works through the gradual layering of climate-linkages into state financial flows, the stimulation of planning activity, the creation of soft bureaucratic incentives, and by funnelling technical capacity to the states at key moments. In the next section, we turn to how states respond to these federal moves.

9.4.3 Political Translation in the States

Some state leaders have managed to elevate the profile of climate-related developmental activities by making a political case for them, thus overcoming low levels of electoral concern about climate change. We use broad-brush examples in this section to show that this process has occurred in a diversity of states, from rich, urban ones like Kerala to poorer agricultural states like Bihar, and across a variety of themes, from air pollution to flooding. This translation process sometimes leads to policy innovations that diffuse vertically to become the standard for national action, giving some credence to the idea of Indian states as laboratories in climate policymaking.

9.4.3.1 Experiments with Co-benefits

Climate change only found mention in the election manifestos of major national parties for the first time in the general election of 2019. In the ruling party's manifesto, it comprised 116 words of 18,327 (0.6 per cent) and was confined to a section on infrastructure. The subject was given only slightly more attention in the principal opposition party's manifesto, occupying 4.6 per cent across multiple sections (Dolsak and Prakash 2019). Parliament has seen little substantive discussion on the issue over the last decade (Dubash 2019).

Operating within the logic of co-benefits, however, states have managed to bridge this political gap well enough to build a noticeable body of climate policy. Alongside the thirty-two adaptation-focused climate action plans mandated by the federal government, states have established at least fifteen solar policies, ten energy conservation building codes, and twenty electric vehicle policies, apart from several LED-village lighting campaigns, energy-efficiency programmes, and afforestation initiatives (Karkun 2021; Kaur and Singh 2019). State policy profiles vary, but leaders exhibit a common inclination towards energy efficiency schemes, which is probably a reflection of high energy prices.

North India's air pollution problem is illustrative of how policies can gain from local political concerns. Air pollution is a complex federal environmental issue because it affects a large swathe of the country's north, including the national capital, and is partially caused by the burning of paddy stubble in the predominantly agricultural states of Punjab and Haryana (Jalan and Dholakia 2019; Sharma and Dikshit 2016). After elevating the issue's profile in campaigning for Delhi's 2020 elections (Sharan 2019), its chief minister unveiled an electric vehicle policy whose primary objective is bringing down pollution (Government of Delhi 2020). The policy contained generous consumer subsidies for Delhi's large urban population and is mostly funded by an 'Air Ambience Fund' built on longstanding diesel taxes. The chief minister's remarks at the launch also positioned the policy as a salve to economic damage caused by Covid-19 lockdowns and laid claim to Delhi's global leadership on the issue.

In nearby Punjab, the state government positioned its response to stubble burning as climate salient as early as 2015 by making a successful proposal to the NSCCC for a technology development programme for the 'gainful utilization' of paddy straw (MEFCC 2017), among a slew of other incentive-based measures (Chaba 2020; Harish and Ghosh 2020) that refrain from exacting costs on the crucial farmer vote bloc. Speaking at a recent national forum, a senior government official from Punjab described these actions as part of a 'climate smart' agricultural strategy (Shekar 2020).

Some state leaders have gone so far as to turn to climate messaging at crucial political moments. In 2018, after the worst floods in Kerala since 1924, the government released a sprawling plan to 'build back better' using climate-first principles. It proposed a major overhaul of infrastructure, institutional coordination, and policies across most areas of governance in service of a 'new Kerala'

(Government of Kerala 2020, 11). The political moment at which this plan emerged gives it a different tenor than previous attempts at climate planning; it came at a crucial and unsteady period in a first-time chief minister's tenure (Padmanabhan 2018). Similarly, the chief minister of Bihar, an agricultural state that experiences frequent and damaging floods, made the unprecedented move of highlighting climate change in his 2020 campaign. This involved a widely covered four-day tour of the state that highlighted new policy measures in water management and agriculture apart from participation in a climate roundtable hosted by the UN Secretary General – which is unusual for an Indian chief minister (Mishra 2019; Press Trust of India 2020).

9.4.3.2 Vertical Diffusion and Institutional Innovation

Such efforts occasionally result in policy innovations that attain national salience. An emblematic recent example is a scheme for solar-powered agricultural pumps in the southern state of Maharashtra. Its ambition of connecting many of its farms to large solar plants could relieve distribution utilities of the burden of supplying subsidized electricity for irrigation (Maharashtra State Electricity Distribution Company Limited 2019). The idea has been eagerly embraced by the central government through a national programme that subsidizes 10 GW of decentralized solar plants (MNRE 2019).

The reasons for Maharashtra's leadership lie in the political economy of its energy development. It has lacked sufficient capacity to meet the demands of its rapidly growing base of small and medium industries in the 1990s, prompting it to implement pioneering wind energy policy (Chaudhary et al. 2014). Maharashtra also established the first clean energy fund in the country (2006), investing in infrastructure and renewable projects through a small tax on commercial and industrial electricity consumers. This was a precursor to the National Clean Energy and Environment Fund, which redirected coal taxes to clean energy projects (Chitnis et al. 2017).

Maharashtra has also established important regulatory precedents. In the months after the restructuring of the Indian electricity sector in 2003, the state electricity regulator put out a 'seminal' order for feed-in-tariffs that was later adopted by the national electricity regulator (Chaudhary et al. 2014, 19). The Maharashtra regulator also established the first Renewable Purchase Standard in the country, an idea subsequently picked up in the National Electricity Policy (2005) and now the primary driver of the renewables transition.

Some states are experimenting with new institutional arrangements, which could constitute an important frontier in experimentation. Climate bodies have emerged in Gujarat, Odisha, Tamil Nadu, and Maharashtra, for example. They are meant to coordinate between state departments and, in some cases, with the central government (Government of Gujarat 2018; Government of Odisha n.d.; Rawal 2021; Sivakumar 2021). The Rebuild Kerala Development Programme, mentioned earlier, is coordinated by a Secretariat (Government of Kerala 2020). Tamil Nadu has established a registered state company to raise funds and implement programmes. The effectiveness of these bodies is, however, unknown.

9.5 Conclusion

The top-heavy nature of Indian federalism sits uneasily with the nature of the climate problem in a large country where consequences of climate change and energy transitions are felt by constituent units foremost. This dissonance between centralized federal institutions and the nature of the climate problem enmeshes the Centre and states in a multi-faceted cooperative relationship, one perhaps more intense than in classical versions of federalism seen elsewhere in this volume.

The modern institutional bias towards the Centre arises from a historical skew in power and resources to the federal government, deliberately crafted in the tumult after Indian independence and just as the global conversation on federalism became more accepting of federal dominance in economic and social policy. Since climate governance is nearly all-encompassing in the scope of actions it demands, the Centre must allocate financial and intellectual resources to stimulate and occasionally supplement action in nearly every area of state jurisdiction. Yet the nature of this top-down force is deeply conditioned by the central government's foreign policy. The government of India has worked assiduously to resist the constriction of its developmental space by international climate change pressures.

The ideas that underlie the federal government's approach to climate change – of necessarily seeking co-benefits to mitigation action and adhering to the principle of common but differentiated responsibilities – filter through to the states and thus establish the normative boundaries for appropriate climate action. This was particularly evident when the SAPCCs were first conceived, and instructions passed down to the states; state governments moulded their actions to a national template and refrained from emphasizing mitigation.

Within this framework, however, state governments have demonstrated a willingness to build political narratives around climate vulnerability at crucial political junctures – as with Kerala and Bihar – which is a precondition to climate policy experimentation. Maharashtra has led in mitigation policy and created a template for several national policies. States, in furtherance of their own developmental and political goals, gently push the boundaries of climate action.

The configuration undoubtedly presents risks. The first of these is inadequate fiscal devolution from the Centre. States have complained vocally that they have not been receiving their fair share in recent years as economic growth has slowed.

A second threat arises from the possibility of a central government that fails to incorporate climate change policy in its programmes and fiscal transfers. The Centre's normative and fiscal power places an upper limit on the depth and pace of state policy. Cumulatively, this configuration could have a chilling effect on state action and rob the multi-level governance system of its promise.

The risks are no less vexing with the states. Large variations in state capacity will become more evident as the impact of climate change grows and calls for drastic mitigation increase. Climate change also threatens to make the equalizing role of the Centre more complex by exacerbating regional inequities. A second state-based threat comes from the absence of a disciplining force from below; climate change's low political salience threatens to result in a patchwork of disconnected and possibly discordant climate initiatives.

This system will face challenges in an era of ambitious, and harder to achieve, carbon targets like net-zero. India's system of cooperation, episodic rebalancing of resources, and bursts of climate policymaking will have to yield to a more predictable, consensual model of action. Moves away from the co-benefits approach will require a new federal compact where states have access to institutions for collective decision making on targets, pace, resources, and policy. Unlike both the USA and Germany, the two countries at opposite ends of our notional spectrum, the voices of Indian states are muted in the upper house of Parliament. It will also demand new ideas about burden-sharing and equity, and, on the flip side, the Centre's punitive powers to punish free-riders. And not least, it places a much greater focus on the alchemical powers of the states, who will be forced to weave the climate crisis into areas of governance always consumed by other priorities.

Notes

1 2017 constant international dollars.
2 Cooling and industrial gasses with global warming potentials much higher than CO_2.
3 Electricity Act, 36 of 2003 (2003).
4 India has three tiers of government. The Union, or federal, government is led by the prime minister and a cabinet of ministers who run over fifty ministries. Laws are made by a bicameral Parliament with an indirectly elected upper house (a council of states). Schedule VII of the Constitution divides legislative powers between the Union and state governments through three lists: the first demarcating the jurisdiction of the Centre, the second of the states, and the third establishing a concurrent list where the Centre and states share authority. State governments therefore enjoy sole legislative power in some areas. States are led by a chief minister and a cabinet who are part of a unicameral or bicameral legislature. The third tier of elected local government is composed of three nested layers from the district to village levels, and a separate system for cities. Climate governance responsibilities are dispersed across these three levels, from Centre to village, with a crucial role played by over a quarter-million village level governments responsible for grass-roots service delivery and key aspects of resilience and disaster preparedness.
5 Environment (Protection) Act, 29 of 1986 (1986).
6 Electricity Act, 36 of 2003 (2003); Electricity (Supply) Act, 54 of 1948 (1948).

7 In keeping with the co-benefits paradigm, these are policies and programmes with developmental objectives whose emission savings have been quantified rather than policies with up-front emission reduction goals. They range from the National Solar Mission and Energy Efficient Buildings Programme to programmes on micro-irrigation, avoiding crop residue burning and improving the efficiency of streetlights.

8 Funds for NSCCC approved projects are sourced from the National Adaptation Fund for Climate Change implemented through a bank under central government jurisdiction, the National Bank for Agriculture and Rural Development. The size of the fund from 2015 to 2017 was INR 350 crore (approximately USD 47m at 2020 rates) (Parliamentary Committee on Estimates, 2018).

9 Data for this analysis were sourced from Bridge to India's repository of solar capacity auction results. Available at: https://india-re-navigator.com/utility/tender-tracker.

10 Rashmi, R. R. Former Special Secretary, Indian Ministry of Environment Forests and Climate Change. (5 November 2019). Personal interview.; Ramesh, J. Former environment minister. (4 May 2020). Personal interview.

11 Chaturvedi, A. Director – Climate Change, GIZ India (28 January 2020). Personal interview.

12 Chaturvedi, A. op. cit.; Mitra, S. Senior Climate and Environment Advisor (5 February 2020). Personal Interview.; Soni, P. Chief – Climate Change, Resilience and Energy (23 January 2020). Personal Interview.

13 Chaturvedi, A. op. cit.

References

Aiyar, Yamini, and Avani Kapur. 2019. 'The Centralization vs. Decentralization Tug of War and the Emerging Narrative of Fiscal Federalism for Social Policy in India'. *Regional & Federal Studies* 29 (2), 187–217. doi: 10.1080/13597566.2018.1511978.

Aiyar, Yamini, and Louise Tillin. 2020. '"One Nation", BJP, and the Future of Indian Federalism'. *India Review* 19 (2), 117–35. doi: 10.1080/14736489.2020.1744994.

Arora, Bhanvi. 2018. 'India Says Never Targeted 100% Electric Mobility By 2030, Scales Down Aim'. *Bloomberg Quint*, 8 March. www.bloombergquint.com/business/india-says-never-targeted-100-electric-mobility-by-2030-scales-down-aim.

Arora, Balveer, and Nidhi Srivastava. 2019. 'Green Federalism'. *Seminar* 717. www.india-seminar.com/2019/717/717_balveer_and_nidhi.htm.

Atal, Vinit, Gireesh Shrimali, and Vaibhav Pratap Singh. 2018. *Addressing Off-Taker Risk in Renewable Projects in India: A framework for designing a payment security mechanism as a credit enhancement device* [Technical Paper]. Climate Policy Initiative, 8 May. www.climatepolicyinitiative.org/wp-content/uploads/2018/05/A-Framework-for-Designing a Payment-Security-Mechanism-as-a-Credit-Enhancement-Device-4.pdf.

Bhardwaj, Ankit, and Radhika Khosla. 2021. 'Superimposition: How Indian city bureaucracies are responding to climate change'. *Environment and Planning E: Nature and Space* 4 (3), 1139–70. doi: 10.1177/2514848620949096.

Chaba, Anju Agnihotri. 2020. 'Explained: Why Punjab rent waiver on farm machines will not help much'. *Indian Express*, 10 October. https://indianexpress.com/article/explained/punjab-farmers-stubble-burning-machines-on-rent-6717113.

Chakrabarti, P. G. Dhar. 2015. 'Federalism and Environmental Policy in India'. In *Green Federalism: Experiences and practices*, eds. P. G. Dhar Chakrabarti and Nidhi Srivastava, 61–71. New Delhi: The Energy and Resources Institute.

Chakrabarti, P. G. Dhar, and Nidhi Srivastava, eds. 2015. *Green Federalism: Experiences and practices*. New Delhi: The Energy and Resources Institute.

Chaudhary, Ankur, Ankita Narain, Chetan Krishna, and Ambuj Sagar. 2014. *Who Shapes Climate Action in India? Insights from the wind and solar energy sectors* (Evidence Report No. 56). Brighton: Institute of Development Studies. https://opendocs.ids.ac.uk/opendocs/handle/20.500.12413/3537.

Chitnis, Ashwini, Shripad Dharmadhikary, Shantanu Dixit, et al. 2017. *Many Sparks but Little Light: The practice and rhetoric of electricity sector reforms in India*. Prayas Energy Group. www.prayaspune.org/peg/publications/item/332-many-sparks-but-little-light-the-rhetoric-and-practice-of-electricity-sector-reforms-in-india.html.

Chokkakula, Srinivas. 2019. 'Interstate River Water Governance: Shifting the focus from conflict resolution to enabling cooperation'. In *Policy Challenges 2019–2024*, 118–21. Centre for Policy Research. https://cprindia.org/wp-content/uploads/2022/03/Policy-Challenges-2019-2024.pdf.

Conceição, Pedro. 2019. *Human Development Report 2019: Beyond income, beyond averages, beyond today*. New York: United Nations Development Programme. https://hdr.undp.org/en/content/human-development-report-2019.

Dasgupta, Aditya, and Devesh Kapur. 2017. 'The Political Economy of Bureaucratic Overload: Evidence from Rural Development Officials in India'. *SSRN*. doi: 10.2139/ssrn.3057602.

Dasgupta, Chandrashekhar. 2019. 'Present at the Creation: The making of the framework convention on climate change'. In *India in a Warming World: Integrating climate change and development*, ed. Navroz K. Dubash, 142–56. New Delhi: Oxford University Press. doi: 10.1093/oso/9780199498734.001.0001.

Dolsak, Nives, and Aseem Prakash. 2019. 'Are India's Political Parties Ignoring Climate Change?' *Forbes*, 13 April. www.forbes.com/sites/prakashdolsak/2019/04/13/are-indias-political-parties-ignoring-climate-change.

Dubash, Navroz K. 2013. 'The Politics of Climate Change in India: Narratives of equity and cobenefits'. *Wiley Interdisciplinary Reviews: Climate Change* 4 (3), 191–201. doi: 10.1002/wcc.210.

Dubash, Navroz K. 2019. 'An Introduction to India's Evolving Climate Change Debate: From diplomatic insulation to policy integration'. In *India in a Warming World: Integrating climate change and development*, ed. Navroz K. Dubash, 2–28. New Delhi: Oxford University Press. doi: 10.1093/oso/9780199498734.001.0001.

Dubash, Navroz K., and Anu Jogesh. 2014. 'From Margins to Mainstream? State Climate Planning in India as a "Door Opener" to a Sustainable Future'. *Centre for Policy Research*, 18 February. https://cprindia.org/briefsreports/from-margins-to-mainstream-state-climate-change-planning-in-india-as-a-door-opener-to-a-sustainable-future.

Dubash, Navroz K., Radhika Khosla, Ulka Kelkar, and Sharachchandra Lele. 2018. 'India and Climate Change: Evolving ideas and increasing policy engagement'. *Annual Review of Environment and Resources* 43 (1), 395–424. doi: 10.1146/annurev-environ-102017-025809.

Dubash, Navroz K., and Sudhir Chella Rajan. 2001. 'Power Politics: Process of power sector reform in India'. *Economic and Political Weekly* 36 (35), 3367–87, 3389–90. https://www.jstor.org/stable/4411059.

Finance Commission of India. n.d. *Finance Commissions: A Historical Perspective*. https://fincomindia.nic.in/ShowContent.aspx?uid1=2&uid2=1&uid3=0&uid4=0, last accessed 30 June 2021.

Forum of Regulators. 2005. *Minutes of the First Meeting of the Forum of Regulators*. www.forumofregulators.gov.in/Meetings.aspx.

Ghosh, Shibani. 2019. *Indian Environmental Law: Key concepts and principles*. Hyderabad: Orient Black Swan. https://orientblackswan.com/details?id=9789352875795.

Global Carbon Project. 2019. *Global Carbon Budget 2019: Summary Highlights*. www.globalcarbonproject.org/carbonbudget/19/highlights.htm.

Government of Delhi. 2020. *Delhi Electric Vehicles Policy, 2020*. https://transport.delhi.gov.in/sites/default/files/All-PDF/Delhi_Electric_Vehicles_Policy_2020.pdf.

Government of Gujarat. 2018. *Overview*. Climate Change Department. https://ccd.gujarat.gov.in/overview.htm.

Government of India. 2018. *India: Second biennial update report to the United Nations framework convention on climate change*. Ministry of Environment, Forest and Climate Change. http://folk.uio.no/roberan/ind/india_proj201709.shtml.

Government of India. 2021. *India: Third biennial update report to the United Nations framework convention on climate change*. Ministry of Environment, Forest and Climate Change. http://folk.uio.no/roberan/ind/india_proj201709.shtml.

Government of India. 2022. *India's Updated First Nationally Determined Contribution Under Paris Agreement (2021–2030)*. https://unfccc.int/sites/default/files/NDC/2022-08/India%20Updated%20First%20Nationally%20Determined%20Contrib.pdf.

Government of Kerala. 2020. *Rebuild Kerala Development Programme*. https://rebuild.kerala.gov.in/reports/RKDP_Master%2021May2019.pdf.

Government of Odisha. n.d. *Aims and Objectives*. Climate Change Cell, Odisha. http://climatechangecellodisha.org, last accessed 4 November 2020.

Harish, S., and S. Ghosh. 2020. *Pursuing a Clean Air Agenda in India During the COVID Crisis*. Centre for Policy Research. www.cprindia.org/research/reports/pursuing-clean-air-agenda-india-during-covid-crisis.

Harrison, T., and G. Kostka. 2018. 'Bureaucratic Manoeuvres and the Local Politics of Climate Change Mitigation in China and India'. *Development Policy Review*. https://doi.org/10.1111/dpr.12386.

Hoegh-Guldberg, Ove, Daniela Jacob, Michael Taylor, et al. 2018. 'Impacts of 1.5°C Global Warming on Natural and Human Systems'. In *Global Warming of 1.5°C. An IPCC Special Report on the impacts of global warming of 1.5°C above pre-industrial levels and related global greenhouse gas emission pathways, in the context of strengthening the global response to the threat of climate change, sustainable development, and efforts to eradicate poverty*, eds. V. Masson-Delmotte, P. Zhai, H.-O. Pörtner, et al., 175–312. Intergovernmental Panel on Climate Change. www.ipcc.ch/sr15.

Huang, Jing, and Shreekant Gupta, eds. 2014. *Environmental Policies in Asia: Perspectives from seven Asian countries*. Singapore: World Scientific. doi: 10.1142/9109.

Hueglin, T. O., and A. Fenna. 2015. *Comparative Federalism: A systematic inquiry*, 2nd ed. Toronto: University of Toronto Press.

India Re Navigator. 2020. *Utility Scale Solar Tenders*. Bridge to India. https://india-re-navigator.com/utility/tender-tracker.

Jalan, Ishita, and Hem H. Dholakia. 2019. *What is Polluting Delhi's Air? Understanding uncertainties in emissions inventories*. New Delhi: Council on Energy, Environment and Water. www.ceew.in/publications/what-is-polluting-delhi-air.

Jörgensen, Kirsten, Arabinda Mishra, and Gopal K. Sarangi. 2015. 'Multi-level Climate Governance in India: The role of the states in climate action planning and renewable energies'. *Journal of Integrative Environmental Sciences* 12 (4), 267–83. doi: 10.1080/1943815X.2015.1093507.

Karkun, Arunesh. 2021. *Can EVs Solve India's Air Pollution Crisis?* New Delhi: Centre for Policy Research, Environmentality. https://environmentality.cprindia.org/blog/can-evs-solve-the-air-pollution-crisis.

Kaur, Nehmat, and Jarnail Singh. 2019. *Driving Climate Action: State Leadership in India*. The Climate Group, 17 May. www.indiaenvironmentportal.org.in/files/file/india_report.pdf.

Krishnan, R., J. Sanjay, and Chellappan Gnanaseelan, eds. 2020. *Assessment of Climate Change over the Indian Region: A report of the Ministry of Earth Sciences (MoES), Government of India*. Singapore: Springer Open. doi: 10.1007/978-981-15-4327-2.

Kumar, Vineet. 2018. *Coping with Climate Change: An analysis of India's state action plans on climate change*. New Delhi: Centre for Science and Environment.

Maharashtra State Electricity Distribution Company Limited. 2019. *Mukhyamantri Saur Krushi Pump Yojana*. Mahavitaran. https://www.mahadiscom.in/solar/index.html>.

MEFCC (Ministry of Environment, Forests and Climate Change, Govt. of India). 2014. *Guidelines for Funding State Action Plan on Climate Change (SAPCC) under Climate Change Action Program (CCAP)*. https://dste.py.gov.in/sites/default/files/guidelinesforfundingsapcc.pdf.

MEFCC. 2017. *Right to Information Request MOENF/R/2017/51130/1*.

Ministry of Finance, Govt. of India. 2020. *Central Sector Schemes*. www.indiabudget.gov.in/expenditure_profile.php.

Ministry of Power, Govt. of India. 2018. *Minutes of the Conference of Power and NRE Minister's of States/UTs held on 3rd July, 2018 at Shimla, Himachal Pradesh*. https://powermin.nic.in/sites/default/files/webform/notices/Final_approved_minutes.pdf.

Mishra, D. 2019. 'What's Climate Change Got to Do with Bihar politics'. *The Print*, 4 December. https://theprint.in/politics/whats-climate-change-got-to-do-with-bihar-politics-nitish-kumar-to-explain-on-his-yatra/330049.

MNRE (Ministry of New and Renewable Energy, Govt. of India). 2019. *Guidelines for Implementation of Pradhan Mantri Kisan Urja Suraksha evem Utthan Mahabhiyan (PM KUSUM) Scheme*.

Mueller, S., and A. Fenna. 2022. 'Dual Versus Administrative Federalism: Origins and evolution of two models'. *Publius: The Journal of Federalism* 52 (4): 525–52. https://doi.org/10.1093/publius/pjac008.

Padmanabhan, A. 2018. 'Can Kerala Floods be a Defining Moment for Pinarayi Vijayan?' *Mint*, 27 August. www.livemint.com/Opinion/3LRdCBlwkZSIC21e2uVfBP/Can-Kerala-floods-be-a-defining-moment-for-Pinarayi-Vijayan.html.

Parikh, S., and B. R. Weingast. 1997. 'Comparative Theory of Federalism: India'. *Virginia Law Review* 83, 1593.

Parliamentary Committee on Estimates. 2018. *Performance of the National Action Plan on Climate Change (*No. 30*)*.

Pillai, A. V., and N. K. Dubash. 2021. 'The Limits of Opportunism: The uneven emergence of climate institutions in India'. *Environmental Politics* 30 (sup1), 93–117. doi: 10.1080/09644016.2021.1933800.

Press Trust of India. 2020. 'Nitish Kumar Shares Bihar's Sustainable Development Efforts at UN Climate Meet'. *NDTV.Com*, September. www.ndtv.com/india-news/nitish-kumar-shares-bihars-sustainable-development-efforts-at-un-climate-meet-2300781.

Ramesh, J. 2017. *Indira Gandhi: A life in nature*. Noida: Simon & Schuster India.

Rao, M. G., and N. Singh. 2004. *Asymmetric Federalism in India (UC Santa Cruz International Economics Working Paper* No. 04-08). doi: 10.2139/ssrn.537782.

Rawal, S. 2021. 'Maharashtra Forms Council for Climate Change to Prepare Action Plan'. *Hindustan Times*, 2 September. www.hindustantimes.com/cities/mumbai-news/maharashtra-forms-council-for-climate-change-to-prepare-action-plan-101630528745993.html.

Sáez, L. 2002. *Federalism without a Center: The impact of political and economic reform on India's federal system*. Thousand Oaks, CA: Sage Publications.
Sarkaria, R. S., B. Sivaraman, and S. R. Sen. 1988. *Report of the Sarkaria Commission*. New Delhi: Ministry of Home Affairs.
Sengupta, S. 2013. 'Defending "Differentiation": India's foreign policy on climate change from Rio to Copenhagen'. In *India's Foreign Policy: A reader*, 1st ed., eds. Kanti P. Bajpai, and Harsh V. Pant. New Delhi: Oxford University Press.
Sharan, D. 2019. 'Kejriwal Claims Pollution in Delhi Down 25% in Four Years'. *Mint*, September. www.livemint.com/politics/news/kejriwal-claims-pollution-in-delhi-down-25-in-four-years-1567760284899.html.
Sharma, M., and O. Dikshit. 2016. *Comprehensive Study on Air Pollution and Green House Gases (GHGs) in Delhi*. Kanpur: Indian Institute of Technology Kanpur.
Shekar, A. 2020. *Punjab: Climate smart agriculture/air pollution*. The Climate Group - State Climate Leadership Forum, Webinar, 24 September.
Sivakumar, B. 2021. 'TN Govt Forms Green Climate Company to Conserve Forests and Wetlands'. *The Federal*, 6 November. https://thefederal.com/states/south/tamil-nadu/tn-govt-forms-green-climate-company-to-conserve-forests-and-wetlands.
Subramaniam, A., and U. Kumar. 2012. 'Growth in India's States in the First Decade of the 21st Century: Four facts'. *Economic and Political Weekly* 47 (3), 7–8.
Tillin, L. 2019. *Indian Federalism*. New Delhi: Oxford University Press India.
Vembadi, Shiv, Nikita Das, and Ashwin Gambhir. 2018. *175 GW Renewables by 2022: A September 2018 update*. Prayas Energy Group. https://prayaspune.org/peg/publications/item/405-175-gw-renewables-by-2022-a-september-2018-update.html.
Wheare, K. C. 1964. *Federal Government*, 4th ed. New York: Oxford University Press.
World Bank. 2020. *GDP per capita, PPP (constant 2017 international $)—India*. https://data.worldbank.org/indicator/NY.GDP.PCAP.PP.KD?locations=IN.
XIV Finance Commission. 2013. *Report of the 14th Finance Commission*. XIV Finance Commission. https://fincomindia.nic.in/ShowContent.aspx?uid1=3&uid2=0&uid3=0&uid4=0.
XV Finance Commission. 2019. *Report for the Year 2020–21*. XV Finance Commission.
XV Finance Commission. 2020. *Finance Commission in Covid Times: Report for 2021–26*. XIV Finance Commission. http://finance.cg.gov.in/15%20Finance%20Commission/Report/XVFC-Complete_Report-E.pdf.

10

Climate Governance and Decentralization in Indonesia

MONICA DI GREGORIO AND MOIRA MOELIONO

10.1 Introduction

Indonesia represents an interesting case for analysis of the relationship between multi-level governance and climate governance for three main reasons. It is a highly decentralized country; it is a major contributor to land-based greenhouse gas emissions; and it is extremely vulnerable to climate change. The chapter first provides a broad overview on Indonesia's climate governance in the context of decentralization, and then focuses on sub-national governance of climate change mitigation in the land use sector, the largest contributor to greenhouse gas emissions in the country.

Indonesia illustrates key advantages of highly decentralized polity structures. Political autonomy has facilitated sub-national climate action through direct engagement of provinces with transnational climate initiatives, and the multiplicity of forums for policymaking has allowed certain provinces to champion sub-national engagement in climate change policy. Decentralization has also facilitated experimental policies in the form of innovative sub-national jurisdictional approaches to climate action in the land use sector. At the same time, though, peculiarities of the decentralization approach in the land use sector have led to perverse incentives that hamper forest-based climate change mitigation action. Drawing on in-country expertise and interviews with provincial government officials from ten highly forested provinces, this chapter explores the relations between national and provincial governments in the processes of forest-based climate change mitigation.

10.2 Climate Change and Land Use Emissions in Indonesia

As a tropical archipelago with a large population chiefly dependent on agriculture, Indonesia is particularly vulnerable to climate change, due to both sea level rises and worsening of extreme weather events leading to the increased frequency and intensity of floods, droughts, and landslides (ICCSR 2009). El-Niño events and carbon rich peatland compound risks of increased emissions from forest fires due to droughts and to human- and climate-induced soil disturbance (Sloan et al. 2017). Climate change has already led to increased food insecurity (Boer and Subbiah 2005).

In 2014 Indonesia became the sixth largest greenhouse gas emitter in the world, and the largest forest-based emitter, due largely to conversion of forest into agriculture (WRI 2020). Since the 1990s, land-use change and forestry has emitted three to four times as much greenhouse gases as the energy sector. Rainforests remove atmospheric carbon and are major stores of carbon, so disturbances lead to release of carbon. Indonesia still contains the third largest tropical rainforest area after Brazil and the Democratic Republic of Congo, but it had one of the highest rates of deforestation worldwide of around 0.7 per cent between 1990 and 2015. Since 2016, primary forest loss has been declining (Global Forest Watch 2020). Forest conversion is the highest priority for mitigation action and requires an integrated approach across forestry and agriculture (Di Gregorio et al. 2017).

Carbon emissions by province reveal differences in levels and sources suggesting the need for distinct jurisdictional approaches to reduce emission reductions. The two very highest emitting provinces are in Sumatra (North Sumatra and Riau) and their emission are more than double over any other Indonesian province. They are followed by East Java and Central Kalimantan and a number of other outer island provinces (Utami et al. 2016). The main source of emissions also differs across provinces. In the more extensive but less populated outer islands, land use change and forestry largely outstrip any other source. In highly populated regions, such as Java, energy production tops the charts.

Forest conversion into agriculture is driven largely by oil palm expansion. Indonesia supplies nearly 50 per cent of palm oil worldwide; demand is predicted to increase and will contribute to drive deforestation in the absence of improved sustainable practices (Schebek et al. 2018). The pressure on forests and the extent and rate of conversion differs across regions. The highest emitting provinces in Sumatra contain the most extensive areas of oil palm plantations. Central and West Kalimantan follow, with some of the highest rates of deforestation and contain a mix of extensive oil palm areas closer to the coast and large tracts of natural forest inland. Finally, Papua and West Papua are among the lowest emitting provinces, contain the most intact areas of primary forest, and are at the early stages of deforestation.

10.2.1 Climate Change Mitigation Commitments in the Land Use Sector

Indonesia developed its first National Action Plan Addressing Climate Change (RAN-PI) in 2007, the same year it hosted the 13th UNFCCC Conference of

Parties (COP) in Bali. COP13 led to the inclusion of avoided deforestation as a carbon sink approach to account for greenhouse gas reductions. The resulting incentive mechanism of REDD+ (Reducing Emissions from Deforestation and forest Degradation) entails that developed countries compensate developing countries for carbon sequestration in tropical forests. In 2009 President Yudhoyono announced Indonesia's pledge to reduce greenhouse gas emissions by 26 per cent from business-as-usual (BAU) by 2020 and up to 41 per cent subject with international support.

In 2011 Indonesia passed a moratorium on deforestation in primary forest and peatland, which became permanent in 2019 (InPres6/2017). The main National Action Plan for the Reduction of Greenhouse Gas Emissions (RAN-GRK), which indicates forests and peatland as main target for emission reductions, was also released in 2011 (PerPres 61/2011). Provincial level plans (RAD-GRK) followed. The RAN-GRK Secretariat at the Ministry for National Development liaises with the provinces on greenhouse gas emission reductions, follows reviews, monitoring and reporting, and in 2019 developed the guidelines for the Provincial Low Carbon Development Plans.

The 2016 Indonesia's National Determined Contribution (NDC) extended the deadline and revised the mitigation commitment to 29 per cent emission reductions and 41 per cent with international support, against a 2030 BAU scenario of 2.87 GtCO2e (GOI 2016; Wijaya et al. 2017). The 2020 NDC confirms the target and indicates 2060 as the net zero target date. It revised the per cent contribution coming from forest emission reductions to 24.5, and the per cent from the energy sector to 15.5. It also committed to forestry becoming a net carbon sink by 2030 (Ibun Aqil 2021). Sectoral mitigation targets are clearly specified in various national regulations. They include the restoration of 5.5 Mha of degraded land between 2015 and 2019; the restoration of 2.4 Mha of peatlands in seven priority provinces in Sumatra, Kalimantan, and Papua; and the allocation of 12.7 Mha to community social forestry by 2020 (BAPPENAS 2014; InPres 6/2017 2017; PerPres 1/2016, 2016).

By December 2021, official MoEF figures suggested that 4.7 M of social forestry areas had been allocated (MoEF 2021). Expected emission reductions in forest and land use by the 2030 deadline for BAU is 650 MtCO_2. They are expected to cost 77 billion IDR – much cheaper than energy-based reductions, and deliver 1.5 times the emission savings. There is little question that, at least on paper, forestry and land use are the low hanging fruit to reach emission reduction at the least possible cost (MoEF 2018a). In 2018 in a bid to leverage private climate finance, the Indonesian government introduced the first green bonds to carbon markets in 2018, and a draft of a presidential decree on carbon pricing was released in 2021. At the same time, the targets need to be understood within the context of major economic development plans, which include a planned expansion of oil palm and forest plantations and mining in Kalimantan (ROI 2011).

10.3 Changing Features of Decentralization in Indonesia: From Decentralization to Recentralization

Indonesia has undergone distinct phases in terms of decentralization and recentralization processes and the level of devolution is not uniform across sectors. Below we present the major shift since democratization. At the time of President Suharto Indonesia was considered one of the most centralized of nations (Butt 2010). After Suharto's fall in 1998, Indonesia embarked on a extensive democratization and decentralization process, which has been labelled as quasi-federalist (Bertrand 2007). Overall, there are five levels of government in Indonesia, consisting of the national government; thirty-four Provinces (*propinsi*); close to 500 districts (*kabupaten/kota*); followed by sub-districts (*kecamatan*); and villages (*desa/kelurahan*). Because of concerns that devolution to provinces might strengthen pre-existing secessionist movements in Aceh and Papua,[1] the 1999 regional autonomy law devolved most powers to the then-292 district governments. This resulted in a very unusual type of decentralization, where 3rd tier district governments enjoyed much greater autonomy than 2nd tier provincial governments (Ferrazzi 2000). A scramble for the creation of new districts ensued, leading to a 50 per cent increase within a decade (Pierskalla 2016).

Regional autonomy brought direct elections of heads of district government, the creation of local legislative assemblies, increasing fiscal transfers and increased local responsibility for public services (Lewis 2013). Political decentralization is extensive with the first three tiers of government all having elected heads of government and legislative assemblies. Second and third tier governments have broad autonomy to legislate in matters not reserved for central government and are responsible for public administration and investments, policing, infrastructure, health, education, the labour force, small and medium enterprises, development planning, agriculture and land management, and the environment.[2] However, whenever a regional law conflicts with a national law, the latter prevails. Central government can also invalidate laws on national interest grounds and decisions on local budgets, and taxes and spatial planning require central approval (Butt 2010).

Fiscal decentralization has resulted in an overall increase in public investment spending towards social goods at the local level (Pal and Wahhaj 2017), although being somewhat limited as revenue collection remain centrally managed. Funds are transferred to 2nd and 3rd tier governments through a 'general allocation grant' indexed according to population and poverty levels. In practice, overlapping and conflicting legislation and practices across governance levels are extremely common (Butt 2010). However, decentralization was neither accompanied by sufficient resources and capacity building for local governments to effectively deliver public goods and substantially promote economic development, nor by sufficient central government capacity to monitor local implementation effectively (Nasution 2016). In practice, however, local government retained a very high level of authority for a unitary state (Bertrand 2007; Shair-Rosenfield et al. 2014).

Major revisions to regional autonomy introduced in 2014 shifted many major functions from district to province. Provinces gained oversight over district governments restoring the first-second-third tier government hierarchy and reversing the decentralization to districts. Environmental responsibilities in the mining, forestry, maritime affairs, and fishery sectors were moved from district to provincial level. At the same time central authorities introduced new administrative penalties for mismanagement leading to the potential dismissal of heads of regional governments, imposing increased central control over all lower tier governments. Provincial governments are only responsible for implementing and assessing policies, expressed as 'educating, supervising, monitoring and evaluating and facilitating' (UU23/2014), while policy formulation in those sectors was recentralized. District level forestry and mining offices have been closed as responsibilities shifted to the province and central ministries. The 2021 law on job creation (Law 11 2021), also called the 'Omnibus Law', underpins further centralization. Intended to facilitate business development at the regional level, it recentralized the allocation of land use permits.

10.4 Climate Governance in the Context of Decentralization

Climate change governance evolved in this changing context of decentralization and recentralization. After outlining the main responsibilities of the central, provincial, and district jurisdictional levels, we present some major instances of horizontal and vertical competition, then illustrate how distinct levels of decentralization in the forestry and agricultural sectors leads to perverse incentives that hamper forest-based climate change mitigation processes.

10.4.1 Central Government Climate Change Architecture in the Land Use Sector

Recentralization tendencies have affected most sectors including climate change. In the central government, a number of departments share responsibilities around climate change policymaking, while provincial governments are involved mainly in implementation. The Ministry of Environment and Forestry (MoEF), and its Directorate General for Climate Change, have a major national climate change mandate. The ministry has a coordination role around environmental matters, but is also a line ministry for forestry, which leads to inconsistencies and diverging interests within the ministry. With respect to land use, most climate related work focuses on forest-based mitigation. While the three highest levels of government have shared mandates on the environment, forestry is much more firmly under central control of the ministry, and even more so since the 2014 recentralization drive. Recentralization processes are particularly relevant in the forestry sector. Recently, these are also driving further deforestation. First, a 2020 MoEF regulation permits clearing of forest for large scale food estates allowing the reassignment of forest areas to 'forest areas for food security'. Second, the 2020 Omnibus Law and associated MoEF regulations remove the requirement to maintain at least 30 per cent of any watershed and island territory as forest lands. In this way, the MoEF reasserts its control over state forest lands – the most extensive land classification in Indonesia. In contrast, the Ministry of Agriculture has a very minor role, being responsible for mainstreaming climate change into agriculture and developing climate-smart solutions. Its main focus is on climate change adaptation.

The coordination Ministry of National Development Planning (BAPPENAS) has the mandate to mainstream climate change into development planning and oversees provincial climate change plans and reporting. It is also responsible for the national level Nationally Appropriate Mitigation Actions, which facilitate access to multi-lateral funding, as well as for national adaptation and mitigation policy plans and their implementation. The ministry has a major climate policy integration role, both horizontally, working primarily on implementation in close collaboration with all sectoral ministries, and vertically, across governance levels. Further, the Ministry of Finance has responsibility for the overall budget and has been claiming a mandate over any form of payments related to climate change, including benefits-sharing mechanisms. It controls the Environment Fund Management Agency (BPDLH), which is responsible for the management of multi- and bi-lateral climate finance (Pham et al., 2021). Infights between the MoEF, BAPPENAS, and the Ministry of Finance on who has jurisdiction on climate change responses has been evident from the start as they compete for control over the climate change agenda. In practice, the former two have overlapping climate mandates, which are not clearly reconciled (Di Gregorio et al. 2017). The Ministry of Foreign Affairs also plays a key role – particularly in relation to global climate change processes under the UNFCCC, because the majority of the climate change related funding comes from international sources.

Given the cross-sectoral nature of climate change, most of the ministries work through multi-sectoral platforms (Di Gregorio et al. 2017). The first national committee on Climate Change and Environment was established in 1992 and included three inter-ministerial working groups led by the Agency for Meteorology, Climatology, and Geophysics, the Ministry of Environment, and the Ministry of Agriculture. Under the Yudhoyono presidency the DNPI included seventeen ministries and seven working groups, as did the REDD+ Task Force and the REDD+ agency. BAPPENAS National Coordination Team on Climate Change has a similar multi-agency set up (Di Gregorio et al. 2015). During Yudhoyono's tenure, the president's Delivery Unit for Development Monitoring (UKP4) had a major supervisory role across all government ministries. Notably, none of the joint committees include sub-national agencies.

10.4.2 Provincial and District Level Climate Change Planning and Implementation

Provincial level and district level governments are involved in local climate change policy development and implementation. By 2013, all thirty-four provinces had developed and enacted, through governor's regulation, their Local Action Plans for Greenhouse Gas Emissions Reductions (RAD-GRK). Such progress was largely achieved first with the support of the National Climate Change Council, and later the guidance of BAPPENAS, as well as with international programmes. Yet, provincial plans remain vague and contain contradictory aims, with mineral and natural resource development plans not being reconciled with climate mitigation aims (Wijaya et al. 2017).

BAPPENAS oversees the local mitigation action plans and its regional branches are responsible for implementation. Provinces have also developed REDD+ Provincial Action Plans. The earlier REDD+ district level strategies were repealed following the 2014 regional autonomy changes, effectively ending any district autonomy on REDD+ policies and shifting control over climate change to the provinces. Aceh, West Sumatra, Jambi, South Sumatra, West Kalimantan, East Kalimantan, Central Kalimantan, Papua, and West Papua developed their REDD+ plans with the support of the REDD+ Agency. By 2019, eleven out of thirty-four Provinces had provincial REDD+ strategies, and five had established ad hoc provincial levels REDD+ Working Groups (Papua, Riau, East Kalimantan, and South Sumatra) or Joint Secretariat in South Kalimantan (Ekawati et al. 2019). For REDD+ implementation there are four key policy features to be developed in addition to the provincial REDD+ strategy itself: provincial reference levels, monitoring, reporting, and verification, safeguards, and benefit sharing mechanisms. So far, most provinces have only developed their reference level. Yet, monitoring, reporting, and verification is necessary for carbon accounting, and safeguards and benefit-sharing institutions are crucial to reduce trade-offs between climate change mitigation and livelihoods and to enhance transparency (Ekawati et al. 2019).

10.4.3 Decentralization and Perverse Land Use Incentives Undermining Forest-Based Mitigation

Policy actors responsible for land use decisions play a key role in climate change policy outcomes in Indonesia, given that land use change is the main source of greenhouse gas emissions and the main target for emission reductions. Two-thirds of the Indonesian territory, 120.6 million ha, was classified as 'forest area' in 2017 – designated as 'permanent forest', although 28 per cent of state forest land is actually not forested and 8 per cent of forest is located outside of this classification. State forest land is divided in production (68.8 mill ha), protection (29.7 mill ha), and conservation forest (22.1 mill ha in 2017) (MoEF 2018b).

In 2014 regional autonomy law shifted responsibility for the management of production and protection forest from district to provincial government and retained conservation forest management under central government authority. As a consequence, district governments lost any authority over forestry decisions, and provincial governments now manage forests through the Forest Management Units (FMUs) established by the ministry. FMUs have long been the MoEF's preferred approach to manage the state forest estate, because they maintain substantial central control over allocation and uses of forest lands. This leaves local government mainly responsible for residual planning and implementation (Sahide et al. 2016). Indeed, FMUs are the main instrument the ministry uses to secure rights to forest areas (Nugroho 2014), and are therefore fundamental means of state territorialization (Peluso and Lund 2011). In practice, however, only fifty-three FMUs had an approved Long Term Forestry Management Plan in 2016 (Santoso et al. 2019). Only a small number of FMUs are fully operational in terms of staff and activities, and many face institutional and capacity constraints. Overlapping claims across levels of government and between the state and local farmers remain largely unresolved (Jodoin 2017). Thus, many FMUs remain such on paper only, with social forestry and REDD+ projects operating in ignorance of existing FMU areas.

Social forestry represents a very small percentage of official management schemes, and in 2017 only 4.1 per cent of the forest estate was in fact managed by local communities (Damarjati 2018). The MoEF simplified social forestry projects' application processes in 2016, with the aim of facilitating achieving social forestry targets. Issuing of permits remains centralized under the Directorate General for Social Forestry, with FMUs having only a supporting role. Thus, the acceleration of social forestry programmes is occurring with limited devolution to forest users. While the MoEF interacts with NGOs implementing the schemes, local community engagement lags behind, resulting in communities benefiting only limitedly from social forestry designations (Suharjito and Wulandari 2019). One of the social forestry schemes, *hutan desa*, has contributed slightly to avoided deforestation between 2010 and 2015; however, deforestation rates fluctuated from positive to negative across the years (Santika et al. 2017). In addition, the total area of *hutan desa* and other social forestry schemes remains a very small fraction of the state forest lands. Finally, new MoEF regulations in line with the 2021 'Omnibus Law on Job Creation' (Law 11/2020) allow for the first time the participation of the private sector in social forestry schemes, which would in practice privatize the management of some schemes. There is also evidence of alliances between MoEF and districts against further devolution, such as resistance against the establishment of customary forests (*hutan adat*), which fall outside state forest areas (Sahide et al. 2016). Further alliances have emerged between provinces and central government against decentralization of forest management to districts. This also means that districts experiencing overlapping claims to forest land have no authority to solve related conflicts. In practice, the MoEF uses tactics reminiscent of a divide-and-rule approach to retain central control over forestry (Sahide et al. 2016).

That said, the level of decentralization or recentralization is not uniform, as districts try to maintain the autonomy they enjoyed in previous decades and political alliances across government levels lead to a whole variety of outcomes locally. Further, the Indonesian indigenous movement, which supports devolution of forests to indigenous communities, has become a major political player in decentralization processes and has achieved significant legal victories in the constitutional court (Sahide and Giessen 2015).

Land outside forest lands remains administered by the Land Agency. Much of this land, classified as 'land for other uses' (APL) is devoted to agricultural use and district governments have extensive control over these areas. They are responsible for issuing land use licences, while provinces are responsible for areas that span across more than one district (Irawan et al. 2019). Consequently, district governments have very strong incentives to lobby the MoEF to release land from the state forest estate to 'land for other uses'. This combination of centralized control over state forest land and decentralized control over agricultural land, leads to perverse incentives whereby local government has a strong interest in accelerating conversion of forests into agriculture, which drives deforestation. Thus, district governments play a crucial role in emissions from forest conversion, yet they are hardly involved in climate change decisions. It is within this context of shifting autonomy from districts to provinces and attempts by the MoEF to retain central control of forests that the innovative sub-national developments to forest-based mitigation have emerged. Below we present the analysis of jurisdictional approaches to REDD+ and assess the complex national–local relations that underpin them.

10.5 Decentralization, Forest-Based Mitigation, and Jurisdictional Approaches to Climate Change

10.5.1 REDD+ and the Rationale for Jurisdictional Approaches

At the 2005 UNFCCC COP meeting in Montreal, Papua and Aceh – two Indonesian provinces that have special regional autonomy status – pushed for the introduction of a new global forest-based mitigation mechanism, later known as REDD+. Other forest-rich provinces supported the mechanism in the hope that it would help to finance forest conservation policies and low-carbon emission development.

To be effective, REDD+ requires a nested approach – in other words a substantial integration of activities and monitoring across levels of governance. In particular, a nested approach facilitates verification of carbon accounting to avoid double counting of emission reductions (Pedroni et al. 2009; Wertz-Kanounnikoff and Angelsen 2009; Wunder et al. 2020). A jurisdictional approach to REDD+ integrates efforts within subnational jurisdictions to deliver emission reductions and co-benefits across the whole territorial boundaries (Boyd et al. 2018). In practice, jurisdictional approaches are led by sub-national governments – province or district in Indonesia – and should include integrated land use plans, and carbon monitoring, reporting, and verification at the scale of the jurisdiction. It is considered a useful step to facilitate climate policy integration within jurisdictions and make it easier to control leakage of carbon emission, which occurs when greenhouse gases are displaced elsewhere instead of being suppressed (Irawan et al. 2019). On paper, a decentralized political structure should facilitate nested approaches leading to effective climate governance. Below, we investigate both opportunities and challenges in REDD+ implementation from the perspectives of leading provinces and assess them within the context of Indonesia's decentralized governance system. After presenting the main developments in jurisdictional approaches, we investigate GCFTF's role in facilitating provincial level jurisdictional approaches to REDD+ in the next sections. The analysis is based on interviews with GCFTF's delegates from Indonesian provinces, its secretariat, and supporting organizations undertaken between 2017–18 and the analysis of climate change policy documents.

10.5.2 Jurisdictional REDD+ in Indonesia

In Indonesia jurisdictional approaches to REDD+ were introduced in 2008. The World Bank and The Nature Conservancy were the first to support district level REDD+ jurisdictional approaches (Fishbein and Lee 2015), while forest-rich provincial governments started to engage in provincial level jurisdictional approaches through the Governors Climate and Forest Task Force (GCFTF). GCFTF is a transnational climate change governance initiative that brings together sub-national governments interested in soliciting international funding for jurisdictional implementation of REDD+ and associated Low Emission Development (LED). Seven forest-rich Indonesian provinces – West, Central, East and North Kalimantan, Papua and West Papua, and Aceh – are part of the thirty-eight-member transnational network (Di Gregorio et al. 2017). Funding for jurisdictional approaches comes primarily from multi-lateral sources and to a much smaller extent from the private sector. In 2009, The Nature Conservancy established the first district level jurisdictional REDD+ project in Berau district in East Kalimantan. In 2014, after the revision of the regional autonomy law, it started to build stronger linkages at provincial level, supporting the East Kalimantan Green Growth Compact, which brings together 150 partners to tackle landscape challenges, and collaborating with the Provincial Council on Climate Change (Hovani et al. 2018). The REDD+ national strategy discusses the role of pilot provinces, but not specifically jurisdictional approaches. Central Kalimantan became the first REDD+ pilot province in 2011, with efforts concentrating at provincial level. In 2020 the World Banks' Forest Carbon Partnership Facility agreed to support jurisdictional REDD+ in the province of East Kalimantan through its Carbon Fund, while the Bio Carbon Fund supported the jurisdictional scheme in Jambi province. District level REDD+ jurisdictional schemes have been underway in Kapuas Hulu and Kubu Raya in West Kalimantan with the support of GIZ, NICFI, and UNDP. Finally, Unilever 's private scheme supports sustainable sourcing of palm oil in Central Kalimantan (Seymour et al. 2020).

The drive towards jurisdictional approaches to address climate change started largely outside of the national climate change policy processes through the collaboration between sub-national governments and international and domestic non-state actors. Over time, the discourse shifted from purely REDD+ jurisdictional approaches to broader Low Emission Development (LED) approaches – although in practice the focus remains largely confined to forest-based mitigation (Di Gregorio et al. 2020; Seymour et al. 2020). GCFTF supports primarily provincial level jurisdictional approaches, which fit well the latest legal provisions on regional autonomy. That said, district level jurisdictional initiatives can be accommodated within these.

After just over a decade of REDD+ readiness activities in Indonesia, the first performance-based payments for emissions reduction were agreed in 2020. In May that year Norway approved a payment of US$56 million for emissions reductions achieved between 2016 and 2017. In August the Green Climate Fund approved a further US$103.8 million for the years 2014–16. The funds are managed centrally through the Environmental Fund Management Agency (BPLDH) established in 2019, which is also responsible for evaluating and approving submission of proposals for funding, including from sub-national governments or non-state actors. A substantial amount is earmarked to extend and enhance social forestry (US$47 million) and FMUs (US$47 million), while the rest supports Indonesia's broader REDD+ architecture (Yong 2020). As social forestry and FMUs are subject to substantial central control by the MoEF and are managed and implemented through its provincial offices, the MoEF is likely to have access to substantial funding. Still, at present it remains uncertain how exactly the funds will be disbursed across levels of governance, across different actors, and how they will contribute to the funding of jurisdictional REDD+ schemes at provincial or district level. We do know, though, that part of the performance-based REDD+ payments will support three pilot provinces of Aceh, West, and Central Kalimantan to implement deforestation-free agriculture (GCFTF 2021; Seymour et al. 2020).

10.5.3 The Scramble for Control Over the REDD+ Policy Mandate at National Level

Within the national government, REDD+ is considered a 'national plan with regional implementation' (Ekawati et al. 2019). It is thus similar to the largely recentralized sectoral approach in forestry, and reflects attempts on the part of the MoEF to retain control of climate change policy decisions related to land use and associated budget lines. A clear indication on the part of the MoEF to claim the mandate to control forest-based mitigation policy has been evident since the very beginning. The MoEF used its power to challenge and ultimately change the organizational climate change and land use policy architecture.

Under the Yudhoyono presidency, and in line with Norway's pressure, the climate change policy mandate fell under the semi-independent entities of the National Council on Climate Change (DNPI), the REDD+ Task Force, and later the REDD+ Agency, who held the mandate for climate change and land use policy development. The office of the president had strong oversight on integration of REDD+ in ministerial policies and action plans through the president's Delivery Unit for Development Monitoring (UKP4). The ministries, in particular the Ministry of Forestry, strongly contested being side-lined from major REDD+ policy decisions and lobbied for control over the climate change policy mandate. In practice, the Ministry of Forestry already had a very strong influence on the national REDD+ policy domain (Brockhaus and Di Gregorio 2014) through its control over forest land. And it used FMUs and social forestry as a means of further strengthening its territorial control. With the election of Widodo to president in 2014, the tables turned and ministries regained full control over climate policy, as the semi-independent REDD+ agencies and UKP4 were dismantled. The merger of the Ministry of Forestry and Ministry of Environment the same year consolidated the control under the newly established Directorate General of Climate Change of the MoEF.

The increased interest of the Government of Indonesia in carbon finance denotes a major intention to raise international private funds to fund REDD+ activities. The national government aims to control carbon finance centrally, and in 2017 the MoEF contacted forest licence holders indicating that they could not independently engage in carbon trading activities (Pham et al. 2021). Further, the 2020 presidential decree draft states that only designated organizations (the BPDLH, the Steering Committee of Carbon Pricing, the MoEF, and connected agencies) would be authorized to engage and manage carbon markets. There are already organizations, including Ecosystem Restoration Concessionaires, that engage in voluntary carbon markets, and if the decree is adopted it will affect their ability to directly engage in such transactions (MMIA 2020; Pham et al. 2021). At the same time, there are also important countervailing tendencies that push for a more decentralized approach to climate policies, such as the push on the part of the GCFTF for jurisdictional approaches to be largely under the control of provincial governments.

10.5.4 The Governor's Climate and Forests Task Force and Jurisdictional REDD+

For the first decade of its existence GCFTF was largely a transnational network that facilitated information sharing, capacity building, and target setting among its thirty-eight sub-national jurisdictions across ten countries. Its main aim was to solicit international funding for provincial level jurisdictional REDD+ as well as broader Low Emission Development. California's membership denotes the attempt to link REDD+ to future sub-national carbon markets. At present, however, REDD + funding opportunities are mainly realized through overseas development aid (Angelsen 2017). In 2020, Norway agreed to fund the implementation of jurisdictional approaches pledging 25$ million to be managed by the UNDP (Di Gregorio et al. 2020). This infusion of funds has increased the relevance of GCFTF to national climate change interests.

According to Indonesia's provincial delegates, the GCFTF enhances the opportunities for provinces to engage in climate change action in a number of ways. It provides opportunities for provincial government to pursue a bottom-up governance approach to climate action that draws on the vision of governors themselves, and it helps both to put and to keep the climate change and forest on the provincial policy agenda. It also strengthens the visibility of provinces as climate change leaders, raising the profiles of provinces at the national level. Finally, GCFTF also facilitates interactions among the various Indonesian provinces that are engaged in forest-based climate change mitigation. All Indonesian provincial GCFTF member governments value the platform's contributions around these functions.

10.5.5 Provincial Governments' Visions on Climate and Forest

Provinces' visions on climate and forests revolve largely around achieving sustainable development outcomes through green growth, illustrating a clear ecological modernization approach, in which forest conservation and greenhouse gas emission reductions go hand in hand with economic development opportunities. The final aim is to enhance economic benefits and improve local standards of living. Among the ten provinces, only West Kalimantan included the aim of reducing greenhouse gas emissions themselves at the core of its vision for climate and forests. East Kalimantan and West Papua put more emphasis on the practical aim of leveraging carbon offset finance. Only two provinces, Aceh and West Papua, highlighted that the final aim of improving forest management and reducing emissions is to provide benefits for local communities. Thus, there is a clear discrepancy between the global REDD+ discourse that aims to reduce emissions, and the visions of provinces, which put much more emphasis on economic co-benefits. At the global level, REDD+ climate discourse includes safeguards that are limited to avoid detrimental effects on livelihoods, while the REDD+ national strategy includes the creation of additional benefits for local people's welfare in the main scope of REDD+ alongside emission reductions. Thus, the global, and to some extent also the dominant national REDD+, discourse differs from that of the provinces, which have a stronger focus on local economic development objectives.

According to delegates, the main challenge that provinces face in achieving their vision for climate and forest is the pressure from the drivers of land use change in terms of the conversion of forest into agriculture – in particular plantation agriculture – and to a smaller extent mining. They suggest that the latter in particular is largely driven by powerful national level actors. The new Omnibus Law and the MoEF regulation on food estate in forest areas also suggest that national level drivers are becoming more dominant. Papua and West Papua are the only provinces to mention poverty as being a major driver of deforestation, and lack of institutional capacity and weak community participation as major challenges. Thus, despite the general adoption of ecological modernization ideas, provincial governments are much more aware and concerned than national and global actors about real trade-offs between achieving economic development and environmental sustainability.

10.5.6 Relations between the National and Provincial Governments

Although predominantly a global platform, delegates indicate that GCFTF creates visibility for provincial governments at national level. As an example, with GCFTF's support West Papua was able to organize a meeting bringing together all seven Indonesian member provinces and the MoEF, providing an opportunity to showcase provincial interests and efforts, enhance the visibility, and consequently the influence of provinces on national climate action. The main challenges mentioned by provincial delegates in terms of national–provincial relations are national level resistance to jurisdictional approaches; major bureaucratic burdens imposed by national government; and the misalignment of policy goals between national and provincial visions. National resistance takes shape in many different forms.

Provincial delegates – in particular those from the provinces with special status – denounce the lack of autonomy in relation to institutions around climate change as a major challenge. Others, however, also indicated that a higher level of autonomy, such as would exist in a federal system, might translate in more competition among federated entities and diverging policy agendas that might lead to lack of alignment across jurisdictional approaches. A main challenge to develop and implement jurisdictional REDD+ approaches at provincial level was uniformly identified by all delegates as insufficient funding. Limited resources, in particular extremely limited environmental budgets, constrain the ability of provinces to take climate action.[3] Any large-scale funding has to go through central government institutions, and delegates talked about the bureaucratic burden of the disbursement process, and the high levels of uncertainty about the level of funding for provinces. In Indonesia, as in many other REDD+ countries, there is an institutional vacuum that is reflected in the lack of rules on the distribution of REDD+ benefits across jurisdictional levels, which fuels uncertainty and leaves decisions largely at the discretion of central government.

It also seems that GCFTF fills functions left vacant by national government in a number of areas. First, it facilitates linkages between provinces and international donors. In a well-functioning multi-level governance system, central government should facilitate such linkages. Instead, several Indonesia provinces indicated that GCFTF played an important role in facilitating direct contact with Norway and the World Bank, which are major funders and supporters of jurisdictional approaches. Second, it facilitates interactions across the various Indonesian provinces. GCFTF has assigned coordinators for countries such as Indonesia, that have a number of member provinces. This also comes with a budget for joint activities as prioritized by the provinces. Delegates suggest that such joint activities build and strengthen in-country blocks of like-minded jurisdictions, which enhances their power in subsequent interactions with national government. Third, GCFTF has helped provinces connect to private sector actors willing to funds jurisdictional initiatives. For example, the GCFTF Indonesia coordinator facilitated Unilever's connection with Central Kalimantan government, which is the first public–private climate initiative on smallholder oil palm certification, operating in two districts in the province.

10.6 Conclusion

In a country as diverse as Indonesia, it would be expected that decentralization supports and facilitates climate action – although it might also create coordination challenges between central and provincial governments. Our evidence showed how provinces are attempting to design climate plans that cater to their specific contexts and needs, but are limited in their ability to experiment. In the land use sector, the institutions and processes of decentralization have created some serious obstacles that hamper forest-based mitigation action. First, a legacy of limited decentralization of the forestry sector in an otherwise highly politically decentralized polity, have created a set of institutional legacies leading to perverse incentives that fuel further deforestation and reduce the ability of provinces to lead forest-based climate mitigation action. Further, districts have largely been excluded from climate change decision making, although they might host major climate mitigation projects. As the sector is attracting substantial international climate finance for mitigation action, recentralizing tendencies of forestry bureaucrats have become more pronounced, as has the competition among sectoral ministries for the control of the climate policy agenda. Districts, and some of the provinces, perceive these developments as a loss in regional autonomy, and an institutional failure in fully adopting the subsidiarity principle. But these recentralizing tendencies do not remain unchallenged.

Sub-national governments have been able to facilitate policy innovation and diffusion, but largely with the help in international processes. New transnational climate governance initiatives support collective action institutions linking provincial governments, facilitating learning and socialization of climate action across provinces. Such support also enhances the visibility of provinces in the national climate change domain. Provincial governments have been using these platforms to develop and disseminate their own ideas and visions for climate and forests. Provinces are increasingly leading jurisdictional approaches to REDD+ and Low Emission Development, despite evidence of resistance by the central government to devolve resources and decision-making power. As implementation of these approaches is just past the pilot phase, it remains to be seen how effectively they will contribute to emission reductions. Constraints on regional autonomy, institutional bureaucratic burdens, limited and uncertain access to funding, and misalignment of national policies with local needs remain some of the key challenges that provinces face vis-à-vis the central government.

At the same time, both national and sub-national governments experience high, although distinct, pressures from private forestry and agribusiness interests driving deforestation that historically contributed to economic development to the detriment of the environment. National and provincial governments will only overcome such pressures if they collaborate more effectively. Instead of working together, bureaucratic national interests seem to be competing with provincial governments in a scramble for control of land, forests, and climate change mandates. Cooperation is further hampered by the distinct *ideas* on the future of climate and forests between national and provincial interests. More inclusive national climate change institutions willing to devolve resources and decision-making power to localities would be more conducive not just to global and national climate change emission reduction targets, but also to important sustainable development targets that are central to the visions and ideas of localities.

There seem to be major differences between decentralized polities, such as Indonesia, and federal systems in the governance of climate change. In decentralized systems, institutions and policies underpinning devolution are more likely to be in flux, and change in response to changes in government, changes in policy agendas, and in the constellation of power across governance levels. This is particularly true in emerging policy domains, such as forest-based climate change mitigation. Decentralization in Indonesia is subject to ongoing political negotiations between the centre and the periphery, and competition over the climate agenda and the associated uncertainty hamper effective forest-based climate change mitigation. Whether a more extensive form of devolution in the climate change arena would translate in enhanced emission reductions remains an open question. It would, however, likely reduce competition across governance levels, which currently hampers effective climate action.

Notes

1 Aceh and Papua gained special autonomy in 2001.

2 Central government is responsible for foreign affairs, defence, national monetary and fiscal matters, and religion. The law, however, also states that central government retains authority to legislate on any area not mentioned in the law.

3 Although interviews were done before the 2020 disbursement of funding, at the time delegates knew that Norway had committed 25 $Mill to support jurisdictional REDD+ implementation at provincial level.

References

Angelsen, A. 2017. 'REDD+ as Result-Based Aid: General lessons and bilateral agreements of Norway'. *Review of Development Economics* 21 (2): 237–64.

BAPPENAS (Ministry of National Development Planning). 2014. *National Medium-Term Development Plan (Rencana Pembangunan Jangka Menengah Nasional (RPJMN 2015–2019)*. Jakarta: Ministry of National Development Planning.

Bertrand, J. 2007. 'Indonesia's Quasi-federalist Approach: Accommodation amid strong integrationist tendencies'. *International Journal of Constitutional Law* 5: 576–605.

Boer, R., and A. R. Subbiah. 2005 'Twenty-Six Agricultural Drought in Indonesia'. In: *Monitoring and Predicting Agricultural Drought: A global study*, eds. V. K. Boken, A. P. Cracknell, and R. L. Heathcote, 330–44. Oxford: Oxford University Press.

Boyd, W., C. Stickler, A. Duchelle, et al. 2018. *Jurisdictional Approaches to REDD+ and Low Emissions Development: Progress and prospects, World Resources Institute Working Paper*. Washington, DC: World Resources Institute.

Brockhaus, M., and M. Di Gregorio. 2014. 'National REDD+ Policy Networks: From cooperation to conflict'. *Ecology and Society* 19 (4): 14. doi: 10.5751/ES-06643-190414.

Butt, Simon. 2010. 'Regional Autonomy and Legal Disorder: The proliferation of local laws in Indonesia'. *Singapore Journal of Legal Studies*: 1–21. www.jstor.org/stable/24870542.

Damarjati, Danu. 2018. 'Kementerian LHK: 95,76% Hutan Berizin Dikelola Swasta'. *DetikNews*, 3 April. https://news.detik.com/berita/d-3951757/kementerian-lhk-9576-hutan-berizin-dikelola-swasta.

Di Gregorio, Monica, Kate Massarella, Heike Schroeder, Maria Brockhaus, and Thuy Thu Pham. 2020. 'Building Authority and Legitimacy in Transnational Climate Change Governance: Evidence from the governors' Climate and Forests Task Force'. *Global Environmental Change* 64: 102126. doi: 10.1016/j.gloenvcha.2020.102126.

Di Gregorio, M., D. R. Nurrochmat, L. Fatorelli, et al. 2015. *Integrating Mitigation and Adaptation in Climate and Land Use Policies in Indonesia: A policy document analysis*, Sustainability Research Institute Paper no. 90. Leeds: Centre for Climate Change Economics and Policy.

Di Gregorio, Monica, Dodik Ridho Nurrochmat, Jouni Paavola, et al. 2017. 'Climate Policy Integration in the Land Use Sector: Mitigation, adaptation and sustainable development linkages'. *Environmental Science & Policy* 67 (Supplement C): 35–43. doi: 10.1016/j.envsci.2016.11.004.

Ekawati, S., K. Budiningsih, G. K. Sari, and M. Z. Muttaqin. 2019. 'Policies Affecting the Implementation of REDD+ in Indonesia (cases in Papua, Riau and Central Kalimantan)'. *Forest Policy and Economics* 108: 101939.

Ferrazzi, G. 2000. 'Using the "F" Word: Federalism in Indonesia's decentralization discourse'. *Publius: The Journal of Federalism* 30 (2): 63–85.

Fishbein, G., and D. Lee. 2015. *Early Lessons from Jurisdictional REDD+ and Low Emissions Development Programs*. Arlington, VA: The Nature Conservancy, World Bank, & Forest Carbon Partnership Facility.

GCFTF (Governors Climate and Forest Task Force). 2021. *GCF Task Force: Provinces in Indonesia mobilize data for sustainable agriculture, driven by the country's recent success in reducing deforestation*. https://gcftaskforce.exposure.co/gcf-task-force-provinces-in-indonesia-mobilize-data-for-sustainable-agriculture, last accessed 20 July 2021.

Global Forest Watch. 2020. *Forest Monitoring Designed for Action*. www.globalforestwatch.org, last accessed 7 August 2020.

GOI (Government of Indonesia). 2016. *First Nationally Determined Contribution*. Jakarta: Government of Indonesia.

Hovani, L., T. Varns, H. Hartanto, S. Rahman, N. Makinuddin, and R. Cortez. 2018. *Jurisdictional Approaches to Sustainable Landscapes: Berau and East Kalimantan, Indonesia*. Arlington, VA: The Nature Conservancy.

Ibun Aqil, A. M. 2021. 'Indonesia Sets Net-zero Emissions Goal Ahead of COP26'. *Jakarta Post*, 1 August. www.thejakartapost.com/news/2021/08/01/indonesia-sets-net-zero-emissions-goal-ahead-of-cop26.html.

ICCSR. 2009. *Indonesia Climate Change Sectoral Roadmap: Synthesis report*. Jakarta: BAPPENAS.

InPres 6/2017 (President of the Republic of Indonesia). 2017. *Instruksi Presiden Republik Indonesia Nomor 6 Tahun 2017, Tentang Penundaan Pemberian Izin Barudan Penyempurnaan Tata Kelola Hutan Alam Primer dan Lahan Gambut*. Jakarta: Republic of Indonesia.

Irawan, S., T. Widiastomo, L. Tacconi, J. D. Watts, and B. Steni. 2019. 'Exploring the Design of Jurisdictional REDD+: The case of Central Kalimantan, Indonesia'. *Forest Policy and Economics* 108: 101853.

Jodoin, S. 2017. 'Rights and Jurisdictional REDD+ in Indonesia'. In Forest Preservation in a Changing Climate: REDD+ and indigenous and community rights in Indonesia and Tanzania, ed. S. Jodoin, 87–116. Cambridge: Cambridge University Press. doi:10.1017/9781316986882.005.

Lewis, B. 2013. 'Local Government Capital Spending in Indonesia: Impact of intergovernmental fiscal transfers'. *Public Budgeting & Finance* 33 (1): 76–94.

MMIA (Coordinating Ministry for Maritime and Investment Affairs). 2020. *Draft of the 896 Presidential Decree on the Carbon Economic Value Instrument for Achieving NDC and 897 Carbon Emission Control in Development*. Jakarta: MMIA.

MoEF (Ministry of Environment and Forestry). 2018a. *Indonesia: Second biennial update report under the United Nations Framework Convention on Climate Change*. Jakarta: MoEF.

MoEF (Ministry of Environment and Forestry). 2018b. *The State of Indonesia's Forests 2018*. Jakarta: MoEF.

MoEF (Ministry of Environment and Forestry). 2021. 'SP.434/HUMAS/PP/HMS.3/12/2021'. http://ppid.menlhk.go.id/berita/siaran-pers/6330/capaian-tora-dan-perhutanan-sosial-di-tahun-2021, last accessed 10 January 2022.

Nasution, A. 2016. *Government Decentralization Program in Indonesia: ADBI working paper*. Tokyo: Asian Development Bank Institute.

Nugroho, B. 2014. 'Menuju KPH Mandiri: Apa yang harus dilakukan?' In *Strategi pengembangan KPH dan perubahan struktur kehutanan Indonesia*, ed. Sugiharto, 91–114. Jakarta: FORCLIME.

Pal, S., and Z. Wahhaj. 2017. 'Fiscal Decentralisation, Local Institutions and Public Good Provision: Evidence from Indonesia'. *Journal of Comparative Economics* 45 (2): 383–409.

Pedroni, L., M. Dutschke, C. Streck, and M. E. Porrua. 2009. 'Creating Incentives for Avoiding Further Deforestation: The nested approach'. *Climate Policy* 9 (2): 207–20.

Peluso, N. L., and C. Lund. 2011. 'New Frontiers of Land Control: Introduction'. *Journal of Peasant Studies* 38 (4): 667–81.

PerPres 61/2011 (President of the Republic of Indonesia). 2011. *Presidential Regulation of the Republic of Indonesia No. 61 Year 2011 on the National Action Plan for Greenhouse Gas Emission Reduction*. Jakarta: Republic of Indonesia.

PerPres 1/2016 (President of the Republic of Indonesia). 2016. *Peraturan Presiden Republik Indonesia Nomor 1 Tahun 2016, Tentang Badan Restorasi Gambut*. Jakarta: Republic of Indonesia.

Pham, T. T., M. Moeliono, J. Yuwono, B. Dwisatrio, and P. Gallo. 2021. 'REDD+ Finance in Brazil, Indonesia and Vietnam: Stakeholder perspectives between 2009–2019'. *Global Environmental Change* 70: 102330.

Pierskalla, J. H. 2016. 'Splitting the Difference? The politics of district creation in Indonesia'. *Comparative Politics* 48 (2): 249–68.

ROI (Republic of Indonesia). 2011. *Masterplan: Acceleration and expansion of Indonesia economic development 2011–2025 (MP3EI)*. Jakarta: Republic of Indonesia.

Sahide, M. A. K., and L. Giessen. 2015. 'The Fragmented Land Use Administration in Indonesia: Analysing bureaucratic responsibilities influencing tropical rainforest transformation systems'. *Land Use Policy* 43: 96–110.

Sahide, M. A. K., S. Supratman, and A. Maryudi et al. 2016. 'Decentralisation Policy as Recentralisation Strategy: Forest management units and community forestry in Indonesia'. *International Forestry Review* 18 (1): 78–95.

Santika, T., E. Meijaard, and S. Budiharta et al. 2017. 'Community Forest Management in Indonesia: Avoided deforestation in the context of anthropogenic and climate complexities'. *Global Environmental Change* 46: 60–71.

Santoso, S. S., D. R. Nurrochmat, B. Nugroho, and I. Santoso. 2019. 'The Feasibility of the Implementation of Forest Management Units (FMUS) Policy: A case study in FMU Yogyakarta and FMU region IX Panyabungan'. *Jurnal Manajemen Hutan Tropika* 25 (1): 1–1.

Schebek, L., J. T. Mizgajski, R. Schaldach, and F. Wimmer. 2018. 'Land-Use Change and CO_2 Emissions Associated with Oil Palm Expansion in Indonesia by 2020'. In From Science to Society: New Trends in Environmental Informatics, eds. B. Otjacques, P. Hitzelberger, S. Naumann, and V. Wohlgemuth, 49–59. Cham: Springer. doi:10.1007/978-3-319-65687-8.

Seymour, F. J., L. Aurora, and J. Arif. 2020. 'The Jurisdictional Approach in Indonesia: Incentives, actions, and facilitating connections'. *Frontiers in Forests and Global Change* 3: 503326.

Shari-Rosenfield, S., G. Marks, and L. Hooghe. 2014. 'A Comparative Measure of Decentralization for Southeast Asia'. *Journal of East Asian Studies* 14 (1): 85–108.

Sloan, S., B. Locatelli, M.J. Wooster, D.L.A. Gaveau. 2017. 'Fire activity in Borneo driven by industrial land conversion and drought during El Nino periods, 1982-2010'. *Global Environmental Change-Human and Policy Dimensions* 47: 95–109.

Suharjito, D., and C. Wulandari. 2019. 'A Reflection of Social Forestry in 2019: Towards inclusive and collaborative government approaches'. *Forest and Society* 3 (1): 137–40.

Utami, A., R. Juliane, and M. Ge. 2016. *Six Things You Never Knew about Indonesia's Emissions and Local Climate Action*. Washington, DC: WRI.

UU23/2014. (Government of the Republic of Indonesia). 2014. 'Undang-undang (UU) tentang Pemerintahan Daerah'. Jakarta: Indonesia.

Wertz-Kanounnikoff, S., and A. Angelsen. 2009. 'Global and National REDD+ Architecture: Linking institutions and actions'. In Realising REDD+: National strategy and policy options, eds. A. Angelsen with M. Brockhaus and M. Kanninen et al., 13–24. Bogor: Center for International Forestry Research.

Wijaya, A., H. Chrysolite, M. Ge, et al. 2017. *How Can Indonesia Achieve its Climate Change Mitigation Goal? An analysis of potential emissions reductions from energy and land-use policies.* WRI Working Paper. Jakarta: World Resources Institute.

WRI (World Resources Institute). 2020. *CAIT Climate Data Explorer*. http://cait.wri.org, last accessed 8 August 2020.

Wunder, S., A. E. Duchelle, C. D. Sassi, et al. 2020. 'REDD+ in Theory and Practice: How lessons from local projects can inform jurisdictional approaches'. *Frontiers in Forests and Global Change* 3: 11.

Yong, C. K. Y. 2020. 'Indonesia Receives Results-Based Payments from the Green Climate Fund'. *UN-REDD Programme*, 1 December. www.un-redd.org/news/indonesia-receives-results-based-payments-green-climate-fund.

11

Climate Governance and Federalism in Mexico

MARCO HEREDIA AND BEATRIZ CORRAL

11.1 Introduction

The chapter addresses key elements of climate policy in Mexico and implementation within the framework of a highly centralised federal system.

Mexico accounts for 1 per cent of global greenhouse gas (GHG) emissions, primarily from the burning of fossil fuels for transportation and the generation of electricity. Other contributing activities are agriculture, the oil and gas industry, and waste. The country is highly vulnerable to climate change, particularly because it is on the route of the most frequent tropical storms on the continent. Seventeen out of thirty-two Mexican states are coastal and concentrate around 47 per cent of the population.

Responsibility for climate change governance in Mexico is shared between the federal, state, and municipal governments. The federal government is responsible for national strategies and policy, as well as strategic areas based on sectors. Its General Climate Change Law establishes principles, planning, management, information and participation, and intersectoral and intergovernmental coordination mechanisms. Structural challenges still prevail, though, in policy implementation, hampering adaptation and mitigation efforts. The General Climate Change Law provides states with the responsibility to legislate and implement policy for transportation, commercial facilities, residential sources, special waste, agricultural and livestock activities, and others within their jurisdictions. While some of the thirty-two states are highly active and engaged in climate action, others lag behind in identifying climate risks and in implementing policy responses to the ever-present challenges that climate change poses to ecosystems and their population. States and municipalities have the authority to design and implement their own policy instruments with a wide margin of flexibility and innovation for mitigation and adaptation in as much as they are aligned with the national climate policy.

Mexican climate federalism allows subnational governments to tailor national climate policies to their specific needs and capabilities. Progress in climate policy implementation varies across the country, and there are contrasts and disparities in the definition of objectives or goals with the national policy.[1] Therefore, a high degree of coordination is needed among the three levels of government for effective climate action. This also demands a more structured coordinated strategy at the federal level.

The State of Yucatan is a good example of local action that is advancing the climate agenda. Yucatan has recently issued specific climate legislation, and has developed a diagnostic that serves as the basis for mitigation and adaptation action. It has maintained a regional and local approach to mitigation and adaptation incorporating state and municipal authorities, and it has integrated climate change into its state development plan and derived state policy.

11.2 Background and Context

11.2.1 Contributions to Climate Change

Mexico contributes approximately 1 per cent of global GHG emissions. In 2019, total GHG emissions were 736.63 Mt of CO_2e, according to the National Greenhouse Gases and Compounds Emissions Inventory (or INEGyCEI as per its acronym in Spanish). Fossil fuel combustion produced 63.52 per cent of total emissions, with the burning of fossil fuels to generate electricity contributing 23 per cent and the transportation sector 20 per cent.

With 50.3 million vehicles on the streets in 2020, transport is a natural target for mitigation action (INEGI 2021). The National Strategy for Climate Change (2013) considers that sustainable mobility is a sector that should be delivering results in a ten-year period and foresees different objectives, strategies, and actions, including increasing energy efficiency and developing a national electromobility strategy with public transportation projects in each state. However, public transportation is the responsibility of state authorities, and they face difficulties in reforming this sector because of unions' and private car owners' resistance. This often results in a lack of appropriate planning, ineffective governance, and poor clean air and climate benefits.

11.2.2 Impact of Climate Change in Mexico

Due to its geographical location and characteristics Mexico is highly vulnerable to the adverse effects of climate change. It has 11,000 kilometres of coastline and is within the most frequent route of tropical cyclones in the region. This is compounded by social and economic issues, which greatly exacerbates Mexico's vulnerability to climate anomalies. The Global Climate Risk Index for 2018 ranked Mexico 26th in risk and 10th in deaths out of 177 countries analysed

(Germanwatch 2020). It is estimated that 70 per cent of the country's population could be affected by climate change.

In the Special Climate Change Programme 2014–18, the federal government projected changes based on a scenario where the temperature increases between 2.5°C to 4.5°C and precipitation decreases by 5 and 10 per cent. This was projected considering 1961–90 temperature and precipitation averages. These projected changes are foreseen in part by the amount, intensity, and socioeconomic consequences of the storms, and also because of the changes in rain patterns with severe droughts in the north of the country, and floods mainly in the south. However, the impact on agriculture, coasts, ecosystems and biodiversity, and strategic infrastructure is no less important.

The socioeconomic and ecological Mexican trends, including urbanisation and pressures on natural resources due to economic and population growth, suggest that under a business-as-usual scenario, climate risks will exacerbate other problems in the country (SEMARNAT–INECC 2016, 12). An example of this is public health. The Federal Commission for Protection against Health Risks (*Comisión Federal para la Protección contra Riesgos Sanitarios* or COFEPRIS) indicates that protecting human health will represent a challenge under climate change scenarios. Fifteen per cent of the territory, 68 per cent of its population, and 71 per cent of its gross domestic product are highly exposed (COFEPRIS, 2017). Challenges here include differentiated temperature increases, change of precipitation patterns, advance in the occurrence of the hot season, extreme weather events, increase in desertification, deforestation, disappearance of glaciers, increase in the sea level, and health problems.

Between 1999 and 2017, disasters with a climate-related declaration were significantly higher in Mexico. The years 2010 and 2013 stand out, representing the highest expenses in climate-related disasters. The most damaging events have been, overall, the tropical cyclones Manuel and Ingrid in 2013. In general, meteorological events in the country (tropical cyclones, landslides, floods, droughts, and severe storms) classified as 'disasters' by the National Centre for Disaster Prevention (CENAPRED), despite certain oscillations, show an increasing trend along with annual average costs.

To address these occurrences and their costs, the Mexican government has two policy instruments: one preventive, the Fund for the Prevention of Natural Disasters (FOPREDEN); and one reactive, the Fund for Natural Disasters (FONDEN). The burden of social costs associated with adverse effects of climate change is on the local communities or local stakeholders, such as municipalities, state authorities, or private stakeholders. In the period between 2012 and 2015, FONDEN's funding (reactive) were seventy-eight times higher than FOPRE-

DEN's (proactive) funding. FONDEN is utilised mostly for loss and damages where there has been a declaration of a disaster. Therefore, local stakeholders have to undertake remediation actions when hydrometeorological events strike a given territory with little or non-existing capacity to address the direct and spillover effects of these climate related events fully (INECC–SEMARNAT 2018; SEMARNAT 2019).

11.2.3 Commitments in Relation to Climate Change

Mexico is an active participant in international processes related to climate change and sustainable development. The country adopted the United Nations Framework Convention on Climate Change (UNFCCC) in June 1992, and it has been a party to the Kyoto Protocol since 1998 (Edwards et al. 2015). This has supported Mexico's mitigation efforts and its capacity to establish inventories, clean development mechanism projects, and other mitigation actions. In 2010 Mexico hosted and chaired the 16th Conference of the Parties to the UNFCCC in Cancun. Its commitment helped to renew the world's interest in addressing climate change and in seeking innovative and creative means for mitigation and adaptation.

In preparation for the adoption of the Paris Agreement at COP 21, Mexico submitted its INDC (Intended Nationally Determined Contribution) on 30 March 2015, becoming the first developing country to include not only ambitious goals in mitigation, but also adaptation pledges in the social ecosystems and productive systems. In 2016, Mexico signed the Paris Agreement.

11.2.3.1 Nationally Determined Contribution Commitments

Mexico's INDC considered a reduction of 22 per cent of its total GHG emissions and a decrease of 51 per cent of black carbon. Black carbon, or soot, is part of the particulate air pollution (PM2.5) and contributes to climate change. It is one of many particles that are emitted when diesel, coal, and other biomass fuels are burned (Climate & Clean Air Coalition). Under this framework, emissions intensity per unit of GDP are scheduled to be reduced by around 40 per cent between 2013 and 2030. The document foresees actions for communities, *ejidos* (a form of traditional town with rights over a specific territory), unions, non-governmental organisations and other groups of citizens; implementation of an ecosystems-based approach; and adaptation of strategic infrastructure and productive systems. The adaptation actions include ensuring food security and water access, comprehensive watershed management biodiversity and land conservation, increasing adaptive capacity through early warning systems, risk management, hydrometeorological monitoring at all orders of government, reducing by at least 50 per cent the number of municipalities classified as most vulnerable, and reaching zero deforestation by 2030.

The states were not involved in the design of the 2015 INDC pledge. They only became involved after the NDC submission in the context of the National Climate Change System (SINACC). In a survey carried out by SEMARNAT and INECC in 2016, eighteen states responded that key areas of action were within their authority: residential and commercial (25 per cent), transportation (25 per cent), AFOLU (14 per cent), agricultural and livestock management (13 per cent), residues (11 per cent), power generation (5 per cent) and industry (5 per cent).

For the update of the NDC (currently Nationally Determined Contributions) in 2020, the adaptation component was strengthened through five axes: prevention and mitigation of adverse effects, resilience and food security, conservation, restoration and sustainable use of biodiversity, integrated management of water resources, and protection of infrastructure and tangible cultural heritage. Synergies with Agenda 2020 were also identified. Mitigation commitments in the NDC update were identical to those in 2015: 22 per cent reduction in GHG emissions and 51 per cent in black carbon as unconditional reductions by 2030. Not adopting more ambitious targets in its mitigation pledge has been criticised by the national and international community (Climate Action Tracker 2021). However, this pledge was in line with the 2018 reform to the General Law on Climate Change.

11.3 Federalism and Climate Change Policy

Mexico is a federal republic with thirty-two constituent units: thirty-one states and Mexico City. These are divided into a total of 2,458 municipalities and sixteen territorial demarcations in Mexico City. Climate change is a shared responsibility spanning federal, state, and municipal governments. This shared responsibility is further outlined by the General Law on Climate Change. The Federal Ministry of the Environment, otherwise known as SEMARNAT, is responsible for issuing the national climate change policy, implementing the national climate change system, and other climate policy instruments in the information, public participation, standards, and management of an emissions registry and an emissions trading system. The Federal Ministry of Finances is responsible for administering the Carbon Tax. This has been in place since 2014 and it is applicable to the sale and importation of fossil fuels. The amount to be levied is calculated based on the amount of carbon in each fuel (for example, in 2022 propane is taxed with 8.2987 cents [MXN] per litre, this is the minimum amount charged. On the other hand, carbon coke is taxed with 51 pesos [MXN] per ton, other carbon fuels are taxed with 55.8277 pesos [MXN] per ton). Natural gas is exempt from the tax (Special Law for Products and Services, article 2.1 (H), DOF, 2013–22). There is not a special destination for environmental or climate projects stemming from the amounts levied for this tax.

State authorities make policy in urban planning and development, transportation, land use, and waste in accordance with the Federal Constitution. Municipalities are also responsible for making and implementing policy related to municipal waste, transportation, and urban planning, consistent with federal and state policy.

11.3.1 Introduction to the Mexican Federal System

While states are sovereign in their internal regime in accordance with the Federal Constitution (CESOP 2006), Mexico represents a paradoxical case of a federal country with a highly centralised form of government, due mainly to the historic social, economic, and political power concentration in Mexico City.[2]

Domestic implementation of international commitments varies according to the subject matter in question. The Mexican legal system, primarily determined by the Federal Constitution, determines those areas where there are shared, common, or exclusive responsibilities. Environmental protection and ecological equilibrium are shared responsibilities. The Federal Congress has the authority to enact legislation that distributes authority among the three spheres of government. Legislation stemming from such processes is known as a General Law. In this vein, environmental legislation comprises a myriad of general laws where the General Law on Ecologic Equilibrium and Environmental Protection is known as a framework legislation. The General Law on Climate Change (GLCC) is part of the series of laws addressing environmental subject-matter related areas, such as biodiversity, residues, and forestry.

The GLCC (articles 7, 8, and 9) establishes authority and concurrent responsibilities of the federal, state, and municipal authorities. Formulation and management of climate policy, financial resource management, promotion of scientific and technological development, education and climate culture, capacity building, the application of incentives and compliance monitoring are present in each level of government. However, boundaries between the jurisdictions of different orders of government are often blurred and decentralisation policies have accentuated these overlaps. Therefore, it is increasingly clear that the design and implementation of effective public policies require formal and informal intergovernmental coordination, ensuring cooperation between the three orders Flamand 2010).

The federal government is responsible for designing national policy instruments, issuing regulatory provisions and official standards, establishing public consultation processes with society, summoning the states and municipalities for the development of concurrent activities, proposing budget forecasts, and issuing recommendations to states and municipalities. Energy regulation and regional and demographic development are reserved for the federal government. Together, the federal government and the states are responsible for agriculture, education, food safety, and the prevention and care of diseases linked to climate change, education, and research. In states and municipalities, powers are concentrated on the protection of natural resources of their jurisdiction (although permits and authorisations are of federal nature) and on waste (other than hazardous) management.

Provisions included in the GLCC do not acknowledge differences in the degree of human and institutional development found in states across the country. It does call upon states to develop their climate change programme that should address a number of elements from transportation, urban development agriculture, livestock management conservation and natural resources restoration under their scope of authority (e.g., state protected areas), food security, infrastructure, education, land use, urban development, civil protection, and diseases prevention and attention. States are responsible for implementing their own greenhouse gases and compounds inventories and developing and implementing their risk atlases.

11.3.2 Climate Policy and Federal Governance

Congress, comprising the Chamber of Deputies and the Senate, has the responsibility of passing legislation on climate change. Executive authorities are responsible for acting on climate change, environmental and sustainable development. Several ministries are responsible for contributing to the design, implementation, and evaluation of climate policy. The Ministry of Environment and Natural Resources (SEMARNAT) and its technical support agency, the National Institute of Ecology and Climate Change (INECC), play a leading role.

SEMARNAT is responsible for formulating and conducting the national policy on climate change (Diario Oficial de la Federación (DOF)); Ley Orgánica de la Administración Pública Federal (1976); Reformada el 22 de enero de 2020). INECC, on the other hand, is responsible for conducting scientific and technological research and sectorial forecasting analysis, and for participating in the development of strategies, plans, programmes, instruments, and actions related to climate change (INECC – SEMARNAT 2015). INECC supports technical NDC-related work through assessing options to achieve mitigation or adaptation targets by identifying pathways and corresponding costs (Partnership on Transparency, 2019).

States issue legislation implementing the GLCC in sectors such as transportation, forestry, waste management, land use, planning, agricultural and livestock management, education, and health. State legislation is mandatory for municipalities within their territory. Municipalities are responsible for waste management, planning, and land use according to specific legislated authority stemming from national or state provisions.

The Mexican Constitution enshrines the human right to a healthy environment and sustainable development. The Constitution has a clause that contains 'interpretation in conformity' and the *pro personae principle.* Pursuant to these, any international treaty enshrining human rights provisions that is signed and approved by the Senate shall be interpreted in the same legal form as the Constitution. The National Human Rights Commission has authority to address any claim pertaining to violation of human rights in the environmental and climate change realms. There is a human rights commission in each of the thirty-two federal entities and they can also address human rights violations related to environmental issues or climate change.

11.3.3 Laws, Policies, Institutions, and Initiatives Developed for Climate Mitigation and Adaptation

The National Climate Change Policy includes planning, information, management, coordination and implementation, financing, monitoring, and evaluation instruments spanning over the three spheres of government. The most important elements for the case of Mexico are indicated below.

11.3.3.1 Legal Framework

The GLCC, which came into force in October 2012, is the main climate policy instrument in the country. This legislation defines planning and policy instruments, institutional arrangements, and provides general guidance for the implementation of climate policy. It also incorporates a long-term, systemic, decentralised, participatory and integrated approach for adaptation and mitigation. Under the GLCC, the Federal Government is mandated to formulate and guide national climate change policy. The role of subnational government is also clearly specified, including the elaboration of state-level GHG inventories and climate programmes (SEMARNAT-INECC 2016, 12).[3] In the same vein, GLCC determines a series of economic, political, information, education, and research instruments that require the co-responsible participation of society (INECC-SEMARNAT 2015, 60).

Based on articles 8 (sections I and XI) and 11 of the GLCC, states have the authority to formulate, conduct, and evaluate their respective policies on climate change following the national policy framework. State governments issue their own state climate change laws. By November 2021, two states, Campeche and Sinaloa, lagged behind in the responsibility to issue legal provisions aimed at addressing climate change. Thirty states have either adjusted their environmental legislation or issued specific climate legislation.

11.3.3.2 Planning Instruments

The National Climate Change Strategy, issued in May 2013, provides the long-term vision for the country with a time horizon of ten, twenty, and forty years to guide climate change policy. It is the basis for a Long-Term Climate Strategy. The National Strategy also incorporates short-lived climate pollutants emissions reduction into national policy. According to the GLCC, the Strategy must be reviewed at least every ten years in mitigation and every six years in adaptation – addressing any differences between projected estimations and the evaluated results. Likewise, the corresponding scenarios, projections, objectives, and goals can be updated based on best information and evaluations. As of October 2021, the update on adaptation has not been issued.

Aligned to the National Strategy, each federal administration has the mandate to develop its Special Climate Change Programme. These stand out as the flagship planning documents for the administration's six-year term. The programmes must include specific objectives, goals, actions, and means for implementation.

States are also responsible for preparing and implementing their climate change programmes, promoting social participation (article 8, GLCC), considering their specific powers, resources, and relevant state level regulations. The GLCC mandates states to carry out their programme on climate change and to establish criteria and procedures to evaluate and monitor compliance. These programmes establish the strategies, policies, guidelines, objectives, actions, goals, and indicators to be implemented and complied with during the corresponding government period, in accordance with the National Climate Change Strategy and the Special Climate Change Programme. INECC reviewed state programmes in 2019; only twenty-five states had issued a programme on climate change. Six states (Aguascalientes, Guerrero, Puebla, Queretaro, Nuevo Leon, and Zacatecas) had not issued a climate change programme, meaning that their ability to implement climate change action was limited due to a lack of diagnosis on vulnerability and sources of GHG. SINACC could play a role in reviewing state programmes and help states adopt policies incorporating advances in the characterisation of vulnerability and GHG sources.

11.3.3.3 Institutional Arrangements

The National System for Climate Change (SINACC) coordinates government bodies and consults with the public, private, and social sectors on salient climate change issues. SINACC comprises: (i) the Inter-ministerial Commission on Climate Change (CICC), a body of fifteen Federal Ministries; (ii) the National

Institute for Ecology and Climate Change (INECC); (iii) the Congress; (iv) thirty-two Federal States; (v) the national associations of municipal officials; and (vi) the Climate Change Council (C3), which consists of at least fifteen experts from civil society (SEMARNAT–INECC 2016, 13). SINACC serves as a permanent mechanism for communication, collaboration, coordination, and consultation on the national climate change policy. It is also a forum for the promotion and coordination of national policy on climate change in the short, medium, and long terms, with a cross-cutting perspective aiming at the implementation of mitigation, adaptation, and vulnerability reduction policies, programmes, and actions to face the adverse effects of climate change (INECC-SEMARNAT 2015).

States participate in SINACC through their environmental authorities; other sectoral authorities such as urban planning or human development often take part in these sessions. SINACC must meet at least twice a year. However, most SINACC meetings have been limited to information sharing, and very few agreements have been reached to advance climate action (SEMARNAT 2020). It has thus shown limited ability to nurture and promote effective implementation of climate policy. Local governments still lack knowledge about SINACC and about the relevance of their participation therein – a situation that is not helped by the short electoral cycles, especially at the municipal level. Participation of the municipalities in SINACC is through their associations, which dilutes direct responsibility of municipalities in this forum. Because of the relevance of SINACC, more active participation of states and municipalities, and a more efficient agenda could help states and municipalities addressing needs and identifying opportunities to enhance climate action.

11.3.3.4 State Climate Policy

State climate policy areas include planning, information for decision making, coordination, management, public participation, education, research financing and evaluation.

Planning instruments include the design and implementation of a State Programme on Climate Change. A climate programme addresses key elements to identify effects of climate change in state territory. The plan is expected to be aligned with the National Strategy and the Special Programme on climate change. However, outdated plans do not reflect this alignment (ten plans out of twenty-five). Actions in the plan are geared towards preservation, restoration, and sustainable management of ecosystems and hydrological resources in their sphere of competence; food safety; agriculture and livestock management, rural development, fisheries and aquaculture; education; efficient and sustainable transportation; infrastructure; human settlements and development planning; environmental protection and natural resources; special management residues;

civil protection and prevention and attention of diseases linked to climate change. The programme should encompass procedures for public participation, follow-up and compliance monitoring, including impact indicators for mitigation and adaptation action.

In information for decision making, states have the responsibility to issue their GHG inventory. Prepared with technical support from the INECC, the inventory identifies sources of GHG and mitigation actions. By 2019, twenty-eight states had a GHG inventory in place. However, not all of these are updated; methodology for their implementation varies across the board and in very few cases, inventories are comparable or compatible with the National Emissions Inventory.

Eighteen states have intersectoral coordination commissions; not all have a solid work programme in place. The Mexico City Climate Commission has met only once since its creation. The Inter-secretarial Commission in the State of Mexico also only held one meeting, at its onset. There are other examples, such as the state of Guanajuato, where environmental authorities have made strides in cross-sectoral coordination. The State of Veracruz initiated a cross-sectoral work through the office responsible for implementing its sustainable development goals to 2030, which is linked to the office of the state governor.

State risks atlases stem from civil protection legislation, but are recognised as a vehicle for identifying climate risks in the climate legislation; not all states have an updated and comprehensive risk atlas that addresses current and future climate-related risks. INECC issued a National Atlas for Climate Change Vulnerability in 2018. It addresses six areas of climate vulnerability at the territorial level of a municipality. States – and municipalities – are progressively looking at the National Climate Vulnerability Atlas and considering its recommendations for their own domestic climate programmes or risk atlases.

At the federal level, the National Register for GHG includes facilities that emit 25,000 tons of CO_2e or more than that amount of GHG. At the local level, states have not put in place registries for GHG state-jurisdiction sources.

States have the authority to issue specific standards and technical guidance – as long as their own respective legislation provides for it. There are examples of environmental standards in Mexico City regarding renewable or clean energy for certain energy intensive business, standards for volatile organic compounds emissions and others for conservation lands or ecological agriculture in its territory. Transportation, however, is an area where state environmental standards could serve to limit GHG emissions and other pollutants. The emissions verification programme in the metropolitan area of Mexico City is one of such cases where without being labelled as standards, administrative provisions serve as guidelines for internal combustion vehicles emissions and for providing exemptions for electrical and hybrid vehicles.

States are responsible for including public participation in their efforts to design, implement, manage, and evaluate climate policy. This is often implemented through consultations and processes foreseen in state legislation. However, public participation is only as worthwhile as the information available to the public for the nurturing of informed opinion. Regarding education and research, states show different degrees of engagement and development in mitigation and adaptation programmes all across the different levels of instruction from elementary to professional degrees.

States such as Puebla, Tamaulipas, Guanajuato, Yucatan, and more recently Sonora, have shown interest in clean and renewable energy and have started to implement projects. Other states do not have the same degree of interest in advancing the climate agenda. Very few states have expressed their interest in implementing a preventive and proactive agenda for adaptation to climate change (Veracruz, Sinaloa), although most of its economic and social burdens rest with state or municipal authorities.

Financing is a necessary condition for successful climate action. States vary in their efforts and understanding of this basic element of policy management. While there are states that have clearly identified public budgets for climate action, others have incorporated specific trust funds for implementation of mitigation or adaptation actions. Other states have not identified specific funds for climate change and their budgets only include allocation for environmental management under their ministries of the environment. The federal budget does not include a specific section for state climate action. Funds from international cooperation agencies are available to support studies or capacity building efforts which may otherwise not be available.

In 2019, the Supreme Court ruled that states legislatures have the authority to establish environmental taxes, including carbon taxes. Reviewing the specific case of the state of Zacatecas, the Supreme Court ruled that the Constitution allowed states to determine such taxes. By December 2021, the State of Mexico and Baja California established a carbon tax, and Nuevo Leon and Yucatan set taxes on emissions.

Grounded on the constitutional right to a healthy environment for peoples' development and well-being, and on the state's role to serve as its utmost steward of these rights, Zacatecas was the first Mexican state to consider a tax on GHG emissions. This is aimed at charging 250 MXN pesos (12 US dollars) for emissions of tCO_2 and other GHG such as methane, N2O, HFC, PFC, and HF6. This tax is applicable to persons, enterprises, local or foreign that have productive units in the state, and other public entities with federal or state autonomy. GHGs will be taxed according to their global warming potential considering CO_2 in a ratio of 1.

After Zacatecas, the State of Mexico established a tax (fiscal year 2022). In this case, each CO_2 emission is taxed with 43 MXN pesos (approximately 2 US dollars) per ton. As per Decree 18 adopted in January 2022, tax levies will be used to improve the state's environmental condition and not solely for tax collection purposes. Expected tax levies are in the range of 80 million MXN pesos or around 3.8 million US dollars. Baja California´s carbon tax is applicable to persons, legal entities, and economic units that have installations or sources where goods or products that generate emissions into the atmosphere are sold to final consumers within the territory of the state. In this case, the contents of tCO_2 are taxed per fuel: gasoline 2.196 kg/l; diesel 2.47 kg/l; natural gas 2.69 and liquified gas 3.00 kg/l. Each kilogramme is taxed with 0.17 MXN pesos. In Nuevo Leon, a new tax is being imposed on emissions of N20 and particulate matter (PM10 and PM2.5) for every ton that surpasses the Mexican Official Standard (NOM, notably, NOM-043-SEMARNAT-1993 and NOM-085-SEMARNAT-2011). Yucatan, following the model of Zacatecas, is taxing GHG considering their GWP potential. Each CO_2e ton emitted is taxed with 2.70 economic units or UMAs, corresponding to 198 MXN for the 2022 fiscal year.

Environmental taxes are surging as a response to the ever-present need for resources to address environmental degradation, and enhance the structural responses of local authorities to address those to adjust or design mechanisms that effectively address the environmental and climate crisis. Tax authorities will be in charge of enforcing these provisions and environmental authorities will be in charge of designing the mechanisms to utilise these resources in a way that truly advances environmental and climate policy.

A more comprehensive national strategy on tax action addressing greenhouse gases is still pending and most needed. SINACC could serve as the forum to design and implement an environmental and climate tax strategy.

Emissions´ trading regulation is reserved for the federal authority (SEMARNAT). The federal mandatory system covers specific sectors (industry, oil and gas, and electricity generation) and only CO_2. States can participate in this indirectly through fostering compensation projects (that can account up to 10 per cent of each regulated facility's emissions) or by fostering voluntary carbon markets allowing emitters to offset emissions by purchasing carbon credits (SEDEMA 2018). States have also shown interest is in participating in REDD+ projects where there is a need to enhance local public and private participation; these include the states of Jalisco (SEMADET 2017), Oaxaca, Campeche, Quintana Roo and Yucatan (Almanza-Alcalde, 2022; CCPY 2021). The federal government has yet to issue guidance, however, on how these reductions will be considered to attain NDC targets.

In the evaluation of climate policy, INECC and the Coordination of Evaluation play a key role. Evaluations are intended to inform national policy improvement or readjustment (INECC-SEMARNAT 2015). In 2018, a subnational evaluation of six states and eighteen municipalities (three municipalities per state) was carried out. Results showed significant differences in the implementation of climate policy across states and municipalities (INECC 2018a). It also found that few states view climate change as an overall threat to social systems. The assessment showed important mitigation opportunities in power generation, transportation, and other areas such as waste management. It also demonstrated that little coordination effort is being made in these areas. SINACC has not implemented a strategy to coordinate climate policy implementation, which is most needed. This strategy could engage environmental and sectoral stakeholders and support the implementation of Mexico's NDC. Thus, opportunities arise for SINACC to host and advance a strategy that could enhance state action. In the adaptation realm, this evaluation revealed that municipalities hardly possess the understanding, technical abilities, or the necessary human resources to design or implement effective climate adaptation action. In most cases, risks or vulnerability to climate change are managed in a reactive approach.

The 2018–24 federal administration has prioritised the enhancement of public companies operating energy generation, distribution, and management. However, these companies have insufficient financial resources to invest in renewable energies, and are ill-prepared to compete with the private sector using cleaner technologies. On the other hand, state administrations have shown political will and commitments to renewable energy, despite some reluctance from the federal government, which claims that the national electricity grid needs to be revamped for a more efficient energy management and distribution in the grid. Notwithstanding, the federal government does not allocate appropriate resources for that overhaul in its 2022 budget initiative. At the national level, there are opportunities to implement projects to distribute energy generation through small-scale projects that might not need the national grid.

The last update of the progress of subnational climate change policy was made in 2019. INECC created a portal where the progress of the thirty-two states and some municipalities was gathered and disseminated (INECC 2021). INECC reviewed progress made by states and selected municipalities in the development of policy instruments. The review found that 75 per cent of the states have a state law on climate change, 78 per cent have a state plan or programme, 43 per cent have an inter-ministerial commission on climate change, and only 12 per cent make an inventory of emissions.[4]

11.4 Federal Governance and Climate Change: The Case of Yucatan

In Mexico, governance challenges have led to responses that are typically partial and fragmented. Advances in climate policy are concentrated at the federal level, while subnational governments are still making climate change a part of their government agendas, especially among state governments. Even in the most proactive states effective climate governance remains a challenge. Due to the flexibility of the Mexican federal system, and the fact that subnational governments have not been included in the delivery of NDC, most state initiatives are no longer in force or are not clearly aligned with national climate objectives. Mitigation efforts and programmes often are unclear about actual climate benefits. An exception is the state of Yucatan, which has made significant advances in the implementation of its climate policy. Yucatan serves as an example for effective cooperation and coordination between different levels of government looking at attaining national climate objectives.

11.4.1 Flexibility, Experimentation, and Innovation

Yucatan's climate action spans across the state, regional, and municipal levels: (i) it has a robust state policy framework with specific legislation in place; (ii) it is part of a *regional* initiative, in coordination with the two neighbouring state governments in the Yucatan Peninsula (Campeche and Quintana Roo); and (iii) through an *inter-municipal* association named *Pucc*, which addresses climate change among other topics of shared concern.

11.4.1.1 State Perspective: State of Yucatan

After the publication of the GLCC in 2012, Yucatan's State Development Plan 2012–18 included objectives aimed at reducing the vulnerability of productive communities to climate change. Yucatan has an Inter-ministerial Commission on Climate Change established in 2010, and a Special Action Programme on Climate Change was published in 2014 with validity to 2018. The State's Climate Change Law was issued in November 2021.

Yucatan addresses climate change in several legal provisions. The issue is covered in five state laws: the Environmental Protection Law, the Law of Sustainable Rural Development Law, the Education Law, Sustainable Forest Development Law, and Conservation and Development of Urban Trees Law. In the case of the last two statutory bodies, climate change is only mentioned but no further provisions are included.

The Special Action Programme on Climate Change 2014–18 defined a roadmap for the substantial reduction of GHG emissions and strengthening local capacity to increase the resilience of the social, environmental, and economic sectors to the effects of climate change by 2030. The mitigation objective is a low-emissions development, preserving the carbon sinks potential of natural areas in Yucatan. The adaptation objective consists of reducing the vulnerability of the state's social, productive, and environmental sectors by 2030 (to reduce the percentage of state GDP affected by losses derived from extreme weather events). The programme also defines strategies, lines of action, and indicators to monitor progress. In addition to being aligned with Mexico's NDC, the objectives and goals defined by the Yucatan government are also adapted to state priorities or conditions, emphasising the conservation of natural areas and the productivity of agricultural activities and forestry.

The Ecological Planning Programme – a key instrument for the effective implementation of climate policy at the subnational level – dates from 2005, and, as expected, it is not aligned with the national climate policy. The authorities of the three levels of government will need to coordinate to determine what would be the best activities to advance climate change and the origin of the necessary resource.

11.4.1.2 Regional Perspective: Peninsula of Yucatan Agenda

The Yucatan peninsula is in the southeast of the country and is divided into three states: Campeche, Yucatan, and Quintana Roo. This region is one of the most important touristic national and international destinations in the country, and it is also one of the most vulnerable to the effects of climate change.

An important part of the territorial perimeter of the peninsula is a transition zone from the Gulf of Mexico to the Caribbean. It is highly vulnerable to the adverse effects of climate change due to the interaction of elements in its coastal environment – exposure to hydrometeorological phenomena, impact on several species of flora and fauna, a sargassum upwelling on beaches in recent years, and an increased risk for lower productivity in regional agricultural production.

In 2010, at the Sixteenth Conference of the Parties (COP 16) in Cancun, Quintana Roo the states of the Peninsula agreed to join efforts and resources to address Climate Change (Coordinación sobre el Cambio Climático de la Península de Yucatan 2015). This regional alliance represents a governance model for subnational authorities to coordinate efforts and resources to undertake initiatives for mitigation and adaptation. It establishes the institutional cooperation framework to implement public policy, and a regional Commission on Climate Change working closely with the Inter-ministerial Commissions on Climate Change in each state of the Yucatan Peninsula.

The Regional Climate Change Commission foresees the participation of the environmental authorities of the three states, as well as a representative of the Federal Inter-ministerial Climate Change Commission (CICC). Strategies of the

Regional Commission are based on three 'Big Vision Projects' (Coordinación sobre el Cambio Climático de la Península de Yucatan 2015): (1) Reducing Emissions from Deforestation and Forest Degradation (REDD+) from forests and mangroves; (2) implementing the roadmap 'Articulation of policy instruments for adaptation to climate change in the Yucatan peninsula' prepared with the support of UNDP in 2013, which includes a Regional Adaptation Strategy for the Yucatan Peninsula, inter-institutional coordination, capacity building and the preparation of diagnoses and studies with a regional vision; (3) creation of the Yucatan Peninsula Climate Change Fund to obtain and distribute funds to mitigation, eliminating deforestation, and promoting environmental restoration and adaptation actions for ecosystems and local communities (Vallejo and Becerril 2018).

11.4.1.3 Local Perspective: Puuc Inter-Municipal Biocultural Board, Intermunicipal Decentralised Public Agency (JIBIOPUUC)

Yucatan is one of the states with the highest rates of biodiversity in Mexico, especially the *Puuc* (*puuc* in Mayan means hill) region located in the southern part of the state. The area is considered important for the environmental and landscape amenity it provides. In addition, this area has historical and cultural characteristics that come from the time of the ancient Maya with important ceremonial centres.[5] The 'Puuc Biocultural State Reserve' spans five municipalities: Muna, Oxkutzcab, Santa Elena, Tekax, and Ticul. This is an area of 135,848 hectares, according to decree 455 published in the Official Gazette of the Government of the State of Yucatan in November 2011.

In 2013, the five municipalities signed an Inter-municipal Cooperation Alliance for the Integrated Management of the *Puuc* Zone for the conservation and management of natural resources. In 2014, the Agreement for the Creation of the Intercultural Biocultural Board of Inter-Municipal Decentralised Public Organisations of Puuc (JIBIOPUUC) was published in the Official State Gazette. The JIBIOPUUC provides technical support to municipalities for the preparation, management, and implementation of projects and programmes related to the environment, natural resource management, and sustainable rural development applicable in their territories, foreseeing climate change among the issues to be addressed.

Municipalities and other stakeholders such as The Nature Conservancy, the German Agency for International Cooperation (GIZ), the REDD+ Alliance, the state's Secretariat of Urban Development and Environment and the National Forestry Commission, communities and municipalities and other relevant stakeholders have joined efforts to promote and implement projects to mitigate climate change, seeking to reach zero deforestation and environmental degradation, favouring best practices in sustainable production and organic agriculture.

In 2019, municipalities signatory to JIBIOPUUC adhered to the Global Covenant of Mayors for Climate and Energy, an international alliance for local and regional authorities that seeks to disseminate and to support actions to face climate change (Yucatan Ahora! 2019). The regional agreement for the Yucatan Peninsula features data on the region and its strategy on climate change (Portal of the Climate Change Strategy of the Yucatan Peninsula).

The case of Yucatan shows us the enriching experience of a state that manages climate policy with the participation of actors from the federal, state (cross-sectorial), and municipal governments, including international cooperation and non-governmental organisations. However, there was no explicit depiction of coordination for consistency or complementarity in the implementation of policies and programmes. This same pattern could also be identified in the participation of federal actors at the subnational level.

Yucatan has worked recently to integrate climate policy into its planning instruments, as well as to address the issue from different spheres and in coordination with other state, local, and federal actors. This dynamic has occurred thanks to the leadership and political will of this particular state, a situation that is not observed in the rest of the country in the same way. The state's 2018–24 Development Plan considers climate change as an issue that needs to be addressed for sound economic, social, and environmental development (Gobierno del Estado de Yucatan 2019). The State Development Plan builds upon climate policy instruments such as the National Atlas for vulnerability. It also seeks to advance to a low emissions economy by means of mitigation action on the industrial, agricultural, commercial, and services sectors. Currently Yucatan is preparing specific legislation on climate change with support from the United Kingdom Partnership for Accelerated Climate Transitions (UK PACT).

11.5 Conclusion

Emissions in Mexico have grown since the early 1990s, but these emissions, notable in the energy and transportation sectors, have slowed their rate of growth in recent years. Because of its geographical and environmental characteristics, coupled with its social and economic problems, the country is highly vulnerable to the adverse effects of climate change. States and municipalities bear the political, social, and economic burdens and costs of hydrometeorological events. Financing, technical, and logistic capabilities are unevenly distributed across Mexican states.

Mexico has played an important role in the development of the international climate change agenda and in promoting and fostering compliance with the Paris Agreement. This has occurred mainly through the implementation of federal policy instruments. Internally, few states have put in place and maintained updated climate policy instruments such as special plans, strategies, inventories, or risks atlases, and some have even embarked on the implementation of carbon taxes. On the other hand, climate action in many states struggles with lack of political will, limited information, and limited capacities to define strategies, or mid-term and long-term objectives and goals. Although there is a National Climate Change System in place, more strategic and structured coordination could strengthen the efficiency and efficacy of federal, state, and municipal climate policy.

Mexico's highly centralised federalism has fostered the design and implementation of nation-wide climate instruments. However, there are gaps for a more appropriate regional and local climate action. The General Law on Climate Change assigns an important role to subnational governments; it does allow some flexibility to adapt this climate policy to their needs as well as for innovation. On the other hand, this statutory body does not recognise territorial differences, nor does it account for the differences in development and capacities among the different regions across the country. Furthermore, the federal government has not significantly integrated subnational actions in climate policymaking processes, such as in the definition of the 2015 and 2020 NDC objectives and goals, and in the allocation of federal funds for climate action. Being relatively aside from NDC design in 2015 and their update in 2020, it remains to be seen how subnational action could contribute to enhancing or raising national ambition. Currently, the progress required throughout the national territory to meet the mitigation and adaptation objectives established in the NDC is not clear.

Despite structural challenges, state responsibilities outlined by the General Law on Climate Change have favoured the emergence of experiences such as the one in the Yucatan peninsula, and by the state of Yucatan, by adopting a regional subnational initiative for the planning and implementation of a climate policy that provides a valuable space for exploiting its mitigation capacity and to identify and address its adaptation needs. Although this sort of initiative shows a gradual progress and enhances governance, several challenges remain in a highly centralised federalism that struggles to make welfare accessible to people and ecosystems across the nation.

Notes

1 This is more widely reviewed and addressed in the Strategic evaluation on the Subnational Climate Change Policy Implementation performed by the Coordination of Evaluation and INECC (INECC, 2018a).

2 Mexico's federal system has been widely studied. See: Victoria Rodriguez. 1998. 'Recasting Federalism in Mexico'. *Publius* 28(1): 235–54.

3 Mexico was the second country to have a national climate change law, after the UK, and the Regulation of the Law was published on 28 October 2014.

4 Cf. notes 1, 4.

5 Notably Oxkintok, Uxmal, Kabah, Sayil, Labna, Xlapak, and Chacmultun.

References

Almanza-Alcalde, Horacio, Poshendra Satyal, and Esteve Corbera, et al., 2021. 'Participatory Injustice in Mexico´s Readinesss Process to Reduce Emissions from Deforestation and Forest Degradation (REDD +), in Human Ecology'. *Human Ecology* 50: 79–90.

Centro de Estudios Sociales y de Opinión Pública (CESOP). 2006. 'Definición'. *Federalismo.* http://archivos.diputados.gob.mx/Centros_Estudio/Cesop/Eje_tematico/federalismo.html.

Climate Action Tracker. 'Mexico'. 15 September 2021. https://climateactiontracker.org/countries/mexico.

Climate and Clean Air Coalition. 'Black Carbon', viewed 30 October 2022. www.ccacoalition.org/en/slcps/black-carbon.

Comisión Federal para la Protección contra Riesgos Sanitarios. 2017. 'Impactos del cambio climático en la salud'. www.gob.mx/cofepris/acciones-y-programas/impactos-del-cambio-climatico-en-la-salud.

Commission for Environmental Cooperation (CEC). 2005. 'Hacia el futuro: Plan estratégico de la Comisión para la Cooperación Ambiental 2005–2010'. Montreal, QC: Commission for Environmental Cooperation. www.cec.org/sites/default/files/documents/planes_estrategicos/9634_2005-2010-strategic-plan_es.pdf.

Commission for Environmental Cooperation (CEC). 2008. 'Cambio Climático'. In *El mosaico de América del Norte: Panorama de los problemas ambientales más relevantes.* 7–10. Montreal, QC: Commission for Environmental Cooperation. www3.cec.org/islandora/es/item/2349-north-american-mosaic-overview-key-environmental-issues.

Commission for Environmental Cooperation (CEC). 2010. 'Plan estratégico 2010–2015 de la Comisión para la Cooperación Ambiental'. Montreal, QC: Commission for Environmental Cooperation. www.cec.org/es/sites/default/files/documents/planes_estrategicos/15040_strategic_plan_2011_es.pdf.

Commission for Environmental Cooperation (CEC). 2015. 'Plan estratégico de la Comisión para la Cooperación Ambiental 2015–2020'. Montreal, QC: Commission for Environmental Cooperation. www.cec.org/es/sites/default/files/documents/planes_estrategicos/strategic_plan_2015_2020_sp.pdf.

Coordinación sobre el Cambio Climático de la Península de Yucatan (CCPY). 2015. 'Agenda de la Península de Yucatan. Secretaría de Desarrollo Urbano y Medio Ambiente de Yucatan, de Campeche, Secretaría de Ecología y Medio Ambiente de Quintana Roo'. www.ccpy.gob.mx/agenda-regional.

Coordinación sobre el Cambio Climático de la Península de Yucatan (CCPY). 'Cambio Climático en la Península de Yucatan'. www.ccpy.gob.mx, viewed 30 October 2022.

Edwards, G., J. T. Roberts, M. Araya, and C. Retamal. 2015. *A New Global Agreement Can Catalyse Climate Action in Latin America.* Washington, DC: The Brookings Institution.

Fickling, M., and J. Schott. 2011. *NAFTA and Climate Change*. Washington, DC: Peterson Institute for International Economics.

Flamand, L. 2010. 'Sistema federal y autonomía de los gobiernos estatales: avances y retrocesos en el caso mexicano'. In *Los grandes problemas de México, Volume XIII*, ed. José Luis Méndez, 495–522. CDMX, Mexico: Colegio de México.

Germanwatch Climate Risk Index. 2020. www.germanwatch.org/en/cri.

Gobierno del Estado de Yucatan. 2019. 'Plan Estatal de Desarrollo 2018–2024, Gobierno del Estado de Yucatan'. www.yucatan.gob.mx/docs/transparencia/ped/2018_2024/2019-03-30_2.pdf.

INECC. 2021. 'Information on the Implementation of the Subnational Climate Policy'. https://cambioclimatico.gob.mx/estadosymunicipios.

INECC. 2018a. 'Acciones de mitigación propuestas hacia 2030'. Ciudad de México: Instituto Nacional de Ecología y Cambio Climático, México. https://cambioclimatico.gob.mx/sexta-comunicacion/material/Mitigacion.pdf.

INECC. 2018b. 'Costos de las Contribuciones Nacionalmente Determinadas de México. Medidas Sectoriales No Condicionadas. Informe final'. Ciudad de México: Instituto Nacional de Ecología y Cambio Climático, México. www.gob.mx/cms/uploads/attachment/file/330857/Costos_de_las_contribuciones_nacionalmente_determinadas_de_M_xico__dobles_p_ginas_.pdf.

INECC. 2018c. 'Evaluación Estratégica del Avance Subnacional de la Política Nacional de Cambio Climático. Coordinación de Evaluación, 26 de noviembre de 2018'. Ciudad de México: Instituto Nacional de Ecología y Cambio Climático, México. www.gob.mx/cms/uploads/attachment/file/418239/EVALUACION_ESTRATEGICA_AVANCE_SUBNACIONAL_PNCC.pdf.

INECC-SEMARNAT. 2015. 'First Biennial Report to the United Nations Framework Convention on Climate Change'. Ciudad de México: INECC/SEMARNAT. https://unfccc.int/files/national_reports/non-annex_i_parties/biennial_update_reports/application/pdf/executive_summary.pdf.

INECC-SEMARNAT. 2017. 'Estimación de impactos en la salud por contaminación atmosférica en la región centro del país y alternativas de gestión. Contaminación y Salud Ambiental'. Ciudad de México: INECC/SEMARNAT. www.gob.mx/cms/uploads/attachment/file/217935/INECC_CAME_Final_14022017.pdf.

INECC-SEMARNAT. 2018. 'Sixth National Communication of Mexico to the United Nations Framework Convention on Climate Change'. https://cambioclimatico.gob.mx/sexta-comunicacion.

Instituto Nacional de Estadística y Geografía (INEGI). 2021. 'Information on Vehicles in Circulation in Mexico'. www.inegi.org.mx/sistemas/olap/proyectos/bd/continuas/transporte/vehiculos.asp?s=est#.

Mexican Government. 2015. *Intended Nationally Determined Contribution*. www.semarnat.gob.mx/sites/default/files/documentos/mexico_indc.pdf.

Partnership on Transparency. 2019. 'Estimating the Costs of Unconditional Mitigation Measures to Support the Implementation of Mexico´s NDC'. Bonn and Eschborn: GIZ. www.transparency-partnership.net/system/files/document/200114_GPD_Mexico_NDC_RZ.pdf.

Portal of the Climate Change Strategy of the Yucatan Peninsula. www.ccpy.gob.mx accessed November 2021.

Ramirez, N., I. Valenzuela, and J. Hernandez. 2013. 'Mexico's Glaciers and their Close Disappearance: A precise thermometer of the global warming advance on a global scale'. *Current World Environment* 8(3): 391–94. www.cwejournal.org/pdf/vol8no3/CWEV08I03P391-394.pdf.

Rodriguez, Victoria. 1998. 'Recasting Federalism in Mexico'. *Publius* 28(1): 235–54.

Rosenzweig, C., W. D. Solecki, and P. Romero-Lankao. 2018. *Climate Change and Cities: Second assessment report of the Urban Climate Change Research Network*. New York: Cambridge University Press.

Santín del Río, Leticia. 2013. 'Las intermunicipalidades y los retos estratégicos para el desarrollo sustentable de los municipios'. *Revista RC ET Ratio, Año IV*, No. 7. http://contraloriadelpoderlegislativo.gob.mx/Revista_Rc_et_Ratio /Rc_et_Ratio_7/Rc7_1_Leticia_Santin_Del_Rio.pdf.

Secretaría de Hacienda y Crédito Público. 2022. *Proyecto de Presupuesto de Egresos de la Federación (Proposal for the Federal Budget).* www.ppef.hacienda.gob.mx.

SEDEMA. August 2018. Secretaría de Medio Ambiente del Gobierno de la Ciudad de México, Lanza GCDMX primer Bono de Carbono Forestal. www.sedema.cdmx.gob.mx/comunicacion/nota/lanza-gcdmx-primer-bono-de-carbono-forestal-de-la-capital.

SEMADET. 2017. Secretaría de Medio Ambiente y Desarrollo Territorial, Gobierno del Estado de Jalisco, REDD+ Información y sub temas, Romero, Alma. https://semadet.jalisco.gob.mx/medio-ambiente/reddmexico/redd-informacion-y-sub-temas (viewed 30 October 2022).

SEMARNAT. 2013. *Bosques, cambio climático y REDD+ en México : Guía Básica.* Segunda edición. Jalisco, México. www.conafor.gob.mx:8080/documentos/docs/35/4034Gu%C3%ADa%20B%C3%A1sica%20de%20Bosques,%20Cambio%20Clim%C3%A1tico%20y%20REDD_%20.pdf.

SEMARNAT. 2014a. *Programa Especial de Cambio Climático 2014–2018. Gobierno de la República*, publicado el 28 de abril de 2014. Retrieved from https://dof.gob.mx/nota_detalle.php?codigo=5342492&fecha=28/04/2014.

SEMARNAT. 2014b. 'El Medio Ambiente en México 2013–2014'. https://apps1.semarnat.gob.mx:8443/dgeia/informe_resumen14/05_atmosfera/5_2_3.html.

SEMARNAT. 2016. 'Informe de la situación del medio ambiente en México 2015: Compendio de Estadísticas Ambientales, Indicadores Clave, de Desempeño Ambiental y de Crecimiento Verde'. Sistema Nacional de Información Ambiental y de Recursos Naturales. https://apps1.semarnat.gob.mx:8443/dgeia/informe15/index.html.

SEMARNAT. 2018. 'Fichas técnicas de las 32 Entidades Federativas en materia de Cambio Climático'. *Dirección General Adjunta para Proyectos de Cambio Climático.* http://iki-alliance.mx/wp-content/uploads/SEMARNAT-2018-Fichas-Estados-politicas-clim.pdf.

SEMARNAT. 2019. 'Informe de la situación del medio ambiente en México 2018: Compendio de Estadísticas Ambientales, Indicadores Clave, de Desempeño Ambiental y de Crecimiento Verde'. Sistema Nacional de Información Ambiental y de Recursos Naturales. https://apps1.semarnat.gob.mx:8443/dgeia/informe18/index.html.

SEMARNAT. 2020. 'Base de datos estadísticos BADESNIARN: Consulta temática'. Sistema Nacional de Información Ambiental y de Recursos Naturales. http://dgeiawf.semarnat.gob.mx:8080/approot/dgeia_mce/html/mce_index.html.

SEMARNAT-INECC. 2016. 'Mexico's Climate Change Mid-Century Strategy'. Mexico City: SEMARNAT/INECC. https://unfccc.int/files/focus/long-term_strategies/application/pdf/mexico_mcs_final_cop22nov16_red.pdf.

SEMARNAT-SINACC. 2016. *Minuta de la Primera Reunión 2016 del Sistema Nacional de Cambio Climático.* 8 December 2016. www.gob.mx/cms/uploads/attachment/file/312024/Minuta_Primera_Reuni_n_2016_SINACC.PDF.

Sims, R., V. Gorsevski, and S. Anenberg. 2015. *Black Carbon Mitigation and the Role of the Global Environment Facility: A STAP advisory document.* Washington, DC: Global Environment Facility.

Sosa-Rodríguez, Fabiola. 2015. 'Política del cambio climático en México: avances, obstáculos y retos. Realidad, Datos y Espacio'. *Revista Internacional de Estadísticas y Geografía* 6(2): 4–23.

UK PACT. 4 November 2021. 'Elaboración de la ley de Cambio Climático del Estado de Yucatan, Periodico Oficial del Estado de Yucatan'. www.yucatan.gob.mx/docs/diario_oficial/diarios/2021/2021-11-04_2.pdf.

Vallejo, R., and J. Becerril. 2018. 'Case of Study: Financial innovation for climate action in the Yucatan Peninsula'. Under2 Coalition. www.under2coalition.org/sites/default/files/under2_coalition_case_study_yucatan.pdf.

Yucatan Ahora! 2019. 'Yucatan sienta bases para reducir efectos del cambio climático'. https://yucatanahora.mx/yucatan-sienta-bases-para-reducir-efectos-del-cambio-climatico.

12

Climate Governance and Federalism in South Africa

JAAP DE VISSER AND ANÉL DU PLESSIS

12.1 Introduction

South Africa adopted a quasi-federal system more than two decades ago. The ability of different organs of state to collaborate is put to the test by, among other things, the demands of an adequate scientifically backed response to climate change and established practice of *centralised* governance. National policy asserts that national, provincial and local authorities are all critical role-players in the country's pursuit of climate-resilient development. Yet, the multilevel system is complex, owing to factors such as the absence of geographical and administrative borders for climate change and the vulnerabilities this causes: a legacy of spatial and environmental injustices; persistent high poverty levels; turf protection combined with competency and skills deficits in government; and a blurred division of authority over environmental matters and related sectors. Recent political changes in some of South Africa's largest cities make for new dynamics in a government dominated by the African National Congress (ANC). Despite these complexities, all three spheres of government have started experimenting ad hoc with climate change law and policy.

The chapter gives an overview of South Africa's quasi-federal system, the country's climate change profile, the way the federal system links up with the demands of climate change governance, and a case study of tangible decentralised climate governance practices and developments. We show that, despite the urgency for climate action in South Africa, its quasi-federal government has not yet embarked on a consolidated process of identifying and implementing consistent and locally tailored solutions. The country's climate change law and policy framework is still in its infancy and can at best be described as patchy, as some line functionaries in the national and local spheres of government scramble to merge climate science, competing socio-ecological demands and short-term political imperatives.

12.2 Climate Change in South Africa

12.2.1 Challenges, Contributions and Impact

Climate change affects South Africa's economy, terrestrial and other ecosystems, water, human health, human settlements in urban and rural areas, agriculture, forestry, its disaster risk profile and the coastal zone (NCCRP; NDP; South Africa's Initial Communication Under the UNFCCC 2004). Current stressors emanate from increased temperatures and rainfall variability (mean annual temperatures have increased by at least 1.5 times the observed global average of 0.65 degrees over the past five decades and extreme rainfall events have increased in frequency (Ziervogel et al. 2014, 605). The stressors include land-use change as a result of land degradation and overuse; water stress in the face of high-water demand; natural disasters such as droughts and flooding; housing deficits; poor infrastructure and municipal service delivery; and the quadruple burden of disease (DEA 2011, 13–14; DEA (Department of Environmental Affairs) 2018). It follows that the country's socio-economic and environmental condition aggravates its exposure to climate risks in as far as the country has to: (a) stimulate a struggling national economy that is still heavily dependent on coal; (b) radically reduce high levels of inequality, poverty and spatial and environmental injustice; (c) address poor spatial and developmental planning and energy infrastructure maintenance; while (d) improving on its public administration, service delivery and governance efforts (Parnell and Walawege 2014, 36–42; Swilling and Annecke 2012, 224–45).

South Africa is the world's seventh largest producer of coal and the fourteenth largest emitter of greenhouse gases (GHGs) and its CO_2 emissions are principally due to a coal-dependent economy. The energy sector contributes approximately 80 per cent of the country's total GHG emissions, of which 50 per cent are from electricity generation and liquid- and solid-fuel production (McSweeney and Timperley 2018). Other contributing sectors include road transportation, manufacturing, construction industries and iron and steel production. Notably, some structural dysfunctions of South Africa's current economic model affect the objectives of 'climate-resilient and low-carbon patterns of development and developmental challenges' (Gulati et al. 2016, 36).

12.2.2 Climate Change Commitments

South Africa is one of the few countries that specify an absolute emission reduction target following a peak–plateau–decline trajectory range. The country's GHG reduction targets are captured in a combination of the National Development Plan (NDP), the Conference of Parties (2009) and South Africa's Nationally Determined Contribution under the Paris Climate Agreement (NDC). In summary, the aim is to achieve a peak–plateau–decline trajectory for GHG emissions, with the peak around 2025; to have an entrenched economy-wide carbon price by 2030; to have zero emission building standards by 2030; and to have absolute reductions in the total volume of waste disposal to landfill each year. In its first NDC of 2016, South Africa committed to:

> [W]orking with others to ensure temperature increases are kept below 2 degrees Celsius above pre-industrial levels, which could include a further revision of the temperature goal to below 1.5 degrees Celsius in light of emerging science, noting that global average temperature increase of 2 degrees Celsius translates to up to 4 degrees Celsius for South Africa by the end of the century. (NDC 2016, 1)

The 2020/2021 updated NDC confirms a deepening commitment to mitigation and shows significant progress in terms of the targets that are set. The upper end of the target range for 2025 has been reduced by 17 per cent, the upper end of the target range for 2030 has been reduced by 32 per cent, and the lower range by 12 per cent. The range between upper and lower bounds narrows significantly from 216 Mt in 2025 and 70 Mt CO_2 – eq in 2030. The government itself admits that '[m]eeting these targets will require South Africa to implement a range of policies and measures . . .' (NDC 2020/2021, 15–16).

The first NDC envisioned the country addressing climate change adaptation through six goals underpinned by adaptation planning, costing of adaptation investments, equity and means of implementation (NDC 2016, 3). The national Department of Environment, Forestry and Fisheries (DEFF) further approved its ten-year National Climate Change Adaptation Strategy 2020 (NCCAS), which articulates the government's vision for adaptation and climate resilience with a focus on water, health, human settlements, agriculture and commercial forestry, biodiversity and ecosystems and disaster risk reduction and management. The Adaptation Strategy is aligned with the updated NDC, which contains South Africa's first adaptation communication with emphasis on the constitutional right that everyone in the country has to an environment not detrimental to human health or well-being and the extent of climate change-associated risk and vulnerability (NDC 2020/2021, 5–7).

12.3 Federalism in South Africa

12.3.1 Introduction and Historical Context

South Africa has a national government, nine provinces and 257 municipalities. Provinces and municipalities have their own, locally elected, provincial legislatures and municipal councils and are headed by indirectly elected premiers and mayors respectively. The local government sphere consists of eight metropolitan municipalities, forty-four district municipalities and, within them, 205 local municipalities. While the hallmarks of a federal structure are present, most design elements, and certainly the federal practice, point towards a centralised or 'quasi-federal' system.

The constitutional arrangement is the product of negotiations held in the early 1990s between liberation movements (most notably the ANC) and the outgoing apartheid government, led by the National Party (NP). While the ANC favoured a unitary state, the NP had become a 'recent convert' to the idea of regional autonomy, more as a break on the imminent ANC hegemony than as a protection of regional ethnic interests (Steytler and Mettler 2001). Vehement arguments for ethnic federalism came from a regional movement in KwaZulu–Natal (the Inkatha Freedom Party). The compromise, ultimately laid down in the 1996 Constitution, was a quasi-federal state with strong unitary elements, which allocates significant powers to local government (Leon 2013; Murray and Simeon 2001). This is important for understanding multilevel government in South Africa, and how it influences subnational action on climate change. It shows that the quasi-federal structure was a compromise, reluctantly agreed to by the ANC, which subsequently went on to control virtually all the levers of power throughout the three spheres of government.

12.3.2 The Distribution of Power

In line with South Africa's reluctance towards federalism, the Constitution describes the system as 'cooperative government'. The central government has plenary powers to make and implement laws with respect to any matter, excluding the matters reserved exclusively for provinces. This authority extends to matters over which it exercises authority concurrently with provinces (see below); matters specifically mentioned for national government in the Constitution; and residual matters not listed anywhere in the Constitution (s. 44.1.a.ii). The exclusive national powers are substantial and include matters such as land, policing and regulating extractive industries and the energy sector. The national government collects most taxes, and distributes these annually across the three levels of government (ss. 214, 228,1 and 229.1).

The national and provincial governments have concurrent legislative and executive authority over matters listed in Schedule 4 to the Constitution (ss. 44.1.a. ii and 104.1.b.i). The list of concurrent powers includes matters such as housing, primary and secondary education, agriculture, environment, trade and health services. Conflicts between national and provincial laws on the same matter are ultimately resolved by the Constitutional Court in terms of a constitutional override clause. The practice of national–provincial concurrency is that the national government legislates, and the provincial governments implement. Provinces are responsible for implementing not only major social functions such as public health, housing, primary and secondary education, but also regulatory functions such as the environment, agriculture and disaster management. They do so within national legislative frameworks and pass very few provincial laws. This is because the national government has taken up most of the legislative space on these major functions. The constitutional override clause itself also makes it difficult to argue for constitutionally permissible provincial deviation from national laws on concurrent matters. The Constitutional Court has not yet been called upon to adjudicate a conflict between national and provincial legislation over concurrent matters. Perhaps most critically, provincial governments are almost entirely reliant on intergovernmental transfers. The Constitution does not allocate any significant revenue-raising powers to provinces, and the national government has not assigned any to them.

Municipalities are responsible for the delivery of basic services, such as water, sanitation, waste management, streets and the delivery of electricity to end-users. They are also responsible for environmental health services and town planning. National and provincial governments may regulate these local government matters, but only by means of standard setting (ss. 155.6 and 155.7). In perhaps the most significant expression of local government autonomy, the Constitution empowers municipalities to raise their own revenue via property rates and service fees (s. 229.1.a).

The local government sphere is made up of metropolitan, district and local municipalities. The metropolitan municipalities are the City of Johannesburg, Tshwane, Ekurhuleni, City of Cape Town, eThekwini, Mangaung, Nelson Mandela Bay and Buffalo City. They are not city states, but single-tiered local governments that, in theory, operate under the broad oversight of a provincial government.

Provinces receive transfers in the form of an annual, formula-based unconditional grant, complemented by conditional grants. Municipalities, on the other hand, are expected to raise much of their own revenue through property rates and fees for services. However, they do receive intergovernmental funding in the form of an unconditional equitable share and limited conditional grants. The overall picture is that cities and larger urban municipalities generally raise significant revenue, but rural municipalities are for the most part grant-dependent.

In addition to the above institutional and financial features, the practice of federalism is strongly influenced by the political reality that the ANC is the dominant political party across all three spheres of government. It controls the central government and eight of the nine provinces with outright majorities. In many of the country's municipalities, it controls the municipal council. However, the 2016 and 2021 local government elections saw a considerable loss of support for the ANC in major cities, such as Johannesburg, Tshwane and Nelson Mandela Bay that are now governed by coalitions.

In conclusion, South Africa practises an 'hourglass' model of federalism: a strong national government, relatively weak provinces and a mix of strong cities and weak rural municipalities in local government (Steytler 2017).

12.4 Climate Change and the Practice of Federalism in South Africa

Navigating South Africa's climate change response through the lens of the principles, ideals and realities of federalism is a tall order. As would be true for many countries, South Africa's climate change governance effort hinges on an intricate combination of government, NGO and private sector actions, a law and policy framework dealing with mitigation, adaptation and non-specific issues and various government actors situated in three, often overlapping, spheres and branches of government. The following discussion flags three key issues in the relationship between climate change governance and federalism in South Africa: governmental siloism; institutional, policy and legal fragmentation; and the absence of a direct relationship between national and local government.

12.4.1 Climate Policy and South Africa's Quasi-federal System

Key taxation and regulatory powers reside with the national government. The regulation and taxation of trade and industry, for example, including critical industries such as car manufacturing, are national powers (s. 44.1.a.ii). This alone makes the national government a key player in climate change mitigation. Furthermore, the national government manages water resources (though not the treatment and sale of water to end-users), biodiversity and electricity generation (though not the sale of electricity to end-users).

Subnational authority with respect to climate change mitigation and adaptation can be summarised, as revolving around municipal planning, the environment and the governance of urban spaces. Provinces play a muted role while the role of local governments and cities is significant.

12.4.2 A Fragmented Law and Policy Framework

Over the past decade, the South African national government adopted several laws and policies relevant to climate change adaptation and mitigation. Provinces have followed suit, albeit with uneven enthusiasm and intensity. Municipal policies on climate change have also come onstream, particularly in cities. The result is a patchwork of policy initiatives and programmes, the key elements of which are sketched below.

12.4.2.1 Legislative and Policy Developments at National Level

In 2008, the national government adopted its 'Vision, Strategic Direction and Framework for Climate Policy', followed in 2011 by the NCCRP. The latter sets two main objectives: (a) to manage expected climate change effects through interventions that build and sustain South Africa's social, economic and environmental resilience and emergency response capacity; and (b) to make a fair contribution to the global effort to stabilise greenhouse gas concentrations in the atmosphere at a level that avoids dangerous anthropogenic interference with the climate system within a timeframe that enables economic, social and environmental development to proceed in a sustainable manner (NCCRP [DEFF} 2011, 11).

In 2015, the government adopted its Climate Change Mitigation System. At the time of writing, the country is awaiting the promulgation of its first national climate change Act. Its forerunner, the *Climate Change Bill* (B9-2022) was published in 2022. The *Bill* covers aspects such the institutional arrangements for climate governance and sectoral emission targets (*Bill* ch. 2 and s. 22). The *Carbon Tax Act* (15 of 2019) became operational on 1 June 2019 and provides for the imposition of a tax on the carbon dioxide (CO_2) equivalent of greenhouse gas emissions.

The national government published an Integrated Energy Plan in 2016, providing an energy plan for liquid fuels, gas and electricity, and an Integrated Resources Plan (IRP), of which the most recent version was published in 2019. The IRP is an electricity infrastructure development plan based on a least-cost electricity supply-and-demand balance, considering security of supply and the environment (minimising negative emissions and water usage). The National Green Transport Strategy (2018–50) addresses the significant contribution of transport to national GHG emissions. The 2020 NCCAS consolidates and prioritises local, provincial and sectoral adaptation options and initiatives with a focus on the following sectors: water, agriculture and commercial forestry, health, biodiversity and ecosystems, human settlements (urban, rural and coastal), disaster risk reduction and management, transportation and infrastructure, energy, mining and oceans and coasts. The NCCAS is a high-level national adaptation policy that is intended to be implemented in functional and operational terms via provincial and local adaptation plans (see below).

The national legislature also responded with sector-specific law reform (for a detailed discussion see Du Plessis and Kotzé 2014). The *Integrated Coastal Management Act* (24 of 2008), for example, requires that coastal provinces and municipalities develop management programmes that consider potential climate change effects in all coastal planning and management. The *Disaster Management Act* (57 of 2002) requires disaster management plans to be developed at national, provincial and local levels. These must include expected climate change effects and risks and disaster risk reduction and climate change adaptation measures (*Disaster Management Act* ss. 14 and 20). Regulations have been published in terms of national air quality legislation and building regulations law to make provision for GHG reporting and energy-efficiency in support of climate mitigation. Despite South Africa's abundance of sunlight, the national government has not yet passed regulations or explicit national policies that commit to solar energy generation as a pillar in the country's transition to a less carbon-dependent electricity grid.

In addition, a comprehensive suite of national local government policy and legislation, mostly administered by the Department of Cooperative Governance and Traditional Affairs, governs municipal affairs. These policies and laws regulate local governance in minute detail and effectively serve to restrain local climate change response measures, especially those with budgetary implications.

12.4.2.2 Policy Development at Provincial and Municipal Level

Local governments, especially metropolitan municipalities, have been rolling out local climate change response policies, plans, strategies and projects. Two of these include the Ekurhuleni Climate Change and Energy Strategy (2007) with a focus on the reduction of the harmful effects of energy use (e.g., pollution and global warming) by promoting cleaner and renewable energy sources, and the Durban Climate Change Strategy (2014) complemented by an Implementation Framework and separate theme reports on biodiversity, food security, health, sustainable energy, transport, and so on. In what is perhaps the most outstanding example of local initiative, the City of Cape Town has a long history of local energy and climate planning and action. It was the first African city to complete a State of Energy Report (2001; updated in 2007, 2011 and 2015) and adopt the Energy and Climate Change Strategy (2006). The City also adopted an Energy and Climate Action Plan (2010) and a Climate Change Policy (2017), and has made associated institutional changes. In 2015, the City of Cape Town adopted the Cape Town Energy 2040 Goal (2015) with its associated energy and carbon-reduction targets. This project, which models a more resilient, resource-efficient and equitable future for Cape Town, commits the City to diversifying Cape Town's energy supply, becoming significantly more energy efficient and reducing carbon emissions. Most recently, the City of Cape Town adopted its Climate Change Strategy (2021) with a vision for the City to become climate resilient, resource efficient and carbon neutral (CoCT 2021, 15). At the time of writing, the City of Cape Town as well as eThekwini, Johannesburg, Tshwane and Ekurhuleni metropolitan municipalities are rolling out plans and timelines to procure electricity from independent power producers (IPPs) and to integrate IPP-generated electricity into the cities' supply networks. IPPs typically rely on wind, solar, biomass and small hydro projects. In 2020, the *Electricity Regulations on New Generation Capacity* of the *Electricity Regulation Act* (4 of 2006) was amended to enable municipalities in good financial standing to procure new generation capacity in accordance with the national IRP.

12.4.2.3 Does the Division of Powers Support Subnational Climate Action?

Given the above patchwork of national, provincial and municipal commitments, policies and programmes, the next question is: what leverage do the subnational governments, that is, provinces and municipalities, have to follow through on these? Where are the opportunities and where are the bottlenecks? Earlier, it was argued that South Africa has tremendous opportunities for climate action in 'electricity supply, urban passenger transport and residential buildings'. In the same vein, the main points of leverage for subnational governments were later identified as planning, the environment and the management of urban spaces. A closer interrogation of the division of powers in these critical sectors reveals a muted provincial role and a municipal role that is, potentially at least, much stronger.

Spatial planning and land-use management powers are strewn across the three spheres of government by the Constitution, resulting in much uncertainty about the roles of each sphere (Berrisford 2011). After a series of Constitutional Court judgments and a new national law, the *Spatial Planning and Land Use Management Act* (SPLUMA) of 2013, the role definitions settled. Municipalities conduct land-use management such as zoning, and decide on land-use applications (De Visser 2016). This is a source of considerable tension, particularly in the context of climate change mitigation (De Visser 2015). Municipalities are regularly accused of allowing urban sprawl. Furthermore, the municipal revenue model, reliant as it is on property rates and service fees, has the propensity to direct municipal attention away from mitigating climate change to maximising revenue from urban development (Steytler 2009, 444) and the provision of services such as (coal-based) electricity. Provincial governments regulate and oversee municipal land-use management and adopt provincial spatial policies, called provincial spatial development frameworks (PSDFs). Given their regional focus, these are potentially critical for adapting to, and mitigating climate change. However, PSDFs are not binding on municipalities (SPLUMA s. 22(3)). The national government adopts a national spatial development framework (NSDF) that directs development nationally, a power that can be used for climate action (SPLUMA ss. 13–14). Again, the NSDF is not binding on municipalities.

In real terms, therefore, municipalities have considerable autonomy to decide on land use and critically influence climate change action in that respect. This has made it difficult for national and provincial governments to use the land-use system to impose constraints on any urban eagerness to develop. However, there are many other constitutional levers that national and provincial governments can use.

One such a lever is the fact that the national and provincial governments share authority with respect to the 'environment' (Schedule 4, Part A). The national government's suite of environmental management legislation on air quality, waste management and environmental development control may be applied, with agreement, by provincial and local enforcement officers (see *National Environmental Management Act* s. 31C). Provinces have passed little or no environmental legislation, but play a critical role in implementing national legislation. National environmental management legislation, for example, empowers provincial governments to demand and approve environmental impact assessments (EIAs), required for developments that trigger an EIA. So, while a development could be approved by a local government, a provincial EIA could be required. A landmark court ruling on the controversial approval of a coal-fired power station determined that the climate change effects of such developments must be considered in the environmental assessment (*Earthlife Africa Johannesburg v Minister of Environmental Affairs and Others* [2017] 2 All SA 519; see also Du Plessis 2018, 11–16; Kotzé and Du Plessis 2020, 634–43). Aside from the important victory for climate change action, this judgment may in future influence the above-mentioned intergovernmental dynamics between provincial and municipal governments on climate change.

Public transport is another sector with tremendous potential for climate action. Urban dwellers, if they have the means, overwhelmingly rely on private car-based transportation (IUDF 2016, 52). For the majority of South Africans, there are minibus-taxis, buses and some trains, all of which are generally inefficient, unsafe and/or expensive. Combined with the urban sprawl in cities (a stubborn result of apartheid spatial design and poor planning), the result is that low-income households spend more than 20 per cent of their monthly income per capita on public transport (StatsSA 2015, 55). Furthermore, the Integrated Urban Development Framework (IUDF) of 2016 reported that, out of forty countries surveyed, South Africans spend the longest time in daily commutes to and from work (IUDF 2016, 52).

The Constitution distributes public transport functions across the three spheres of government. In practice, the public transport sector is very fragmented and riddled with historical inefficiencies (IUDF 2016, 53). Commuter rail (even within urban boundaries) is managed by a national utility, interprovincial transport is regulated nationally, and most road-based public transport (such as taxis and buses) is regulated and licensed provincially. Subsidy arrangements are uneven and disjointed across spheres of government, resulting in 'separate systems for different public transport modes hav[ing] become embedded' (IUDF 2016, 54).

Cities have long clamoured for greater authority to manage and integrate the various urban-based public transport systems, critical to streamline services, enable people to choose public transport instead of private cars and thus ultimately reduce emissions. At a policy level, the national government agrees. In the IUDF, it committed supporting the devolution/assignment of functions for various public transport modes to local government, based on the premise that local government is the sphere most able to manage and integrate public transport with other infrastructure and services (IUDF 2016, 55). The *National Land Transport Act* (5 of 2009) indeed envisages greater decentralisation of public transport functions to local government and even contains something akin to a subsidiarity principle (*National Land Transport Act* s. 11(3) Palmer, Moodley and Parnell 2017, 219). However, this decentralisation has not gained much momentum. The ANC-led national government is reluctant to cede power and budget to the cities, where voters are increasingly interested in voting for opposition parties (see, Siddle and Koelble 2012, 197–204).

12.4.2.4 Institutional Actors, Support and Monitoring, and Evaluation

The national policy intention is for cross-cutting climate change responses to be included in national, provincial, and local planning, law, and policy regimes such as the national Industrial Policy Action Plan, the national Integrated Resource Plan for Electricity Generation, the Provincial Growth and Development Plans and municipal Integrated Development Plans (IDPs) (NCCRP [DEFF} 2011, 14–15). The Department of Performance Monitoring and Evaluation in the Presidency has been tasked with ensuring that the implementation of climate change adaptation measures is properly integrated across government levels. Reporting on climate responsibilities and adaptation measures must be integrated into the Programme of Action and the ministerial delivery agreements as well as the quarterly reporting requirements of government at all levels (NCCRP [DEFF] 2011, 47–8).

Specific national government bodies responsible for climate change response action include the International and Domestic Reporting Unit and the Climate Change and Air Quality Branch, both in the DEFF. A total of thirty-two organs of state and universities are involved in the preparation of the country's National Communications to the United Nations Framework Convention on Climate Change, for example.

In the provincial sphere it is mostly provincial departments responsible for environmental affairs that take the initiative regarding climate change adaptation and mitigation but few provinces have dedicated provincial structures for this. Several municipalities have reshuffled their structures in recent years to make provision for directorates and departments dedicated to climate change governance. The eThekwini Municipality, for example, has an Environmental Planning and Climate Protection Department with a staff complement of over thirty.

Capacity building, the development of technical skills and knowledge generation on climate change response action happen across the three spheres of government in a fragmented and ad hoc fashion. In one example, the National Treasury has a Cities Support Programme on the Provision of Technical and Strategic Recommendations to Mainstream Climate Responsiveness into City Plans, Budgets and Grant Conditions. The programme aims to develop tools for the mainstreaming of climate responsiveness into city planning, budgeting and projects, as a low-cost approach to efficient city management and fiscal sustainability (National Treasury 2018, i).

12.4.3 Intergovernmental Coordination and Planning

Given the patchwork of policies and law across the three spheres, the inefficiencies in the distribution of powers and the multiple institutional actors involved, what is then the intergovernmental glue that brings these seemingly disparate strands together? The importance of coordination for the government's climate action project is well-recorded (Ziervogel and Parnell 2014, 59).

12.4.3.1 Intergovernmental Relations

South Africa's multilevel system is held together by an array of intergovernmental mechanisms for dialogue, coordination and information sharing. Much of the country's intergovernmental mechanisms are expected to support the coordination of climate change action across the three spheres of government. The Constitution itself contains 'principles of cooperative governance' that call on organs of government to share information, consult, refrain from litigation, and so on. The *Intergovernmental Relations Framework Act* (13 of 2015) provides a general framework for intergovernmental relations between national, provincial and municipal levels, including intergovernmental forums.

12.4.3.2 Intergovernmental Forums

The President's Coordinating Council brings together the presidency, key national ministers, provincial premiers and a representative of the South African Local Government Association (SALGA) (accredited as local government's voice in intergovernmental relations). National ministers, in turn, regularly meet with their provincial counterparts. In practice, these standing intergovernmental bodies are important vehicles for national–provincial coordination. Most of them include SALGA representatives. In the context of climate change, the MINMEC: Environment is particularly noteworthy. It meets quarterly and brings together the national minister of environmental affairs, all nine members of the provincial Executive Councils (MECs) responsible for environmental management functions and a representative of SALGA.

The Forum of South African Directors is the technical equivalent of the President's Coordinating Council, comprising the most senior government officials of the national and provincial governments as well as SALGA senior management. Much of this architecture is repeated at the provincial level, where provincial executives meet with municipal mayors. There are also several coordinating bodies that focus specifically on climate change, such as the Inter-Ministerial Committee on Climate Change (IMCCC) and the Intergovernmental Committee on Climate Change (IGCCC).

There is certainly no shortage of forums and structures for coordination. However, it is not always clear whether these forums go beyond information sharing, and into more programmatic alignment and coordination. The intergovernmental forums neatly follow the 'hierarchy' of national, provincial and local government. In addition, intergovernmental forums tend to be dominated by the most senior sphere of government involved, thereby hollowing out their integrative potential. Steytler (2011, 420) argues that instead of focusing on common issues, the intergovernmental relations forums are mostly used by the national government to monitor the performance of provinces.

The greatest flaw in this architecture is that cities are not directly connected to the national government. Given the pivotal role of cities, this rather diminishes South Africa's ability to coordinate climate change action. Other than the President's Coordinating Council (which meets sporadically and is far removed from the sites of pragmatic policy coordination), the *Intergovernmental Relations Framework Act* does not provide for an intergovernmental forum that connects cities directly to the national government. As stated earlier, MINMECs generally include local government but only through SALGA. The assumption is thus that cities connect with the national government through one representative of organised local government at the President's Coordinating Council or a MINMEC, or via their provincial governments. This does not satisfy the need for city–national intergovernmental relations, particularly in the context of climate change. In practice, municipal–national intergovernmental relations thus take place mostly outside the generic intergovernmental relations framework. The National Treasury has, for example, established dedicated structures and programmes to engage with subnational authorities, such as the Cities Support Programme referred to earlier.

12.4.3.3 Integrated Development Planning

Local government legislation provides for an intricate framework of integrated development planning as a strategic planning and budgeting framework that brings together the plans and programmes of all three spheres of government. Each municipality adopts a five-year strategic plan, which is expected to function as the 'landing strip' for all government planning in the municipality's jurisdiction. The key entry point for this 'all of government' approach are the eight metropolitan and forty-four district municipalities. The spatial development frameworks (SDFs) mentioned earlier are key components in this framework. As explained, they must be developed at municipal, provincial and national level. Potentially at least, they are important instruments for climate change adaptation and mitigation. First, these SDFs are expected to 'identify long-term risks of spatial patterns and contain measures to mitigate those risks' (SPLUMA s. 12(1)(j)). Second, while they are not binding, they must inform the exercise of discretion in relation to land use and development. This is most pronounced at municipal level. Proposed developments that contradict a municipal SDF must be specifically motivated (SPLUMA s. 22 (1)). Arguably, therefore, if climate change mitigation and adaptation principles are articulated in these SDFs at all three levels, this ought to influence the direction of development.

12.5 Opportunities for Greater Climate Change Resilience

12.5.1 Provincial Government

The extent to which provinces and municipalities in South African engage in climate change is informed by the degree of devolved autonomy. As alluded to earlier, provincial autonomy is limited both because of the constitutional design – including the absence of fiscal autonomy for provinces – and a highly centralised political practice.

Provincial governments have the authority to make significant contributions, but mainly in the regulatory sphere. They have little financial leverage: provinces are almost completely grant dependent; do not provide much grant funding to local governments; and most of their spending power is in social services, namely education, health care and housing. Of those three, housing is the most closely associated with climate change action. However, the provincial role is primarily focused on the project management of subsidy housing projects. Here, the imperative is to produce as many housing opportunities as possible, given the large housing backlogs. Housing subsidy rules are determined nationally and, as discussed earlier, the location of new housing projects is not determined by provincial governments. So, there is relatively little scope for provinces to invest in climate change mitigation or adaptation programmes or influence public funding streams towards these goals.

Provincial governments also constrain themselves by not making full use of the policy and legislative space afforded to them by the Constitution. Provinces adopt very few provincial laws and do not compete with the national government over concurrent powers (De Visser 2017, 229), including those most closely related to climate change. This is borne out by the absence of any Constitutional Court jurisprudence on the application of the constitutional override provision with respect to concurrent powers. As a result, all critical sectors are regulated nationally. Furthermore, even when exclusive provincial powers offer some leverage for climate action, there is little uptake. Provinces are reluctant to do so for fear of trampling on municipal autonomy, which is vigorously asserted by cities.

12.5.2 Local Government

Municipalities, metropolitan municipalities in particular, have greater leverage for two reasons. The first relates to the constitutional design and the functions allocated to local government. While the 'big ticket' provincial functions are social services, the municipal functions relate closely to the built environment, where a significant part of climate change mitigation and adaptation is located. Municipalities conduct town planning and thus determine the direction and location of new developments. They approve building plans, determine urban design and have the authority to develop their own (green) building codes. They deliver electricity, water and sanitation services and control municipal roads and traffic. They also see to waste management and can implement waste-to-energy projects. Second, municipalities (in particular those governing cities) have greater budget autonomy and command revenue sources that can be used to mitigate or adapt to climate change. Electricity and water sales, as well as property rates and development charges are critical revenue sources that reduce a city's grant dependency. Cities can also use their taxation and service tariff policies to influence behaviour towards climate action (De Visser 2012). Lastly, unlike provinces, cities actively borrow on national and international capital markets and some cities issue city bonds (Khumalo et al. 2016, 210).

However, cities are also constrained by a lack of autonomy. Climate action at the city level is subject to a dense national legal framework, designed to reign in errant local governments, that works mostly to constrain much-needed innovation. Long-term public–private partnerships, procuring renewable energy and other essential tools for urban climate action are so tightly regulated that many municipalities shy away from them (De Visser 2012). Furthermore, as already alluded to, cities lack critical aspects of the broad built environment function that enable them to contribute to climate change action: while they control town planning, they lack direct control of housing subsidies (Palmer, Moodley and Parnell 2017, 251) and urban public transport. Similarly, whereas they control (most of the) electricity reticulation function, they are prohibited from generating their own electricity without national permission. Without national permission, they are even prohibited from purchasing electricity from any other supplier than the national power utility, Eskom, which delivers predominantly brown energy.

On all three issues, namely public transport, housing and electricity generation, there is no shortage of policy, rhetoric and promises to increase city powers. However, progress has been slow, mainly due to a lack of political will at the centre to devolve powers. The difficulties with respect to public transport were outlined earlier (see Section 12.4.2.4). With respect to housing, greater decentralisation of housing funds to cities commenced enthusiastically around 2010. However, as more cities turned into battlegrounds for opposition politics, the national government got cold feet and this has all but stalled.

Electricity generation has been in an outright crisis for over a decade with regular episodes of 'load shedding' (scheduled blackouts) crippling the economy, and seemingly unbridled air and water pollution by South Africa's power utility, Eskom. Eskom holds a firm monopoly on power generation and the management of the power grid, protected by national legislation and the assurance that electricity generation lies entirely within national jurisdiction (see Section 12.2.2). It is, however, not able to meet the country's electricity demands. This is due to persistent political interference by national politicians, large scale corruption, poor planning and the reliance on a dated model of highly centralised electricity generation, predominantly from South Africa's abundant coal reserves. This situation stresses how timely it is for municipalities in good financial standing to be able to turn to IPPs in terms of the 2020 *Electricity Regulations on New Generation Capacity* referred to earlier.

South Africa must thus urgently make a just transition to green energy. The reality is that cities are chomping at the bit to contribute. However, they are held back by the restraints on their autonomy over electricity generation and a national government that is reluctant to loosen the reins. However, in a major policy shift in 2020, the national government amended the Electricity Regulations, enabling municipalities in good financial standing to develop their own power generation projects. This signals a careful but important first move to transform the energy sector.

Once the *Climate Change Bill* is enacted, provincial and local authorities will be expected to undertake 'climate change needs and response assessments' to be reviewed every five years, as well as 'climate change response implementation plans', informed by the mentioned assessments (*Climate Change Bill* (B9-2022) ch. 3). These assessments will have to be aligned with national sectoral emission targets and should address adaptation considerations and options as well as risks and vulnerabilities, such as the impact of climate change on ecosystems and households (*Climate Change Bill* (B9-2022), s. 15).

Prior to the national government's adoption of the NCCRP, the *Climate Change Bill* and the *Carbon Tax Act*, it was mostly individual municipalities (especially metropolitan municipalities) that have shown an interest and leadership in climate governance. Notably, the National Climate Change Response Database by 2016 listed 125 climate change adaptation projects implemented since 2011. Government implemented around half of these projects with metropolitan municipalities at the forefront, said to have been responsible for approximately thirty of the government-implemented projects (TNC 2018, 177; see also Ziervogel et al. 2014, 610–612), followed by the NGO sector (TNC 2018, 178).

By 2016, all eight metropolitan municipalities in South Africa had a climate change plan or strategy, either completed or in process, as well as plans and practices that integrated measures for adaptation (DEA 2018, 177). A few smaller municipalities have shown remarkable initiative and have been quite resourceful as far as their local climate actions are concerned. In 2018, the National Treasury published a detailed report on climate mainstreaming in South African cities. The project probed whether climate change responsiveness is reflected in the language of the city planning instruments. The findings indicated that most metropolitan municipalities highlight climate responsiveness in their IDPs, but the integration remains largely at a high level with very few specifics (National Treasury 2018, 42). Many of these municipalities have started to reflect climate responsiveness in their SDFs but the coverage is 'patchy and inconsistent'; when included, it is again at a relatively high level (National Treasury 2018, 42). The mainstreaming of climate change responsiveness in municipal operations did not form part of the study and the valid concern has been raised that the effectiveness of local initiatives lies in the ability of municipalities to translate climate policy and planning into action in the face of institutional complexity and human capacity constraints (Ziervogel et al. 2014, 612).

12.6 Conclusion

South Africa is said to have the most advanced research, observation and climate modelling programme on the African continent (Ziervogel et al. 2014. 606). Yet, despite the constitutional emphasis on cooperative government and intergovernmental relations and the unequivocal local effects of climate change, South Africa's quasi-federal government has not yet embarked on a consolidated *process* of deliberation, cooperation and intergovernmental learning and planning concerning climate governance. The reasons for this range from climate change not yet being high enough on the political agenda, to the diversity of climate change impacts in the country, institutional complexity, a lack of resources and political problems.

South Africa has a national policy compass for subnational action, in the *National Development Plan* and the *National Climate Change Response White Paper*. Yet, many of the details required for climate change response planning, budgeting, implementation and reporting are spread across line functions situated in the national and local government spheres. While some cities have eagerly taken up possibilities for being involved in local climate change governance by way of local plans, policies and programmes, most municipalities are struggling to make ends meet and to see to the delivery of the most basic of municipal services. Climate change thus serves to put the spotlight on the country's hourglass model of federalism: a strong national government (particularly active in the climate change policy, planning and legislative arenas); relatively weak provinces (mostly publishing provincial climate change policies and plans); and a continuum of very strong metropolitan cities, weakening secondary cities and very weak rural municipalities in the local government sphere. Despite significant constraints on their autonomy, the strong municipalities tend to be involved in the initiatives of global city networks and transnational climate change partnerships, and are changing structures, adopting strategic climate change plans, amending spatial decision-making orientations and development objectives, as well as making bold commitments to reduce their emissions.

The muted role for provinces in climate change is perhaps the logical result of their orientation towards social services, the absence of fiscal autonomy and the backdrop of South Africa's reluctant entry into the family of federations. However, the constraints on cities to leverage their control over the built environment and make very meaningful contributions to climate action are not in keeping with the constitutional and policy commitment to devolution. Instead, they seem informed by the political economy of decentralisation in South Africa, that is, the reluctance on the part of the national government to accept strong cities. This said, the government architecture of South Africa is such that it holds huge potential benefit for a coherent yet diversified and context specific policy, law and programmatic response to climate change. Much of the possibilities sit in the multifarious existing financial, strategic, environmental and disaster-risk related planning and management instruments of all three spheres of government, and in the nuts and bolts of a law and policy framework that is all for a government working closely together to the benefit of the health and well-being of the present and future generations.

References

Berrisford, Stephen. 2011. 'Unravelling Apartheid Spatial Planning Legislation in South Africa: A case study'. *Urban Forum* 22 (3): 247–63.

Climate Action Tracker. 2018. 'South Africa'. *Climate Action Tracker*. https://climateactiontracker.org/countries/south-africa/2018-11-30.

CoCT (City of Cape Town). 2006. *Energy and Climate Change Strategy*. CoCT, Environmental Planning Department. https://resource.capetown.gov.za/documentcentre/Documents/City%20strategies,%20plans%20and%20frameworks/Energy_+_Climate_Change_Strategy_2_-_10_2007_301020079335_465.pdf.

CoCT (City of Cape Town). 2010. *Climate Change Action Plan*. CoCT. https://resource.capetown.gov.za/documentcentre/Documents/City%20strategies%2C%20plans%20and%20frameworks/CCT_Climate_Change_Action_Plan.pdf.

CoCT (City of Cape Town). 2015. *Cape Town Energy 2040*. CoCT. https://savingelectricity.org.za/wp-content/uploads/2018/01/2040_energy_vision_cct_brochure.pdf.

CoCT (City of Cape Town). 2015. State of Energy Report. https://resource.capetown.gov.za/documentcentre/Documents/City%20research%20reports%20and%20review/State_of_Energy_Report_2015_2015–09.pdf.

CoCT (City of Cape Town). 2017. *Climate Change Policy*. CoCT. https://savingelectricity.org.za/wp-content/uploads/2018/01/climate_change_policy.pdf.

CoCT (City of Cape Town). City of Cape Town. 2021. Climate Change Strategy. https://resource.capetown.gov.za/documentcentre/Documents/City%20strategies,%20plans%20and%20frameworks/Climate_Change_Strategy.pdf.

COGTA (Department of Cooperative Governance and Traditional Affairs). 1998. *White Paper on Local Government*. COGTA. www.cogta.gov.za/cgta_2016/wp-content/uploads/2016/06/whitepaper-on-loca-gov.pdf.

COGTA (Department of Cooperative Governance and Traditional Affairs) and IUDF (Integrated Urban Development Framework). 2016. *Integrated Urban Development Framework: A new deal for South African cities and towns*. IUDF. https://iudf.co.za/pdf_downloads/2016-integrated-urban-development-framework.

Copenhagen Accord. 18 December 2009. FCCC/CP/2009/L.7.

De Visser, Jaap. 2012. 'Cities and Climate Change: *ex abundanti cautela* – From an excess of caution?' In *Climate Change at the City Scale: Impacts, Mitigation and Adaptation in Cape Town,* eds. Anton Cartwright, Susan Parnell, Gregg Oelofse and Sarah Ward, 122–46. London: Routledge.

De Visser, Jaap. 2015. 'Addressing Climate Change through Provincial Planning in South Africa: A case study of the Western Cape'. Presented at Strenghtening Green Federalism – Sharing Best Practices 2015.

De Visser, Jaap. 2016. 'Devolution by Court Injunction: The case of land use planning and management in South Africa'. *Ugandan Journal of Management and Public Policy Studies* 10 (1): 84–100.

De Visser, Jaap. 2017. 'Concurrency in South Africa'. In *Concurrent Powers in Federal Systems: Meaning, making and managing*, ed. Nico Steytler, 222–40. Leiden: Brill.

Department of Economic Development and Environmental Affairs. 2011. *Eastern Cape Climate Change Response Strategy*. Urban Energy Support. www.cityenergy.org.za/uploads/resource_182.pdf.

Department of Economic Development, Environment and Tourism. 2013. *Limpopo Green Economy Plan Including Provincial Climate Change Response*. DEFF. www.environment.gov.za/sites/default/files/docs/limpopogreen_economyplan.pdf.

Department of Agriculture and Environmental Affairs. 2013. *Report to the NCCRP Workshop: Progress report on climate change activities of the KZN*. DEFF. www.environment.gov.za/sites/default/files/docs/kzn_nccrp_progress_report.pdf.

DEA (Department of Environmental Affairs). 2000. *Initial Communication under the United Nations Framework Convention on Climate Change*. UNFCCC. https://unfccc.int/resource/docs/natc/zafnc01.pdf.

DEA (Department of Environmental Affairs). 2011. *South Africa's Second National Communication under the United Nations Framework Convention on Climate Change*. UNFCCC. https://unfccc.int/resource/docs/natc/zafnc02.pdf.

DEA (Department of Environmental Affairs). 2015. *The National Climate Change Response Monitoring and Evaluation System Framework*. DEA. www.environment.gov.za/sites/default/files/reports/nationalclimatechangeresponse_MESF.pdf.

DEA (Department of Environmental Affairs). 2018. *South Africa's Third National Communication under the United Nations Framework Convention on Climate Change*. UNFCCC.
https://unfccc.int/sites/default/files/resource/South%20African%20TNC%20Report%20%20to%20the%20UNFCCC_31%20Aug.pdf.

Department of Environmental Affairs and Development Planning. 2014. *Western Cape Climate Change Response Strategy*. Western Cape Government. www.westerncape.gov.za/text/2015/march/western_cape_climate_change_response_strategy_2014.pdf.

DEFF (Department of Environment, Forestry and Fisheries). 2011. *National Climate Change Response White Paper*. DEFF. www.environment.gov.za/sites/default/files/legislations/national_climatechange_response_whitepaper.pdf.

DEFF (Department of Environment, Forestry and Fisheries). 2019. *National Climate Change Adaptation Strategy*. DEFF. www.environment.gov.za/sites/default/files/docs/nationalclimatechange_adaptationstrategy_ue10november2019.pdf.

Department of Mineral Resources and Energy. 2019. *Integrated Resource Plan 2019*. Department of Mineral Resources and Energy. www.energy.gov.za/IRP/2019/IRP-2019.pdf.

Department of Transport. 2018. *Green Transport Strategy for South Africa: (2018–2050). Department of Transport.* www.transport.gov.za/documents/11623/89294/Green_Transport_Strategy_2018_2050_onlineversion.pdf/71e19f1d-259e-4c55-9b27-30db418f105a.

Du Plessis, Anél. 2018. 'The Promise of Well-being in Section 24 in the Constitution of South Africa'. *South African Journal on Human Rights* 34 (2): 1–18.

Du Plessis, Anél, and Louis J. Kotzé. 2014. 'The Heat is On: Local government and climate governance in South Africa'. *Journal of African Law* 58 (1): 145–74.

Ekurhuleni Metropolitan Municipality. 2007. *Energy and Climate Change Strategy*. Ekurhuleni Metropolitan Municipality. www.ekurhuleni.gov.za/council/reports/environmental-reports/468-energy-and-climate-change-strategy-2007/file.html.

Environmental Planning and Climate Protection Department. 2014. *Durban Climate Change Strategy*. eThekwini Municipality. www.durban.gov.za/City_Services/energyoffice/Documents/DCCS_Final.pdf.

Environmental Planning and Climate Protection Department. 2019. *Durban Climate Change Strategy*. eThekwini Municipality. www.durban.gov.za/City_Services/development_planning_management/environmental_planning_climate_protection/CAP/Documents/Durban%20CAP.pdf.

Gulati, Manisha, Louise Scholtz, and Saliem Fakir. 'Climate Change and the Low Carbon Transition'. In *Greening the South African Economy: Scoping the issues, challenges and Opportunities*, eds. Josephine K. Musango, Mark Swilling, and Jeremy Wakeford, 36–49. Claremont: UCT Press.

Khumalo, Bongani, Ghalieb Dawood, and Jugal Mahabir. 2016. 'South Africa's Intergovernmental Fiscal Relations System'. In *Kenyan-South African Dialogue on Devolution*, eds. Nico Steytler and Yash Pal Ghai, 201–27. Cape Town: Juta.

Kotzé, Louis J., and Anél du Plessis. 2020. 'Putting Africa on the Stand: A bird's eye view of climate change litigation on the continent'. *Environmental Law* 50 (3): 615–63.

Leon, Tony. 2013. 'Ten Thoughts on Provinces'. *Politicsweb*. www.politicsweb.co.za/archive/ten-thoughts-on-the-provinces–tony-leon.

McSweeney, Robert, and Jocelyn Timperley. 2018. 'The Carbon Brief Profile: South Africa'. *The Carbon Brief*. www.carbonbrief.org/the-carbon-brief-profile-south-africa.

Murray, Christina, and Richard Simeon. 2001. 'Multi-sphere Governance in South Africa: An interim assessment'. *Publius* 13 (4): 65–92.

National Treasury. 2018. *Mainstreaming Climate Responsiveness into City Plans, Budgets and Grant Conditions*. National Treasury. https://csp.treasury.gov.za/csp/DocumentsProjects/A%20Review%20of%20Infrastructure%20Grant%20Conditions,%20Outputs%20and%20Outcomes.pdf.

NDC. Republic of South Africa Nationally Determined Contribution. 2016. www4.unfccc.int/sites/ndcstaging/PublishedDocuments/South%20Africa%20First/South%20Africa.pdf.

NDC. Republic of South Africa Nationally Determined Contribution. 2020/2021. www4.unfccc.int/sites/ndcstaging/PublishedDocuments/South%20Africa%20First/South%20Africa%20updated%20first%20NDC%20September%202021.pdf.

NPC (National Planning Commission). 2012. *National Development Plan 2030: Our future – make it work*. South African Government. www.gov.za/sites/default/files/gcis_document/201409/ndp-2030-our-future-make-it-workr.pdf.

Palmer, Ian, Nishendra Moodly, and Susan Parnell. 2017. *Building a Capable State: Service delivery in post-apartheid South Africa*. London: Zedbooks.

Parnell, Susan, and Sophie Oldfield, eds. 2014. *The Routledge Handbook on Cities of the Global South*. New York: Routledge.

Parnell, Susan, and Walawege Ruwani. 2014. 'Sub-Saharan African Urbanisation and Global Environmental Change'. In *Africa's Urban Revolution*, eds. Susan Parnell and Edgar Pieterse, 35–59. London: Zedbooks.

StatsSA. Statistics South Africa. 2015. www.statssa.gov.za/publications/P0318/P03182015.pdf.

Steytler, Nico. 2009. 'The Decisions in *Wary Holdings (Pty) Ltd v Stalwo (Pty) Ltd and Another 2009* (1) SA 337 (CC): Be aary of these holdings'. *Constitutional Court Review* 2: 429–31.

Steytler, Nico. 2011. 'Cooperative and Coercive Models of Intergovernmental Relations: A South African case study'. In *The Federal Idea: Essays in honour of Ronald L. Watts*, eds. Thomas J. Courchene, John R. Allan, Christian Leuprecht, and Nadia Verrelli, 413–28. Montreal: McGill-Queen's University Press.

Steytler, Nico. 2017. 'The Constitutional Court of South Africa: Reinforcing an hourglass system of multi-level government'. In *Courts in Federal Countries: Federalists or unitarists*, eds. Nicholas Aroney and John Kincaid, 328–66. Toronto: University of Toronto Press.

Steytler, Nico, and Johann Mettler. 2001. 'Federal Arrangements as a Peace-making Device during South Africa's Transition to Democracy'. *Publius* 31 (4): 93–106.

Siddle, Andrew, and Koelble, Thomas. A, 2012. *The Failure of Decentralisation in South African Local Government: Complexity and unanticipated consequences*. Claremont: UCT Press.

Swilling, Mark, and Eve Annecke. 2012. *Just Transitions: Explorations of sustainability in an unfair world.* Claremont: UCT Press.

Taviv, Rina. 2012. *Implementation of Gauteng Climate Change Response Strategy (GCCRS).* DEFF. www.environment.gov.za/sites/default/files/docs/gauteng_climate_change_strategy.pdf.

Ziervogel, Gina, Mark New, and Emma Archer van Garderen et al. 2014. 'Climate Change Impacts and Adaptation in South Africa'. *Wiley Interdisciplinary Reviews: Climate Change* 5 (5): 605–50.

Ziervogel, Gina, and Susan Parnell. 2014. 'Tackling Barriers to Climate Change Adaptation in South African Coastal Cities'. In *Adapting to Climate Change: Lessons from natural hazards planning,* eds. Bruce C. Glavovic and Gavin P. Smith: 57–73. Dordrecht: Springer.

Ziervogel, Gina, Emma Archer van Garderen, and Penny Price. 2016. 'Strengthening the Knowledge–Policy Interface through Co-production of a Climate Adaptation Plan: Leveraging opportunities in Bergrivier Municipality, South Africa'. *Environment and Urbanisation* 28 (2): 455–74.

13

Climate Governance and Federalism in Spain

SUSANA GALERA RODRIGO, MERCEDES ALDA-FERNANDEZ,
AND MARIO KÖLLING

13.1 Introduction

Spain's geographical location strongly exposes it to global environmental challenges. The minimum temperature has increased over the last century by around 1.7°C and annual rainfall has been declining, while also becoming increasingly torrential (Greenpeace 2018). The most important economic sectors are considered potentially problematic for the environment: industry, tourism, transport, energy, and agriculture. The emergence of environmental policy as an independent policy area has been closely linked to an 'environmental consciousness' within public opinion that has been increasingly widespread since the 1980s due to United Nations summits, the rise of environmental NGOs, and serious environmental problems around the world.

The legal framework for climate policies derives from international treaties and European legislation as well as from the Spanish Constitution and the Statues of the Autonomous Communities (ACs). In this regard the role of the central government in Spanish climate change governance has been characterised as falling between supranational demands for EU convergence and domestic regional divergence (Pérez Gabaldó 2013). Although the decentralisation process in Spain has been quite successful, for years now, experts have been calling for a revision of the Constitution in order to adapt the text to the current reality of the territorial model and to establish a framework granting unity and diversity and an equilibrium between shared rule and self-rule. In fact, due to regulatory overlaps and coordination deficits, economic policy, environmental protection, and energy policy are long-standing sources of tension between the central and AC governments and have been challenges to effective climate action over the past decades. Moreover, the last decade in Spain has been particularly turbulent in many aspects. Since 2015, due to the fragmentation and polarisation of the party system, no party has been able to form a stable governmental majority after elections. The fragmentation of the party system has intensified since the 2019 general elections when twenty-two parties obtained representatives at the Congress of Deputies, the lower house of the Spanish Parliament (Ramos and Alda 2020, 32).

Despite this, in recent years more systematic climate change strategies have been adopted in Spain at all levels. The first National Plan for Adaptation (NPACC) was adopted in 2006 closely aligned with the EU regulatory framework. The ACs have also adapted strategies related to climate change within their jurisdiction on environmental issues and climate-related sectors. Finally, several plans and programmes have been implemented at the local level.

This chapter provides a brief overview of challenges and commitments in the field of climate change in Spain, analysing climate governance in the context of Spain's federal system. It aims to show that, despite the numerous climate strategies and policies adapted at all territorial levels over the last decade, there are no formal vertical channels of influence for AC or local actors on policy formulation at the federal level, which limit climate policy diffusion at the vertical dimension. Moreover, the coordination mechanisms for the implementation of climate policies are not very effective and do not consider the early phases of public action nor the participation of all involved authorities. However, the ACs and local governments have ample room for policy experimentalism in adapting their climate change strategies to the peculiarities of their different territories. In this regard the Spanish State of Autonomies has provided a favourable context for dynamic processes of climate policy diffusion between ACs and local entities.

13.2 Climate Change in Spain

Spain is distinguished by its high level of biodiversity, terrestrial and maritime protected areas, low degree of population density, and high concentration in urban areas (69 per cent population in areas with more than 50,000 inhabitants) (European Commission n.d.). But Spain is also severely challenged by climate change risks.

13.2.1 Climate Change Challenges

By EU standards, Spain has an extraordinarily high level of biodiversity. The country is ranked as the largest ecological reserve in Europe, with around 8,000 plant species, 540 bird species, 95 mammal species and 80 fish species. Spain has the largest terrestrial protected area in the EU – 138,000 square kilometres, which represents 27 per cent of the total area of the country (European Commission 2021). It also has a remarkable climatic variety, ranging from humid Atlantic conditions, with annual rainfall of more than 2,000 mm, to large semi-arid areas, with severe hydrological stress, and cold alpine climates in some isolated areas (AEMET 2018).

Spain is also one of the countries in the EU most vulnerable to climate change. In 2020, the average temperature in the country was 1.7 ºC higher than the average between 1850 and 1900, and the rate of warming has accelerated in the last few decades (AEMET 2021). According to the National Institute of Meteorology the years 2017 and 2020 were the hottest since the 1960s and summers are almost five weeks longer than at the beginning of the 1980s (Gobierno de España 2020a). The sea level in Spain has risen 13.5 cm since the 1990s (UNESCO, UN-Water 2020). Seven of the ten water basins with the greatest water stress in Europe are located in Spain, and almost 90 per cent of the extension of glaciers in Spain has disappeared since 1920 (Greenpeace 2018). It is estimated that 74 per cent of the Spanish surface is at risk of desertification (at different levels).

In the near future, key climate-related hazards are predicted to increase their effects and consequences. For example, heat waves, droughts, and extreme winds multiply the risk of wildfires. The area affected by wildfires in Spain in 2019 was 94 per cent higher than the average of the last decade (Greenpeace 2020). Several studies indicate that rising sea temperatures will lead to new and more extreme weather phenomena. Coastal flooding will become more extensive when the sea level rises, and heavy precipitation will increase the risk of landslides.

Over the period 1980–2019, Spain was the EU member state with the fifth highest economic losses in absolute terms caused by climate-related events. According to the NPACC 2021–30, the decrease in water resources due to changing precipitation patterns and longer droughts will have important implications for agriculture and livestock farming, urban supply, hydroelectric production, and ecosystems. Moreover, the spread of invasive species, as a secondary effect of climate change, could also have potential implications for human health (Gobierno de España 2020a). Although there are important national challenges, and the ACs share certain social and institutional characteristics, the ACs are affected quite differently by climate change risks because of their different geographical situations, environmental characteristics, and economic structures. For example, between 2050 and 2100 the risk of flooding could triple in the Basque Country, but Andalusia will suffer more frequent, longer, and more intense heat waves. In 2050, Murcia's own water resources will be 40 per cent less. Most Mediterranean ACs will notice an increased torrential intensity (Gobierno de España 2020a).

13.2.2 Spain's GHG Emission Profile

During recent years there have also been successful efforts to reduce carbon emissions. The emission rates are in line with the EU average (0.1)[1] and below the

Table 13.1 *Spanish national evolution of aggregate emissions*[2]

	1990	2005	2010	2015	2017	2018
CO_2-eq (Kt)	289.383	443.440	358.859	338.245	340.298	334.255
% Variation vs 1990	100%	153.2%	124.0%	116.9%	117.6%	115.5%

Source: *Informe de Inventario Nacional de Gases de Efecto Invernadero* (2020).

average of the OECD countries (0.2). The Spanish greenhouse gas (GHG) emissions inventory estimates gross emissions of 313.5 million tons of CO_2 equivalent for the year 2019. This means a reduction in emissions compared to the previous year of 6.2% + 8.3% compared to 1990 and −29.3% compared to 2005 (Table 13.1).

In 2019, the main greenhouse gases emissions sectors were: transport (29%), industry (20.6%), electricity generation (13.5%), agriculture (12.5%), LULUCF[3] sector (12%), residential, commercial and institutional (RCI) (9%), waste (4.3%), off-road machinery (3.7%), refinery combustion (3.5%), and fluorinated gases (1.5%) (Gobierno de España, 2020b). Taking the year 1990 as a reference with 100%, the variations by sector have been in 2018: waste (138.9%), energy (118.9%), agriculture (107.0%), industrial processes and product use (93.7%).

13.2.3 Spanish International Commitment on Climate Change

The central government represents Spain at the international level as a subject of public international law and is entitled to conclude international treaties on all subject matters. Commitments derived from these international treaties become part of Spanish law and are binding for the ACs. The central government is committed to existing multilateral environmental protection regimes and has ratified all main international and European agreements, strategies, or programmes related to climate change. These include the United Nations Framework Convention on Climate Change; the Paris Agreement on Climate Change; the Convention on Biological Diversity; the United Nations Convention to Combat Desertification; the Sendai Framework for Disaster Risk Reduction (2015–30); the 2030 Agenda for Sustainable Development (UN GA 2015); and the Protocol to the Barcelona Convention on Integrated Coastal Zone Management of 1995.

The main EU Strategies represent a further framework of domestic climate policies – for example, the EU Strategy on adaptation to climate change (European Commission 2013); the European strategic long-term vision for a prosperous, modern, competitive, and climate neutral economy (European Commission 2018);

the European Green Deal (European Commission 2019); and the Governance of the Energy Union and Climate Action.[4]

13.2.4 Climate Objectives in Climate Law and Policies in Spain

Spain's policies regarding sustainability, protection of its exceptionally diverse natural habitats, or general environmental quality have been ineffective for decades. Since 2018 a number of new initiatives have been adopted. Complying with the obligations stated in the EU regulation on Governance of Energy and Climate action, in December 2020 Spain adopted the National Integrated Energy and Climate Plan (ENCP) 2021–30, which includes measures on both mitigation and adaptation. The long-term goal of the plan is to make Spain carbon neutral by 2050; achieve a 90 per cent reduction in GHG emissions from 1990 levels; and, also by 2050, base the electricity system exclusively on renewable sources of generation. Meanwhile, the National Plan for Adaptation to Climate Change (NPACC) for the period 2021–30 was approved in 2020 following a joint agreement with the ACs. The Plan establishes strategic objectives and defines a system of indicators for impacts and adaptation to climate change, as well as requiring the preparation of risk reports.

Following years of work, in 2021 the *Climate Change and Energy Transition Law* was passed (Law 7/2021, of 20 May). This establishes the following minimum national targets for the year 2030 (article 3.1):

- Reduction of greenhouse gases emission by at least 20% compared with 1990.
- Increase in renewable energy to at least 35% of final energy consumption.
- At least 70% of electricity produced from renewables.
- Improving energy efficiency by reducing primary energy consumption by at least 35% from the baseline in accordance with EU regulations.

13.3 Climate Governance and Federalism

The decentralisation process in Spain started at the end of the 1970s, after the Franco dictatorship, in parallel with the transition to democracy, economic development, and administrative modernisation. In 1986, Spain became a fully-fledged member of what was then the European Community. Even though the Constitution of 1978 eschewed the term 'federation', over the past decades the Spanish model of territorial administration, known as the State of Autonomies, has come to exhibit the basic structures and processes typical of federations and can be defined as a federation in practice, if not in name (Watts 2010). The seventeen ACs are the constituent units, in addition to which there are two autonomous cities, fifty provinces, and 8,124 municipalities. All levels have their own legal status; however, the provinces and local level have only administrative, and no legislative, autonomy.

13.3.1 The State of Autonomies and Climate Governance

The Constitution divides powers such that some are exclusive to the central government, while the ACs are able to assume in their Statutes of Autonomy all matters not allocated to the central government, as well as the legislative development of the (central government) framework legislation and the implementation of central government legislation (Tudela and Kölling 2020). In this regard power is shared between both layers of government. This enables the central government to define nationwide standards and the ACs to adjust, at least to some extent, those laws to their own preferences. Over the last forty years, ACs have adopted their own Statutes of Autonomy defining their institutions and powers and they have assumed responsibility for the provision of a wide range of public services of a regional or local nature. There is not a specific constitutional provision for 'climate change' issues, which are considered as 'environmental' matters. While environmental protection is a shared responsibility between the two orders of government, with the central government establishing the legislative framework in which the ACs can legislate according to their own preferences, 'climate change' issues also concern 'electricity, energy market and the general coordination of the economy' clauses, and these are exclusive powers of the central government.

Although article 149.1.1 SC reserves for the central government power in respect of 'international relations', the so-called international clause does not exclude the ACs acting at international or EU level if their own powers are affected. However, the Constitutional Court determined that the ACs' activities at international or EU level have to be within the framework of the central government's policies. According to general constitutional doctrine, either the central government or the ACs may assume responsibility for the implementation and transposition of EU Law. In this regard, responsibilities regarding the domestic implementation of international commitments, environmental issues, energy production, natural resources, public emergencies, or natural disasters are shared between the two orders of government.

Since 1978, the Spanish Constitution has been amended only twice, both times because of external pressure from the European Union. However, practically all seventeen Statutes of Autonomy have been modified and climate change clauses have been included in several statutes.[5] In particular, the so-called Wave of Reforms taking place in 2007[6] has been characterised as an example of emulation also regarding environmental matters, for example, the reformed Statues of Aragón and Castilla-Leon included powers over 'policies that contribute to mitigating climate change'.[7] Over the years, the Statutes have also included environmental powers by adding new categories such as 'ecosystems', 'biological corridors', or 'soil pollution'.

Given the fact that Spain has evolved since the end of the 1970s from a unitary state with a long-standing centralist tradition to a highly decentralised state without constitutional reforms, the Constitution approved in 1978 can be characterised by a certain openness and flexibility. But the 'openness and flexibility' as regards the division of powers and relations between the central government and the ACs and between the ACs themselves have given rise to, and continues to generate, conflict – especially in economic policy, environmental protection, energy, water, and emergency planning (Alberton 2020, 36). As a consequence of this, together with the fact that climate governance involves different powers belonging to both the central government and the ACs, the Constitutional Court assumed a specific role in determining powers and responsibilities in climate governance. Since 1980, environmental issues have been among the most conflictual issues for the Court, reaching a total of 200 appeals. Constant sources of dispute were protected areas, biodiversity, and forests. Other matters, such as water, land planning, and energy, were also highly controversial (Alberton 2021). In the many conflicts over jurisdiction brought before the Constitutional Court, the constitutional provisions on energy and economy coordination have been insistently applied contrary to the environmental clauses invoked by the ACs (Galera Rodrigo 2018). Moreover, implementing EU environmental law, the central government tends to invade AC powers (Nogueira 2012). Based on recent case law, some scholars have thus pointed to a re-centralization of environmental powers (Casado Casado 2018).

13.3.2 The Institutional Framework and Intergovernmental Relations

The Constitution does not establish an institutional framework that would reduce the conflicts of jurisdictions and facilitate continuing dialogue and cooperation between the levels of government. There is neither a permanent institutionalised representation of regional interests at the national level, nor a framework for intergovernmental relations (Tudela and Kölling 2020). Intergovernmental cooperation was only acknowledged in the Spanish legislative framework in 1992. Although the Senate is defined in the Constitution as a chamber of territorial representation (Art. 69), not only does it have only limited legislative power, of its 266 members, 208 are elected by popular vote, and only 58 members are appointed by the regional legislatures. As a consequence, it does not fulfil its ostensible function and does not work as a forum for the participation of the ACs in central government legislation and they have no right of veto over decisions that affect them (Aja 1999). Instead, the first chamber – the Congress of Deputies – has become the central forum for intergovernmental negotiations especially for Autonomous Communities with strong regional parties (Rodríguez López et al. 2018, 230). Since some of the nationalist parties (particularly from Catalonia and the Basque Country) had enough seats in the Congress of Deputies, they became important actors with considerable bargaining power in cases of minority government (Field 2016). However, these parties are concerned overwhelmingly with regional interests, and their preferences for national-wide climate governance tend to be limited.

Nevertheless, the framework for cooperation and coordination between central and regional governments has developed over the past thirty years and vertical cooperation has improved. Since the beginning of the 1980s, cooperation between the central government and the governments of the ACs has been progressively assumed by sectoral conferences (*Conferencias Sectoriales*). These are multilateral cooperation bodies for specific policy sectors (e.g., the environment) and bring together central government and the ACs (Perez Medina 2020). Each sectoral conference has established a specific framework for cooperation at the administrative level, albeit with a very weak organisational structure. Whether sectoral conferences will be convoked, and which topics will be discussed, is a decision taken either by central government or if one-third of ACs convene a meeting (Colino 2021). Horizontal relations between the ACs have traditionally been weak. However, between 2004 and 2020 there was a significant improvement in these relations, at least in terms of formalised mechanisms (Ramos and Alda 2021).

There is no specific Climate Change Sectoral Conference, and those most closely involved are the Environment Sectoral Conference and the Energy Sectoral Conference and, until certain extent, the ones relating to Local Issues and Infrastructure and Land Management. In 2001, an Advisory Council on Environmental Policy for EU Affairs was created within the Environment Conference and was given responsibility for matters concerning environmental issues in EU affairs. However, the number of agreements reached between the parties as a result of negotiations on draft legislation has been very limited compared to other Sectoral Conferences (Alberton 2020). The agreements reached have mainly concerned the distribution of federal subsidies to the ACs in relation to environmental matters, as well as agreements on the transfer of funds for the management or execution of EU environmental measures, plans, and actions. During the preparation of the National Climate and Energy Plan (2018, 2019) the Sectoral Conference on the Environment met only three times; the conference on Energy met just once (Ramos, Alda, and Cicuéndez 2019); and neither the Conferences on Local Issues nor on the Conference on Infrastructure and Land Management was called at all. Meanwhile, Spain's peak intergovernmental meeting, the Conference of Presidents, bringing together the heads of government, has not discussed climate governance so far (Alberton 2020).

Besides the formal cooperation within the Sectoral Conferences, the Ministry for Ecological Transition and Demographic Challenge maintains informal contacts at the technical level with the ACs and the other public administrations. There are two further administrative bodies at the central level which collaborate on an ad hoc basis with the ACs and local authorities. The Spanish Climate Change Office is in charge of drawing up regulatory proposals for climate policy that are consistent with international and European commitments. The Institute for Energy Diversification and Saving is in charge of the preparation and implementation of climate change measures regarding energy, particularly energy efficiency.

Due to the specific features of the decentralisation process, bilateral cooperation between the central and AC governments has been constant since the creation of the Statutes of Autonomy (Ridaura Martínez 2007). Despite the progress made with regard to the multilateral cooperation mechanisms, bilateral cooperation is still preferred by several ACs. In light of this, the amendments to the Statutes of Autonomy during the last decade have institutionalised the bilateral commissions that are intended to enable permanent collaboration between individual ACs and the Spanish government. Only on rare occasions have the ACs and central government adopted regulations or agreements about environmental matters.

Over the past decades, several advisory or functional bodies have been established in order to facilitate collaboration and information sharing on climate change between the levels of government. However, their functions are not clearly defined, and their composition does not allow effective policymaking (Presicce 2020). Consequently, the effectiveness of these organisations has been low. The ACs take part in the National Council of Climate which was created in 1998 and involves the different ministries of the central administration (twenty-four members); the ACs (one representative for each of seventeen ACs); the municipalities and provinces (three members); and research institutions and social actors (twenty members). The Council prepares proposals and recommendations on climate change policies alongside the reports which are legally required in specific cases. They also participate in the Coordination Commission of Climate Change Policies, which ensures the coordination of climate change and adaptation strategies, and the goals on the prevention and reduction of GHG emission within the central administration, ACs, and local authorities, as well as with the National Council of Climate.

Recently, a further framework for cooperation has been aimed by Law 7/2021 on Climate Change and Energy Transition which has created an ostensibly new governance system. Article 37 creates the Committee of Experts on Climate Change and Energy transition, an advisory body that will prepare an annual report to be submitted and discussed in the Parliament and whose membership and working rules are still pending. Meanwhile, under the ambitious title of Inter-Administrative Cooperation on Climate Change and Energy, article 38 requires that ACs provides yearly information on their Climate and Energy Plans to the Coordination Commission of Climate Change Policy. Even though these articles are qualified as 'governance tools', there is little evidence that they represent a significant improvement in multilevel coordination and governance. The Committee of Experts relates more to civil society participation and knowledge diffusion than to governance and inter-territorial relations; regarding the information to be provided by the ACs on their climate planning, it does not include the early phases of elaboration but the post-approval ones, when they can be found in the official journal and websites.

In general, the coordination and intergovernmental relation in Spain, and also for climate policies, are addressed traditionally through collective bodies where ACs are represented – such as the Coordination Commission of Climate Change or the Sectoral Conferences. These provide little scope for real input into policymaking.

13.3.3 Fiscal Federalism and Climate Governance

As a share of GDP, in 2014 Spain had the 14th-lowest environment-related tax revenue among the OECD countries. Environment-related tax revenues amounted to 1.8 per cent of GDP, compared with an average of 2.0 per cent among OECD countries. Recently, the central government has introduced several new environmental taxes, but the ACs had started introducing environmental taxes a decade ago.

The system of revenue assignment between the levels of government in Spain is rather complex because of the marked asymmetry between the financing regime of the two 'charter' (*Foral*) regime ACs and the fifteen Common Regime ACs, and because of the complex variety of sources from which Common Regime ACs draw their revenues (López-Laborda et al. 2023).[8] The Common Regime is based on three pillars: (i) inter-governmental transfers and unconditional equalisation grants; (ii) shared taxes; and (iii) own-source tax revenues (Leon 2015). Own-source taxes at the level of the ACs include environmental taxes such as those on large commercial establishments situated on the outskirts of cities. Four main regional categories of the ACs' energy/environmental taxes can be distinguished: atmospheric emissions, installations and activities that affect the environment, wind energy taxes, and taxes on wastewater. However, these taxes are usually focused on facilities and technologies (infrastructure) and not on damage and consumption, so their capacity to achieve change in environmental behaviour is very limited (Gago et al. 2019, 6). Although revenues from environmental taxes are still very low, the environmental tax regulation allowed ACs to adapt their policies to the peculiarities of their different territories (Lago-Peñas 2019).

13.4 Federal Governance and Climate Change

Although decentralisation in Spain has been quite successful, experts have for years now been calling for a revision of the Constitution in order to adapt the text to the current reality of the territorial model and to establish a federal framework granting unity and diversity and an equilibrium between shared rule and self-rule. The shortcomings of the model can also be evidenced for climate policy. Although the ACs and local governments have ample room for policy experimentalism and for adapting their climate change strategies to the peculiarities of their different territories, there are no formal channels of influence for AC or local actors on policy formulation at the federal level, and processes of climate policy diffusion between governments can only be observed at the horizontal level.

13.4.1 The Central Government

Most recent plans and strategies adopted by the central government have followed the timing and path set by the EU. In this regard, the central government could learn from other Member States' successes and failures in designing and implementing policies. The EU multilevel governance provided opportunities for policy learning by enabling policymakers to meet, communicate, cooperate, and exchange ideas with one another through various forums and institutions. Especially since 2000, national strategies and policies have been adopted in Spain mainly driven by commitments at European and international level and, particularly, by specific obligations stated by the European Directives on emissions, renewable energy, and energy efficiency – for example the approval of Law 1/2005 on permits for emissions trading and the national plan for the allocation of greenhouse gas emission (2005–7) followed the EU Directive 2003/87/EC, and the National Action Plan on Renewable Energies (2011–20) adopted in 2009, followed the EU Directive 2009/28/EC. In 2021, after a decade of negotiations, Spain finally adopted its first Climate Change and Energy Transition Law. Although the law is more ambitious in certain areas than similar laws in other EU countries, the law has been requested by the EU for a long time.

13.4.2 The Autonomous Communities (ACs)

The involvement of ACs in the setting up and implementation of national strategies, plans, or laws on climate change has been minimal, but the ACs did not demand more participation and focused individually on designing and implementing their own regional climate change strategies.

The first NPACC adapted in 2006 was discussed at the Environment Sectoral Conference but the ACs were left aside in the decision-making process, since there are no legal or institutional provisions for them to play any role at all. Moreover, during the preparation of the ENCP, there were no specific meetings organised for the Sectoral Conferences on Environment and on Energy.

There has been further coordination through informal working groups of which we do not know their composition or their rules of operation. In this regard, the current NPACC 2021–30 foresees the participation of the ACs on the implementation phase in several coordination and advisory forums. Most of these working groups reinforce the inter-institutional coordination, both in its intersectoral dimension and its territorial dimension (with special attention paid to the connection between the Central Administration, ACs, and local administrations). For example, the Impact and Adaptation Working Group brings together departments of the Central Administration and the ACs with the general objective of coordinating and integrating the different strategies and adaptation plans. The current NPACC states that its aim is to provide for a high degree of transparency of these bodies by laying down their functioning rules.

But ACs introduced proactively their own policies. Some ACs were actively involved in climate change governance long before the central government (Table 13.2). The ACs of Andalusia and the Basque Country had already adopted climate/environmental strategies in 2002 and Galicia in 2005 long before the first National Plan for Adaptation was adopted in 2006. Five ACs adopted strategies against Climate Change in 2008 and four ACs adapted their strategies in 2009. In this regard, ACs' governments could learn from one another's successes and failures in designing and implementing policies. Over time, ACs with a longer tradition of action on environmental issues developed expertise and infrastructure to implement climate change strategies. These ACs were also better prepared to develop policies related to climate and energy.

Over the past decade all ACs have adopted strategic frameworks, action plans, and/or programmes both on mitigation and on adaptation to climate change (Table 13.2). This trend can be explained by competition in the adaptation of climate policies between the levels of government. However, such competition did not only relate to the aim of adapting the best strategies, but also to the possibility of getting EU funding for regional projects. This may also explain the similarities among the ACs' strategies. Objectives, structures, and scopes of the frameworks have been emulated among ACs. For example, most strategies have common objectives regarding the reduction of GHG emission, but also different targets reflecting local realities such as geographical, economic, and environmental characteristics (Table 13.2). In addition to most strategies having a similar structure, there are also similarities with the structure of the national plans. The majority of ACs adopted plans instead of climate change laws in order to avoid constitutional conflict with central government. Regarding the institutional

Table 13.2 *Main climate change plans and objectives by AC*

	Strategies adopted before 2020	Strategies adopted after 2020	Climate objectives, e.g. reduction of GHG
Andalusia	Andalusian Climate Change Strategy 2002 Andalusian Climate Action Plan 2007–12	Andalusian Climate Action Plan (2020)	ETS[9]: –18% to 2005
Basque Country	Basque environmental strategy for Sustainable Development 2002–20 Basque Plan Against Climate Change 2008–12	Basque Country Climate Change Strategy 2050 (2015)	2030: –40 to 2005 2050: –80% to 2005
Galicia	Galician Climate Change Strategy 2005	Galician Strategy on Climate Change and Energy 2050 Energy and Climate Integrated Regional Plan 2019–23	
Murcia	Regional Strategy against Climate Change 2008–12	Mitigation and Adaptation Strategy for Climate Change (2019)	2030: 26% to 1990 ETS 40% to 1990 total.
Rioja	Regional Strategy against Climate Change 2008–12		2012: no more 37% to 1990
Navarra	Strategy and Action Plan against Climate Change 2008–12 Navarra Strategy of Climate Change 2010–20		2020: –20 to 2005 2030: –45 to 2005 2050: –80% to 2005
C. Valencia	Valencian Strategy on Climate Change 2008–12 Valencian Strategy on Climate Change 2013–20	Valencian Strategy on Climate Change and Energy 2030	
Balearic Islands	Action Plan to fight against Climate Change 2008–12 Balearic Strategy on Climate Change 2013–20 Mitigation Action Plan against Climate Change 2013–20		2030: 40% to 1990 2050: 90% to 1990
Aragon	Aragonese Climate Change and Clean Energy Strategy 2009–18	Aragonese Climate Change Strategy. Horizon 2030	–40% related to 1990 –26% ETS to 2005

Table 13.2 (*cont.*)

	Strategies adopted before 2020	Strategies adopted after 2020	Climate objectives, e.g. reduction of GHG
Asturias	Plan to Monitor greenhouse gas emissions (2009)		
Extremadura	Climate Change Strategy for Extremadura 2009–12 Extremadura Climate Change Strategy 2013–20		
Castilla y León	Regional Strategy of Sustainable Development 2009–14 Regional Strategy on Climate Change 2009–12–20.		
Castilla-La Mancha	Climate Change Regional Strategy. Mitigation and Adaptation 2010–12–20	Strategy of Climate Change. Horizon 2020 and 2030 (2019)	2020: 10%ETS. 21% the rest= –15% 2030: –20% ETS
Madrid	Air Quality and Climate Change Strategy 2013–20. Blu+ Plan		2020: –20% to 2010
Catalonia	Catalonian Climate Change Adaptation Strategy 2013–20	Catalonian Climate Change Adaptation Strategy 2021–30	2030: –40% to 1990 2040: –65% to 1990 2050: –100% to 1990
Canary Islands	Canarian Strategy on Plastic Containers 2014–20	Canarian strategy against Climate Change (2020)	2010: 36.7% to 1990, –3.3 to 2005 2015: 22% to 1990; –13.7% to 2005
Cantabria		Climate Change Action Strategy of Cantabria 2018–30	2020: –10% to 2005 ETS 2030: –26% to 2005 ETS 2050: –80%

Source: Own elaboration based on Autonomous Communities official webs and Official Journals.

framework there is further evidence of emulation. Most ACs created similar specific regional agencies with the objective of coordinating energy and/or climate issues. The ACs also created inter-departmental bodies, or agencies, to coordinate actions relating to climate change within their own territory. Moreover, in line with efforts at the national level, the ACs built up consultative participation bodies that bring together different social, economic, and environmental organisations and scientific institutions.

There has also been a certain divergence between the climate objectives by ACs (Table 13.2) which can, to a certain extent, be explained by elements of partisanship. The presence of co-partisans in central government (vertical partisan congruence) may produce objections to the introduction of innovative climate objectives and/or strategies. Moreover, party ideology is a further factor in explaining the divergence, especially regarding the introduction of environmental taxes. There is a higher probability of environmental taxes being introduced by left-wing governments. In ACs where the conservative Popular Party has traditionally governed (Castilla y León, Galicia, Madrid) less ambitious targets have been set. However, most AC governments are coalitions of different parties which reduces the partisan effect on climate governance.

Only three ACs – Catalonia, the Balearic Islands, and Andalusia – have reinforced their legal framework by enacting their own climate change laws. These laws were enacted even before the central level adopted the general law on climate change in 2021 (Cocciolo 2020). Although these regional laws display different approaches to climate strategy, all of them have created similar administrative bodies for coordination purposes within the public administration and for communication with civil society and committees of experts. In this regard, an 'emulation' process can be identified among the ACs. Furthermore, the constitutional court judgment on the Catalan law of Climate Change determined discussion in the rest of ACs. In this regard, the ACs of Andalusia and the Balearic Islands could learn from Catalonia's failures in designing and preparing climate change law, while other ACs didn't start with the legislative process. Catalonia adopted the first *Law of Climate Change*. The law was challenged in the Constitutional Court which deactivated important parts of the law, since the law not only addressed environmental protection but also the energy and economic sectors – the exclusive powers of the central government.

By contrast, the *Law of Measures Facing Climate Change and the Energy Transition* adopted by the AC of Andalusia excluded from its scope the restrictions of emissions which are within the responsibilities of the central government. Finally, *Law 10/2019 of Climate Change and Energy Transition* adopted by the AC Balearic Islands includes also lessons learnt from the Catalan experience in setting up a framework for energy transition in such way that it does not affect the competences of the central government. In this regard the law was discussed *ex ante* in a bilateral commission between representatives of the central government and the AC in order to prevent a conflict before the Constitutional Court. Currently six others (Aragon, Asturias, the Canary Islands, Valencia, La Rioja, Navarre, and the Basque Country) are in the process of adapting their own climate change laws.

In short, we can state that, with or without specific regional legal frameworks on climate Change, all ACs have adopted climate policies through general and specific strategies, plans, and programmes. The ACs' strategies can go beyond the national climate objectives if they are not interfering with the central government competence on 'energy' or on 'general economic planning', which prevents, for example, that the ACs increase the economic sectors which are submitted to the emission trading system.

13.4.3 The Local Level

The participation of local governments in climate policies can be distinguished in three different stages. First: limited participation of local entities in the decision-making process for climate decision and planning being carried out at state and AC level, mainly during the setting up of the state and ACs' climate strategies. Second: sectoral policies and programmes that account for climate objectives are implemented at the local level, where, for example, urban regeneration programmes, housing policies, or waste management are within its competences. Third: the local level implements specific climate policies based on international agreements if they decide to join voluntarily. The Covenant of Mayors for Climate and Energy is an international agreement that provides common tools and methodological standards – as climate and energy planning templates – for local entities. Even the Covenant is based on a voluntary accession, and it has had high success among the Spanish municipal and provincial authorities.

In 2019, 29 per cent of the municipal and provincial governments in Spain (which account for around 62 per cent of the population) had concrete climate change adaptation plans. However, most large Spanish cities have adopted their own plans and strategies on environmental and climate issues. Some local entities have conducted specific local-scale vulnerability and impact analyses or developed adaptation plans, although the latter are still in the minority (FEMP 2019).

The Spanish Network of Cities for Climate is a thematic network created in 2009 by the Spanish Federation of Municipalities and Provinces and the Spanish Ministry for Ecological Transition for joint action on climate. As of 2019, the network included 316 Spanish local entities. According to Alda and Ramos (2018), local authorities also participate in the main transnational networks of local governments on climate and energy policies, both at European and global levels

Table 13.3 *Spanish non-state governments in transnational networks*

Network	Number of Spanish participants	Types of entities
Local Governments for Sustainability (ICLEI)	10	Municipalities (4); Provincial Councils (1); Municipal Association (*Mancomunidades*), Metropolitan Areas (5)
Carbon Climate Registry	21	Municipalities (20), Provincial Councils (1)
Energy Cities	5	Municipalities (4); Energy agency (1)
Climate Alliance	1	Provincial Council
Climate Group	6	Autonomous Communities
C40	2	Municipalities
Covenant of Mayors*	2.151	Municipalities, Provinces and Local Associations
Urban Development Network[10]	72	Municipalities
Non-State Actor Zone for Climate Action (NAZCA)	365	Municipalities (359) Autonomous Communities (6)
Global Covenant of Mayors	2.151	–

Source: Alda and Ramos (2018).

(Table 13.3). These networks are examples of the dissemination of best practice in climate policy at the local level. For example, based on the Covenant of Mayors, several municipalities adopted harmonised Energy and Climate Change Plans. Some provinces joined the Covenant of Mayors as 'coordinators', also developing provincial climate policy on the basis of the setting up of the local plans. For example, the Province of Tarragona published public tenders for the adoption and implementation of Local Energy and Climate Plans. As a result, from 2013 to 2021, 184 out of 188 municipalities endorsed the Covenant of Mayors, and from 2015 to 2021, 144 municipalities adopted a Local Energy Plan. Similar to the AC level, the adaption of Local Energy and Climate Plans did not only relate to the aim of adapting the best strategies, but also to the possibility of getting funding for local projects. This may also explain the similarities among the municipal strategies.

Finally, the recent Law 7/2021 on Climate Change and Energy Transition requires urban mobility plans for municipalities above 50,000 inhabitants (article 14). Furthermore, in an indirect way it requires some other local policies aligned with Climate Change Strategies as the urban planning (article 21), fair transition (article 28), or public procurement (article 31).

13.5 Conclusions

Climate change governance in Spain is framed between international commitments and EU legislation and encompasses several levels of government and stakeholders. There are many strategies and climate policies at all territorial levels adopted over the past decade. But there are no formal channels of influence for AC or local actors on policy formulation at the federal level, which limits climate policy diffusion at the vertical dimension. Climate change governance encompasses various responsibilities, some of which are the exclusive responsibility of central government, while others are shared between the two levels of government. In addition, climate change governance is determined by EU objectives and policies. But it is only the central government that negotiates the targets set at the EU level and designs and presents the agenda for national climate change governance. The ACs are not involved in this decision-making process but have to implement and fulfil the targets. As we show coordination is not always necessary, and ACs could adopt their climate change plans and targets; however, coordinated action at the vertical and horizonal dimension could lead to a more effective approach.

In this respect, one can also see a certain retarding effect on Spanish climate change policy, as the ACs could not push for a national climate change law that included their preferences. The complexity and transverse character of climate change government requires a reinforced system of coordination and cooperation, but, although there are several coordination bodies, there are no intergovernmental institutions for coordination, decision-making and implementation of the climate change policies in Spain. The constitutional framework was, and is, a source of conflict too. The unclear division of powers not only create controversies when EU climate change measures have to be implemented in Spain, but regulatory overlaps in environmental protection and energy policy are long-standing sources of tension between the central and AC governments and have been challenges to effective climate action over the past decades.

However, the ACs and local governments have ample room for policy experimentalism. Climate change policy initiatives adopted by the ACs and local level entities at the beginning of the 2000s generated opportunities for policy entrepreneurship, and a chance for the ACs and local authorities to disrupt the status quo. The initiatives of the ACs and local governments have also allowed them to adapt their policies to the peculiarities of their different territories, although performance has been quite uneven and there are not official regional reports on implementation and its results. In this regard the Spanish State of Autonomies has provided a favourable context for dynamic horizontal processes of climate policy diffusion.

Notes

1 Kgs./PPP -purchasing power parity of GDP.
2 Numbers in Kilo tonnes CO_2-eq.
3 Land use, land-use change, and forestry.
4 *Regulation (EU) 2018/1999 of the European Parliament and of the Council of 11 December 2018 on the Governance of the Energy Union and Climate Action, amending Regulations (EC) No 663/2009 and (EC) No 715/2009 of the European Parliament and of the Council, Directives 94/22/EC, 98/70/EC, 2009/31/EC, 2009/73/EC, 2010/31/EU, 2012/27/EU and 2013/30/EU of the European Parliament and of the Council, Council Directives 2009/119/EC and (EU) 2015/652 and repealing Regulation (EU) No 525/2013 of the European Parliament and of the Council.* 2018. OJ L 328, 21.12.2018: 1–77. http://data.europa.eu/eli/reg/2018/1999/oj.
5 (1) The Statue of Cataluña, included in article 46, the powers over 'prevention and control of activities that alter the atmospheric and climatic regime'.
(2) The Statue of Andalucía, included in article 57.3, the powers over 'regulation of the authorisation and monitoring regime for greenhouse gas emissions'.
(3) The Statues of Canary Island and Extremadura, included in articles 153.1.ñ and 9.1.33 respectively, the mention of 'climate change' as part of the environmental powers.
6 Between 2006 and 2011, eight out seventeen Statutes of Autonomy were reformed in a significant way adding new regional competences.
7 Articles 71.22 and 70.1.35 respectively.
8 Within the *Foral* ACs, the provinces of the Basque Country and the AC Navarra have the power to establish and regulate their tax systems, including the ability to collect, manage, and inspect all taxes with the exception of import duties and value added tax.
9 ETS: Emissions Trading Scheme.
10 Average number of participants in one of the network's activities, held in 2016. The information is merely indicative.

References

AEMET (Agencia Estatal de Meteorología). 2018. *CLIVAR Exchanges No 73: Volumen especial sobre el clima en la Península Ibérica: una visión científica global y coordinada por el Comité CLIVAR-España*. Madrid: Ministerio para la Transición Ecológica, AEMET. www.aemet.es/documentos/es/conocermas/recursos_en_linea/publicaciones_y_estudios/publicaciones/Clivar_num73_clima_peninsula_iberica/Clivar_num73_clima_peninsula_iberica.pdf.

AEMET. 2021. 'El informe sobre el estado del clima en 2020 en España muestra que fue el año más cálido y se dispararon los récords de calor'. *AEMET*. www.aemet.es/es/noticias/2021/05/informe_estado_clima_2020.

Aja, E. 1999. *El Estado Autonómico: Federalismo y hechos diferenciales*. Madrid: Alianza Editorial.

Alberton, M. 2020. 'La praxis de las relaciones intergubernamentales en España: un examen cuantitativo y cualitativo de la cooperación en materia ambiental'. *Revista Catalana de Dret Ambiental* 11 (2): 1–44. www.raco.cat/index.php/rcda/article/view/378495.

Alberton, M. 2021 'The Swing of Intergovernmental Relations Concerning Environmental Matters through the (Un)balanced Doctrines of the Constitutional Courts in Spain and in Italy'. *Revista d'estudis autonòmics i federals* 33: 15–57. https://dialnet.unirioja.es/servlet/articulo?codigo=8096312.

Alda, M., and J. A. Ramos. 2018. 'La inserción de los gobiernos locales españoles en las redes transnacionales en materia de clima y energía'. In *Conference: IX Congreso Internacional en Gobierno, Administración y Políticas Públicas GIGAPP*, Madrid.

www.researchgate.net/publication/328415866_La_insercion_de_los_gobiernos_locales_espanoles_en_las_redes_transnacionales_en_materia_de_clima_y_energia.

Alzaga Villaamil, O., and I. Alvarez Rodriguez. 2021. *Derecho político español según la Constitución de 1978: Tomo I. Constitución y fuentes del Derecho*, 7th ed. Madrid: Marcial Pons.

Aragón Reyes, M. 2006. 'La Construcción del Estado Autonómico'. *Cuadernos Constitucionales de la Cátedra Fabrique Furió Ceriol* 54 (55): 75–95.

Casado Casado, L. 2018. *La recentralización de competencias en materia de protección del medio ambiente*. Institut d'Estudis de l'Autogovern, Barcelona. https://presidencia.gencat.cat/web/.content/ambits_actuacio/desenvolupament_autogovern/iea/publicacions/01_IEAg/IEAg_arxius-i-vincles/IEAg_07.pdf.

Cocciolo, E. 2020. 'Cambio climático en tiempos de emergencia: Las comunidades autónomas en las veredas del 'federalismo climático' español'. *Revista Catalana de Dret Ambiental* 1. https://revistes.urv.cat/index.php/rcda/article/view/2868/2881.

Colino, C. 2021. 'Los retos de la coordinación y la cooperación en el Estado autonómico'. In *Retos de la gobernanza multinivel y la coordinación del Estado autonómico: de la pandemia al futuro*, ed. C. Colino, 11–33. Madrid: Instituto Nacional de Administración Pública (INAP).

ECA (European Court of Auditors). 2018. *Special Report No 33: Combating desertification in the EU: A growing threat in need of more action*. Luxembourg: ECA. www.eca.europa.eu/Lists/ECADocuments/SR18_33/SR_DESERTIFICATION_EN.pdf.

European Commission. 2013. *Communication from the Commission to the European Parliament, the Council, the European Economic and Social Committee and the Committee of the Regions: An EU Strategy on adaptation to climate change*. Doc COM(2013)216 final.

European Commission. 2018. *Communication from the Commission: A Clean Planet for All: European strategic long-term vision for a prosperous, modern, competitive and climate neutral economy*. Doc COM(2018)773 final.

European Commission. 2019. *Communication from the Commission: The European Green Deal*. Doc COM(2019)640 final.

European Commission. 2021. 'Biodiversity statistics'. *Eurostat*. ISSN: 2443-8219. https://ec.europa.eu/eurostat/statistics-explained/index.php?title=Biodiversity_statistics#Natura_2000_-_the_cornerstone_of_biodiversity_protection_in_the_EU.

European Commission. n.d. *Eurostat*. https://ec.europa.eu/eurostat/web/main.

FEMP (Federación española de municipios y provincias). 2019. *Sexto Informe Sobre políticas locales de lucha contra el cambio climático*. Madrid: FEMP. https://sextoinforme.redciudadesclima.es/sites/default/files/2020-06/Sexto_Informe_sobre_Politicas_Locales_de_lucha_contra_el_Cambio_Climatico.pdf.

Field, B. N. 2016. 'Minority Government Performance and Comparative Lessons'. In *Why Minority Governments Work: Europe in Transition: The NYU European Studies Series*. New York: Palgrave Macmillan. https://doi.org/10.1057/9781137559807_12.

Gago, A., X. Labandeira, J. Labeaga, and X. Lopez-Otero. 2019. 'Impuestos energético-ambientales, cambio climático y federalismo fiscal en España'. ('Energy-Environmental Taxes, Climate Change and Fiscal Federalism in Spain'. With English summary.). *Ekonomiaz*, 95: 274–90. https://labandeira.eu/publicacions/wp0219.pdf.

Galera Rodrigo, S. 2016. 'Changing the Energy Model: Step back on the Europe 2050 strategy?' *European Energy and Environmental Law Review* 25 (2): 65.

Galera Rodrigo, S. 2018. 'Las competencias en materia de clima: La complejidad jurídica del gobierno multinivel'. In *Políticas locales de clima y energía: Teoría y práctica*, eds. S. Galera and M. Gómez, 215–54. Madrid: INAP.

Gobierno de España. 2020a. *Plan Nacional de Adaptación al Cambio Climático 2021–2030*. Madrid: Ministerio para la Transición Ecológica y el Reto Demográfico (MITECO). www.miteco.gob.es/es/cambio-climatico/temas/impactos-vulnerabilidad-y-adaptacion/pnacc-2021-2030_tcm30-512163.pdf.

Gobierno de España. 2020b. *Informe de Inventario Nacional de Gases de Efecto Invernadero*. Madrid: MITECO.

Gobierno de España. 2021. *Informe de Inventario Nacional de Gases de Efecto Invernadero*. Madrid: MITECO.

Greenpeace. 2018. *Imágenes y datos: Así nos afecta el Cambio Climático*. https://es.greenpeace.org/es/wp-content/uploads/sites/3/2018/11/GP-cambio-climatico-LR.pdf.

Greenpeace. 2020. https://es.greenpeace.org/es/trabajamos-en/bosques/incendios-forestales (accessed December 2020).

Lago-Peñas, S. 2019. 'La descentralización tributaria en España: avances significativos, retos pendientes'. *Cuadernos de Información Económica* 272: 45–52.

León, S., ed. 2015. *La financiación autonómica. Claves para entender un (interminable) debate*. Madrid: Alianza Editorial.

López Nieto, L., ed. 2009. *Intergovernmental Relations in Democratic Spain: Interdependence, autonomy, conflict and cooperation*. Madrid: Ed. Dykinson.

López-Laborda, J., J. Martinez-Vazquez, and C. Monasterio. 2021. 'Spain'. In *Fiscal Federalism: Comparative Experiences and Challenges*. Forum of Federations, forthcoming.

López-Laborda, J., J. Martinez-Vazquez, and C. Monasterio. 2023. 'Spain'. In *The Forum of Federations Handbook of Fiscal Federalism* (ed.) Jean-François Tremblay. Palgrave Macmillan

Nogueira López, A. 2012. 'La transposición de Directivas ambientales en el Estado Autonómico'. In *Transposición de directivas y autogobierno: el desarrollo normativo del Derecho de la Unión Europea en el Estado autonómico*, ed. Xabier Arzoz Santisteban, 85, 281–344. Barcelona: Institut d'Estudis Autonòmics. www.gencat.cat/governacio/pub/sum/iea/IEA_85.pdf.

Pérez Gabaldón, M. 2013. *La gestión intergubernamental de la política de cambio climático en España*. Pamplona: Aranzadi.

Pérez Medina, J. M. 2020. 'Dinámica de las conferencias sectoriales: Entre la intergubernamentalidad y la cooperación administrative'. *Revista d'estudis autonòmics i federals*, ISSN 1886–2632, 31: 17–64. https://doi.org/10.2436/20.8080.01.44.

Pérsica, L. 2020. 'Buscando instrumentos de coordinación para la gobernanza climática multinivel en España'. *Actualidad Jurídica Ambiental* 101. www.actualidadjuridicaambiental.com/wp-content/uploads/2020/04/2020_05_04_Presicce_Gobernanza-climatica-multinivel.pdf.

Ramos, J. A., M. Alda, and R. Cicuéndez. 2019. 'The Institutionalized Dimension of Intergovernmental Relations in the Estado Autonómico'. In *Intergovernmental Relations in Democratic Spain: Interdependence, autonomy, conflict and cooperation*, ed. Lourdes López Nieto, 377–96. Madrid: Ed. Dykinson.

Ramos, J. A. and M. Alda. 2021. 'Las relaciones de cooperación entre Comunidades Autónomas: un balance centrado en el periodo 2004–2020'. In *Retos de la gobernanza multinivel y la coordinación en el Estado autonómico: de la Pandemia al Futuro*, ed. C. Colino, 131–48. Madrid: INAP.

Ramos, J. A. and M. Alda. 2020. 'Relaciones intergubernamentales en tiempos revueltos: El Estado autonómico entre la gran recesión y la Covid-19, pasando por el procés'. *RIBRIG* 1: 1–42.

Ridaura Martínez, J. 2007. 'Las comisiones bilaterales de cooperación en el sistema autonómico español'. *Cuadernos Constitucionales de la Cátedra Fadrique Furió Ceriol* 60 (61): 65–84.

Rodríguez López, E., J. B. Harguindéguy, and A. Sánchez Sánchez. 2019. 'The Perfect Storm? How to explain the rise of intergovernmental conflicts in Spain? (1984–2014)'. *Innovation: The European Journal of Social Science Research* 32 (2): 211–30. doi: 10.1080/13511610.2018.1520628.

Tudela, J., and M. Kölling. 2020. 'Spain (Kingdom of Spain)'. In *The Forum of Federations Handbook of Federal Countries 2020*, eds. Ann Griffiths, Rupak Chattopadhyay, John Light, and Carl Stieren, 329–40, London: Palgrave Macmillan.

UNESCO, UN-Water. 2020. *United Nations World Water Development Report 2020: Water and climate change*. Paris: UNESCO. https://en.unesco.org/themes/water-security/wwap/wwdr/2020.

UN GA. 2015. *Transforming our World: The 2030 agenda for sustainable development*. 70th Sess., UN Doc A/RES/70/1.

Velasco Caballero, F., ed. 2021. *Tratado de Derecho Local*. Madrid: Marcial Pons.

Watts, Ronald L. 2010. 'España ¿una federación multinacional encubierta?' In *España y modelos de federalismo*, eds. J. Tudela and F. Knüpling, 1–32. Madrid: CEPC.

14

Climate Governance and Federalism in Switzerland

MARLENE KAMMERER, SEAN MUELLER, KARIN INGOLD, AND MARIA GALLMANN

14.1 Introduction

Switzerland is widely perceived as a climate policy 'pusher' (Liefferink and Wurzel 2017). However, the climate change performance index 2022 ranks Switzerland 15th, behind European countries like Denmark, Sweden, Norway, or the UK (Burck et al. 2021, 7). This comparatively low position reflects the lack of Swiss ambition on renewable energy development; the tendency to compensate CO_2 emissions abroad instead of achieving reductions at home; and the failure to reduce emissions in transport (Kammerer et al. 2021). Although Switzerland's climate policy goes back to the early 1990s and comprises a mix of different measures in all relevant sectors, the 20 per cent reduction in CO_2 (compared to 1990) emission reduction target specified in the 2013 *CO_2-Act* has not yet been achieved, and a more ambitious version of the *CO_2-Act* was rejected in a referendum in June 2021 (Swissvotes 2021). Thus, while Switzerland genuinely pushes for stronger climate policies in international negotiations and during the UN Framework Convention on Climate Change Conferences of the Parties (UNFCCC COP; see Ingold & Pflieger 2016), in terms of domestic regulation it tends to wait for, and align itself with, the positions of the European Union (EU). In fact, rather than a climate 'pusher', Switzerland is more accurately described as a 'follower' of the EU on climate matters (Kammerer et al. 2021).

This chapter aims to unpack a further paradox, namely the one relating to the advantages and disadvantages offered by the Swiss federal structure as both enabling and hindering effective and sustainable climate governance (see also Casado-Asensio and Steurer 2016; Reich 2021). We show that the fragmented nature of the Swiss polity, with its twenty-six constituent units (cantons) and some 2,000 municipal polities with each more or less autonomy in key policy areas, is not in itself an obstacle. However, tackling the complexity arising from such a multilevel structure is time consuming, at best. The short answers to the questions posed in Chapter 1 read as follows:

1. Swiss federalism is very decentralized. This has facilitated *locally tailored solutions and policy innovation*, especially in terms of climate change adaptation, but inter-jurisdictional *learning is limited.* Moreover, the resulting patchwork of regional and local policies does not compensate for the absence of an ambitious climate mitigation policy at national level.
2. The nature of Swiss federalism is such that lower levels of government can indeed *compensate* for the inaction or failures of the next higher level, meaning that municipalities can tackle climate change where 'their' cantons do not, and cantons can similarly act if they perceive the federal government as too slow or lax. However, both lower levels (cantonal and municipal) lack one of the most important instruments to properly address climate change, since all major indirect taxation powers (on fuel or flight tickets, for instance) fall within the jurisdiction of the federal government. In turn, the domestic backseat position of the federal government is not due to federalism, but instead has to do with the many veto points offered by direct democracy and the overall rather conservative preferences of the electorate.
3. Because climate change is not treated as its own policy field but cuts across a number of primarily subnational domains (notably environment, buildings, transport, and spatial planning), there is a conspicuous *lack of coordination* both across levels of government and across policy domains. While there does exist a coordination body, the 'tripartite conference' (meaning the federal government, the cantons, and the municipalities), it has not so far been discussing climate change as such.[1] The full potential of Swiss federalism as a laboratory of ideas and innovation is not, therefore, harnessed.

What is more, federalism not only plays a role through the vertical division of powers (degrees of de/centralization), but also by specifying the operational model. In fact, unlike dual federations such as the USA or Canada, in Switzerland the twenty-six cantons are in charge of implementing (most) federal decisions (administrative federalism; on which, see Mueller and Fenna 2022). If we further distinguish between climate change mitigation and adaptation, the matrix shown in Table 14.1 is obtained.

Consequently, national guidelines are implemented in a variety of ways across the cantons. Focusing on that subnational variety, this chapter asks how that variety is shaped. What factors determine the *cantonal* formulation of climate change *adaptation* strategies (see Wieser 2018), on the one hand, and the *municipal* adoption of climate change *mitigation* policies, on the other – specifically when it comes to the adoption of the 'gold standard' of the Energy Citylabel (see Schmid 2018)? In doing

Table 14.1 Vertical division of climate change powers in Switzerland

	Mitigation	Adaptation
Confederation	*De jure*: legislative powers based on Art. 74 (environmental protection) and 89 (energy policy) of the Federal Constitution *De facto*: *CO_2-Act* with rather unambitious goals and instruments (e.g., import restrictions on certain cars), 2021 revisions rejected in referendum	*De jure* and *de facto*: subsidiary role (e.g., project-specific matching grants for infrastructure, general-purpose fiscal equalization and other transfers) National Climate Change Adaptation Strategy[2] as guiding document but without binding force
26 Cantons	Implementation duties for most federal legislation (except those relating to customs and indirect taxes) Policy design competences deriving from the national CO_2-Act in the energy and building sector (cantons can go beyond, e.g., imposing a ban on oil heating) Ownership over natural resources on their territory (incl. wind, water, geothermal) No major indirect tax powers (minor competences with regards to car registrations, for instance)	Overall climate change adaptation and risk management on their territory Inter-cantonal coordination through Energy, Landscape, Transportation, Planning and Environment Agriculture, Economy (incl. tourism) and Finance conferences (but no cross-cutting 'climate change conference')
ca. 2,000 municipalities	Implementation duties for most federal and cantonal legislation Significant own powers in building and spatial planning (e.g., power to approve renewable energy projects on their territory; building permits)	Primary responsibility for natural disaster prevention (floods, avalanches, landslides) Implementation duties for most cantonal legislation

so, we highlight the role of inter-cantonal conferences (coordinating bodies of cantonal ministers) and their potential to provide opportunities for the diffusion of best practices and joint learning. We also discuss the pitfalls and potentials for climate policymaking in a strongly decentralized system, more generally. In the USA, for example, 'boomerang federalism' (Fisher 2013) and a lack of national commitment has led to stronger climate change regulation at the subnational level, with subsequent spillover effects for national policymaking. In Switzerland, the situation is quite different. While climate change *adaptation* benefits from decentralized and tailor-made solutions at the regional level, for climate *mitigation* a 'healthy competition' among decentralized entities has so far been largely missing. Our case studies thus provide evidence for both the advantages and disadvantages of (Swiss) federalism: experimentation, innovation, and responsiveness to local needs with regards to adaptation; but an incoherent, haphazard, and only slowly evolving national mitigation strategy.

14.2 Climate Change in Switzerland

Switzerland has already experienced some pronounced effects of climate change, due to its alpine geography. Measurements dating back more than 150 years show that near-surface air temperatures have increased in all regions of Switzerland by an average of 2.1°C (MeteoSwiss 2020). That is more than double the average global increase. Moreover, mean temperature deviations across the country reveal a persistent warming over the last thirty years. This warming has already led to an upward shift in the tree line by 300–400 m; a decrease in the alpine glaciers volume by 60 per cent since the 1850s; and up to 50 per cent fewer snow-days in the lower elevation regions (NCCS 2018). Even if the UNFCCC goal of limiting global warming to a 2°C increase is achieved, Switzerland will have experienced a warming of between 2.1 and 3.4°C by the end of the twenty-first century. In an even grimmer scenario (RCP 8.5 model), temperatures could increase by up to 6.9°C (NCCS 2018). When it comes to precipitation, varying regional patterns have been observed, but overall, rain is expected to decrease in the warmer seasons and increase in the colder ones. This presents growing disaster risks as winter precipitation will increasingly be in liquid rather than solid form (leading to flooding, debris fall, rock avalanches, and landslides), while in summer droughts are likely to be more frequent (MeteoSwiss 2020; NCCS 2018).

Even though Switzerland covers only a small land area, it is geographically diverse and different regions face different challenges in coping with climate change. Varying socio-economic conditions and uneven degrees of urbanization add to these diversified effects. Paradoxically, while it is here that federalism offers its most beneficial contribution in permitting locally tailored solutions in adaptation, the regionally varying effects (both in type, time, and intensity) reduce the political pressure for a common mitigation strategy. For instance, the country's largest cities – Zurich, Basel, and Geneva – are the ones most strongly affected by the health risks of more hot days and nights during summertime (a

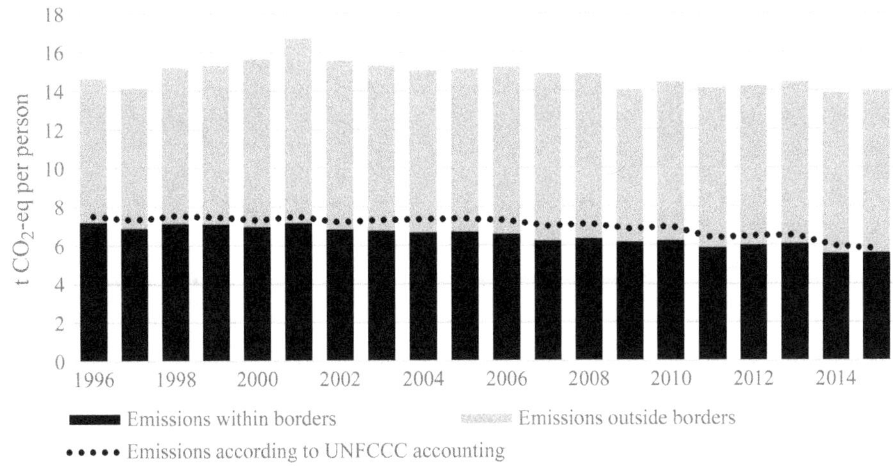

Figure 14.1 Switzerland's overall GHG footprint, 1996–2015.
Source: Own figure with data from FOEN, 2020a+b.

spatial planning problem), while in the alpine regions, climate change heavily affects winter tourism and thus leads to reduced income (an economic problem). Moreover, climate change also presents different opportunities to different regions, such as increasing summer tourism, extended vegetation periods for agriculture, and decreasing winter heating needs (NCCS 2018).

What is more, although Switzerland accounts for less than 0.2 per cent of greenhouse gas (GHG) emissions worldwide, it has comparably high per capita emissions. In 2018, average per capita GHG emissions amounted to 5.4 tonnes of CO_2 equivalents – slightly above the global average, but below that of the OECD (see Figure 14.1).

As is visible from Figure 14.1, GHG emissions have decreased. Switzerland did meet its 8 per cent GHG reduction commitment under the Kyoto Protocol's first commitment period (2008–12). However, only half of this was accounted for by domestic emissions reduction; the rest resulted from the purchase of emission reduction certificates abroad and local forest sinks (FOEN, 2014).[3]

Within Switzerland, by far the largest part of greenhouse gas emissions (77.2 per cent in 2017) stem from transport and heating of buildings (FOEN 2020c). Emissions from domestic electricity generation remain low in comparison to other OECD states as a majority of Switzerland's supply is (still) based on hydro- and nuclear power. Figure 14.2 provides the developments of GHG emissions per specific sector in 2018 since 1990. The main contributors to the decrease in total emissions are industry, buildings, and agriculture – a consequence of improved thermal insulation and energy efficiency, as well as declining livestock and reduced fertiliser use, respectively (FOEN 2020c). Of further note is the fact that

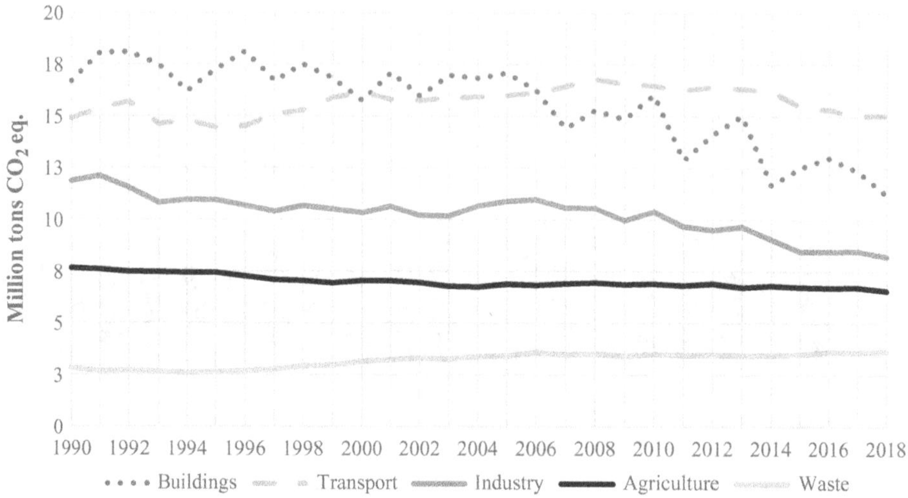

Figure 14.2 Swiss GHG emissions per sector, 1990–2018.
Source: Own figure with data from FOEN (2020a, 2020b).

Switzerland is credited with negative emissions (−3.4 per cent) in the UNFCCC category 'land use, land-use change and forestry' due to its growing forest area (since 1902, a Federal Act has forbidden a reduction in total forest size; see, Dardanelli and Mueller 2019; supplementary data, 27).

In sum, considering its vulnerability, its substantial grey emissions, its high GDP, and its depiction of a role model at the international scale, Switzerland might be expected to invest heavily in climate change mitigation measures. That it fails to do so is not, however, primarily federalism's fault; in fact, federalism has enabled subnational polities – cantons and municipalities alike – to go beyond national targets. However, the cantons, but more so the municipalities as the lowest level, are limited in how far they can go through the lack of indirect tax powers and their own political considerations. Upscaling of ambitious regional and local policies to the national level has been impeded by direct democracy, the strength of certain lobby groups, and the overall rather conservative nature of the Swiss electorate (e.g., Swissvotes 2021).

14.3 Climate Policy in the Swiss Confederation

14.3.1 Swiss Federalism in a Nutshell

In its modern form, the Swiss federation dates to 1848, when after a brief civil war, the constitutional framework was agreed upon as a compromise between liberal centralizers and conservative regional autonomists (Linder and Mueller 2021; Vatter 2018). Although since then the federal government has steadily acquired increased powers, the cantons continue to wield significant legislative, administrative, and especially fiscal authority (Dardanelli and Mueller, 2019). The roughly 2,000 local governments also exercise significant powers and control own-source revenue (Ladner et al. 2019; Mueller 2015). Most importantly, Switzerland has moved strongly towards an administrative division of powers whereby the lower levels generally implement the decisions of the higher level(s) as well, of course, as their own. For this reason, the federal government only disposes of a very small administrative workforce: fewer than 35,000 full-time equivalents (EPA 2019) while, by comparison, the city of Zurich alone has 23,000 (Statistik Stadt Zürich 2017, 289).

Three core principles define the workings of Swiss federalism: symmetry, diversity, and subsidiarity (Vatter 2018). *Symmetry* means that despite immense differences between the cantons and communes in terms of size, resources, and state capacity, all are treated alike by both the federal government and each other. Legally speaking, all cantons are equal, and legislation does not generally distinguish between cities, conurbations, or rural and mountain communes. *Diversity*, in turn, refers to a largely accepted consequence of cantonal and local autonomy, namely that in non-centralized policy areas such as education, energy, or environmental protection, there may well be very different types and levels of public services. The same is true for the degree of local autonomy, which varies from one canton to the other: generally speaking, eastern municipalities have the most, western ones the least amount of autonomy (Mueller 2015).

The third and final principle, *subsidiarity*, amounts to a basic presumption of responsibility in favour of lower levels. In other words, a higher level of government is only entitled to intervene if the lower level cannot – or is no longer willing to – fulfil a public task. That counts as much for the division of powers between local and cantonal governments as for that between cantonal governments and the federation. One consequence of this is the 'enumeration principle': a higher level of government can only legislate in a given policy area once a constitutional clause (at federal level for the Confederation, at cantonal level for the cantons) *explicitly* enables it to do so. One effect of this is slower policy change, since every constitutional change has to be approved in a referendum at the corresponding level; at federal level, constitutional change is even harder since a majority of voters and a majority of cantons must agree. In turn, the principle entails considerable room for experimentation at lower levels of government. The potential for policy experimentation is further enhanced by still substantial degrees of subnational fiscal autonomy, both in terms of own-source revenue (direct taxation mainly occurs at cantonal and local level) and expenditures. However, in Switzerland fiscal autonomy also generally means fiscal responsibility and the need to find the appropriate funding sources, which in turn must pass their own referendum test. In the city of Bern, for instance, the ordinary budget is subject to a mandatory referendum every year.

14.3.2 Swiss Climate Policy in the Federal Context

Regarding climate policies, cantons are not only in charge of implementing national decisions in environment, energy, and transport, but they can also take their own decisions in these and related areas. While climate change *mitigation* policy may well be defined at the national level (though implementation is largely left in the hands of the cantons), *adaptation* policy is even less centralized.[4]

14.3.2.1 Mitigation

Mitigation is strongly affected by international agreements and the Conferences of the Parties (COPs), and the decisions therein, related to the UNFCCC (United Nations Framework Convention on Climate Change). It is thus the central government's responsibility, and the federal parliament's final decision, to comply with international commitments and design domestic policies accordingly (Ingold and Pflieger 2016). For instance, the key legal document, the federal *CO_2-Act* of 2013, defines an overall national CO_2 emissions reduction target and introduces some core instruments for its implementation at national level.[5] These include a CO_2 tax on combustibles, tradable permits (CO_2 certificates), import restrictions for some vehicles, and a technology fund to support clean-tech innovations. To comply with its own nationally determined contribution (NDCS) to the 2015 Paris Agreement, Switzerland needs to revise that Act. A first version contained a tax on flight tickets and further CO_2 reduction measures for the finance sector. However, the reform failed narrowly in a binding referendum in June 2021, so the old *CO_2-Act* is prolonged over the next years with many core policy instruments phasing out in 2022. A new proposal was published by the Swiss government in December 2021, but it will most likely again have to overcome the referendum hurdle once through parliament. In consequence, the next version most likely will pursue a focus on fundings instruments instead of taxes and levies.

However, climate policy is far from completely centralized since the two key sub-sectors most heavily contributing to national CO_2 emissions are almost exclusively in cantonal hands: building and transport (see also Figure 14.2). Together with policy autonomy, huge differences therefore exist across the Swiss territory: each one of the twenty-six cantons defines and implements its own standards relating to building insulation, energy efficiency, heating, and public transport, but also in terms of renewable energy promotion. Moreover, not only the legal and administrative, but also the *fiscal* capacities vary significantly: while Swiss residents pay income and property tax to all three levels of government, the lion's share of direct taxes either stays at subnational level or is transferred back to it. The federal government, in turn, mainly disposes of income from indirect taxation such as VAT, tariffs, alcohol and tobacco duties, and other specialized levies. A sophisticated system of fiscal equalization, last fully revised in 2008, ensures that even the poorest cantons are equipped with exactly 86.5 per cent of the Swiss-wide average of fiscal resources. The system is funded both vertically, by the federal level, and horizontally, by the richer cantons. Most cantons have established similar fiscal equalization systems for their local governments. However, such transfers – although unconditional in nature – are generally used to cover running administrative costs or for debt service and investments into basic infrastructure, so that there is not much left to engage in costly environmental innovation even if beneficiary jurisdictions had the political will to do so.

An important pillar of Swiss climate and energy policy is the support and promotion of renewable energy. Here, differences across the cantons could not be bigger. After the 2011 nuclear catastrophe in Fukushima, the federal government decided to phase out nuclear energy, which at that time produced some 40 per cent of all electricity consumed. To reach this goal, in 2018 a new energy policy entered into force with the aim of shutting down all nuclear power plants within the next few decades, reducing overall energy consumption, and increasing the production from renewable sources. Since Switzerland almost reached the peak of its potential in hydropower production (BFE 2019), the focus is on other renewables such as solar, geothermal, and wind power.

However, the great variation in geographical and political conditions of the cantons has resulted in very different implementation. Some cantons prefer to increase the output from hydropower. But landscape and environmental protection are not always compatible with requests of the new energy strategy. This also holds true for wind power. The potential in Switzerland is not huge and wind park projects often face local opposition by landscape or bird protectionists. Finally, some cantons and municipalities wish to promote solar panels, which need different types of regulation (incentives and promotional measures, through such mechanisms as tax deductions or project grants) than wind or hydropower (a spatial planning problem). In sum, the different cantons face different challenges, like physical power, local opposition, or lacking policy instruments to support the local energy source (Kammermann 2018; Stadelmann-Steffen et al. 2018). Thus, not all cantons would or should rely on the same policy portfolio, as each source or context requires diverse and tailor-made policies and instruments (Stadelmann-Steffen and Dermont 2018).

14.3.2.2 Adaptation

Adaptation, meanwhile, is characterized by strong non-centralization, and most design *and* implementation powers are in the hands of the cantons – or even, depending on the canton, the municipalities. This has both historic and pragmatic reasons. On the one hand, certain climate change adaptation sectors such as flood prevention or landscape protection belong to the oldest regulated fields in Switzerland. Sectoral policies developed long before something 'unifying' called 'climate change' or 'climate change adaptation' even existed. On the other hand, given Switzerland's territorial fragmentation and socio-economic diversity, non-centralization and tailored adaptation are justified by the different needs of the twenty-six cantons in land use and economic development, with tourism being the main factor in the mountains.

Nonetheless, even though climate change adaptation is heterogeneous and non-centralized, the Swiss government released an action plan for 2014–19 that defined specific goals, challenges, and measures. It was updated to a new action plan for 2020–5. To guide the implementation of the strategy at cantonal and local levels, an advisory pilot programme and a guideline for climate-adapted settlement development was launched by the Federal Office for the Environment (FOEN), the designated national authority for climate policy. However, only very few monitoring mechanisms and no sanctions are laid out – which is why the concrete design and introduction of measures as well as their implementation remain at the mercy of lower levels of government (see also Table 14.1).

In principle, a political system composed of some 2,000 local, twenty-six cantonal, and one federal government, all largely disposing of their own powers and resources, might seem prone to end up in a race-to-the bottom competition and excessive policy fragmentation. However, unlike US federalism for instance, competition is accompanied by a strong sense of within- and between-canton solidarity (Linder and Mueller 2021). In addition, the different levels of government generally refrain from encroaching onto each other's policy spheres. Instead, there are a great number of vertical and horizontal cooperation bodies, both general and policy-specific, as well as hundreds of binding inter-cantonal treaties (Vatter 2018). Despite its dual origins, the Swiss federation today is much closer to the German administrative model (Mueller and Fenna 2022).

Furthermore, the federal level typically lacks the political will, legal basis, and/or the revenue to become active in areas already occupied by the cantons. It thus takes a concerted effort by cantonal governments or significant public pressure for the national level, for instance through a popular initiative, to amend the federal constitution or to take away powers from the cantons. At the same time, most cantons are very small – twenty of the twenty-six cantons have fewer than 500,000 inhabitants (BFS 2019) – and equally modest is what they can, or want to, have managed publicly, while spillover effects abound. Political and policy cooperation is also practised within cantons, given that no cantonal executive is formed by a single party (BFS, 2019), and inter- and intra-cantonal cooperation reinforce each other (Bolleyer 2009; Mueller and Hechter 2019). The net effect of all these structural features is, on the one hand, that public action is endowed with great degrees of democratic, bottom-up legitimacy. On the other hand, the need to build consensus and to cooperate across political parties and territorial borders slows down policy innovation and exacerbates differences. The following section explores these differences analytically.

14.4 Janus-Faced Swiss Federalism? Cantonal Adaptation and Municipal Mitigation

To test whether and to what extent Swiss federalism permits local innovation and experimentation, this section compares the Swiss cantons and selected municipalities in view of developing their own climate change policies. The intuitive explanation would be a polity's degree of vulnerability or exposure. Because of topographic or physical reasons, some cantons are more exposed to climate change effects than others, which in turn might explain why they act faster and/or more comprehensively.

At the municipal level, in turn, we investigate climate change mitigation and the adoption of the 'energy city' label, including climate-friendly measures and the promotion of the 2000-Watt society.[6] For both analyses, we rely on two master's theses defended at the University of Bern (Schmid 2018; Wieser 2018) under the supervision of one of the co-authors. Disentangling the socio-economic, political, and institutional factors for differences in subnational climate policy provides us with a look deep inside the actual workings and deficiencies of Swiss federalism. The method used in both studies is QCA (see Box 14.1); the technical details of the results are explained in Boxes 14.2 and 14.3. For an overview of factors studied, see Table 14.2.

Box 14.1
Qualitative comparative assessment

Different from a regression analysis that accounts for the causality between an independent and a dependent variable and maybe some interaction effects or control variables, QCA is strong in explaining outcomes (dependent variables) via *the combination* of conditions (independent variables). It thus follows the Boolean logic and investigates if (a) a factor is a necessary or a sufficient condition for an outcome to occur, and (b) if the presence and absence of certain factors is important for this outcome (see Dusa 2019; also 'SetMethods' in Medzihorsky et al. 2018). In our cases, the outcome is either the adoption of a climate change adaptation strategy at the cantonal level or the adoption of a gold standard energy label at the municipal level. The conditions are the six factors discussed above (Table 14.1).

Table 14.2 Factors potentially explaining cantonal climate adaptation and municipal climate change mitigation

Cantonal level: climate change adaptation		Municipal level: climate change mitigation	
ALPS	High percentage of coverage with Alps	PERFORMANCE	High energy efficiency performance
PERCEPTION	Cantonal citizens perceive climate change as a threat (above and below the Swiss mean)	EXTERNAL CHANGE	Changes in energy policy at the cantonal level
GOAL ADAPTATION	Climate change adaptation mentioned as a legislative goal	GOAL ENERGY	Energy city gold standard defined as a legislative goal
MITIGATION	Climate change mitigation is supported by the canton	ENERGY SUPPLY	City has its own energy supply company
LEFT	Head of department from left or green party	LEFT	Head of department from left or green party
ADMIN	Size of cantonal administration	LEGISLATIVE	Proportion of left-green seats in municipal parliament

14.4.1 Subnational Diversity in Climate Change Adaptation

Twelve cantons currently have an ambitious or even very ambitious climate change adaptation strategy.[7] The other fourteen do not have a proper legislative document, and most of them do not even explicitly aim at adapting to climate change. What are the necessary and sufficient conditions (factors) that lead to the adoption of a climate change adaptation strategy at the cantonal level? To assess this question, we looked at six different conditions: (1) As seen above, cantons with a high percentage of coverage with ALPS tend to be more vulnerable to climate change, rendering adaptation action more likely. (2) Naturally, this is spurred by a high PERCEPTION of climate change being a threat by cantonal citizens.[8] Furthermore, cantons show greater ambition in climate change adaptation if (3) such adaptation is already mentioned as a legislative goal (GOAL ADAPTATION), (4) the federal climate change MITIGATION policy is supported by the canton, and (5) the head of government is from a LEFT or green party.

No single condition was evaluated as necessary (Wieser 2018). The only condition coming close to the degree of necessity (Ragin 2008) is MITIGATION, or the explicit support of national mitigation targets (e.g., CO_2 reduction). In other words, most cantons that support the national mitigation targets also introduce a cantonal adaptation strategy. But, generally, the situation of when and why cantons adopt an adaptation strategy is more complex (see Box 14.2). Some conditions are sometimes present and sometimes absent (like MITIGATION or the support of a LEFT department head), yet still an adaptation strategy was introduced. Nevertheless, we can still infer some generalities. Climate change adaptation anchored as a GOAL in the cantonal legislature, and an explicit support of national MITIGATION targets are never present together at the same time. But we can conclude that both their independent presences can be a condition that leads, in combination with other conditions, to the adoption of climate change adaptation policies. Furthermore, the vulnerability of a canton (ALPS) seems to make a difference, also for the other conditions that need to be present. In alpine regions, GOAL or MITIGATION, the latter together with strong citizens' PERCEPTION of the problem, lead to the adoption of a climate change adaptation strategy. In contrast, in non-alpine regions it is additionally the presence of a LEFT department head that seems decisive. We thus find an interesting mix of the degree of affectedness, politics, and policy that jointly leads to the climate adoption of climate change adaptation policy at the cantonal level. What this means for Swiss federalism is discussed below, after having looked at the municipal level.

Box 14.2

Analysis of cantonal adoption of adaptation strategy

Table 14.3 *Solution pathways to the adoption of a cantonal climate change adaptation strategy*

	Consistency	PRI	Raw coverage	Unique coverage	Cases
*mitigation*GOAL*	1	1	0.325	0.191	AG, BL, ZH, UR, GR
*MITIGATION*LEFT*alps*	1	1	0.189	0.189	BS, SH
*GOAL*left*ALPS*	1	1	0.189	0.054	UR, TI
*MITIGATION*left* ALPS*PERCEPTION*	0.846	0.638	0.209	0.15	VD, VS

Note. Consistency of 1 is a maximal consistency of the result. PRI is the proportional reduction in consistency. Coverage indicates how much this combination of conditions covers in comparison to all possible combinations. This is also related to the number of cases explained.

Continued

Box 14.2 (cont.)

The first column in Table 14.2 indicates the four pathways that lead to an adoption of a cantonal adaptation strategy. Five cantons have a strategy (AG, BL, ZH, UR, and GR) that is explained by the combination of the absence of mitigation support and the presence of an adaptation goal set for the legislature. So, in QCA language, the presence of a condition is always indicated via capitals, and the combination is indicated through a *. In other words, the * reads like an 'and'. The other three pathways combine (a) MITIGATION support at the national level with a left department head and the absence of alpine regions; (b) the climate change adaptation GOAL formulation, the absence of a left department head, and the presence of alpine regions (ALPS) in the canton; and (c) the support of national MITIGATION targets, the absence of a left department head, and the presence of both alpine regions and a high climate perception amongst the cantonal population (ALPS and PERCEPTION).

There are no deviant cases where one of these combinations would also be true for a canton that does not adopt an adaptation strategy. However, two cantons, Solothurn and Geneva, are not explained via either of these consistent pathways.

14.4.2 Climate Change Mitigation at the Municipal Level

Although only a small country, Switzerland has more than 2,000 municipalities that possess considerable autonomy, the capacity to undertake policy actions, and act as innovative entrepreneurs. One example for local innovations are labels such as 'Energy City' (*Energiestadt*). Schmid (2018) investigated the drivers for nineteen medium-size Swiss cities to make a strong commitment to climate change and energy efficiency (gold standard energy label). Ten of them adopted the 'gold standard', nine did not. Box 14.3 summarizes the four pathways that lead to the adoption of a municipal energy label (see Table 14.4). Generally speaking, at the local level a 'race to the top' – to greener and more ambitious climate change mitigation policy – characterized those cities that had already paved their way beforehand. These exhibit remarkable energy performance and have not much to add to fulfil the standard's requirements, receive incentives from the national or other cantonal subsystems, or have already made up their mind in terms of defining their own energy efficiency goals. While this is again good news for Swiss federalism in terms of not holding back the local climate pushers, it does little for the rest for the country and may even act as an excuse for the federal level *not* to get (too) involved (see also Keeler 2007, 354).

More particularly, at the local level political factors play a decisive role. The energy department and/or the municipal parliament in the hands of the left are two conditions present in three out of four pathways explaining ambitious local climate change mitigation policy. Also, external changes and evidence as well as energy goals facilitate the introduction of the label. Finally, and this is a consistent finding with other studies (Kammermann 2018), economic factors and local entrepreneurship induce innovative action. This is why the presence of an own energy supply company in a town also spurs adopting more ambitious energy standards.

Box 14.3
Analysis of municipal climate change mitigation policy

Table 14.4 *Solution pathways to the adoption of a municipal energy label*

	Consistency	PRI	Raw coverage	Unique coverage	Cases
*PERFORMANCE*ENERGY SUPPLY*LEGISLATIVE*	0.922	0.922	0.368	0.251	Uster, Neuchâtel, Schaffhausen, Frauenfeld
*CHANGE*GOAL*LEFT*	1	1	0.349	0.079	Dietikon, Zug, Köniz, Montreux, Frauenfeld
*CHANGE*LEGISLATIVE *LEFT*	1	1	0.313	0.067	Vernier, Köniz, Montreux, Frauenfeld
*PERFORMANCE*CHANGE *GOAL*ENERGY SUPPLY*	1	1	0.199	0.058	Riehen Frauenfeld

Note. Consistency of 1 is a maximal consistency of the result. PRI is the proportional reduction in consistency. Coverage indicates how much this combination of conditions covers in comparison to all possible combinations. This is also related to the number of cases explained.

Again, there is no deviant case, and this time, there is even no case that stays unexplained. So, the ten positive cases (i.e., cities with the gold standard energy label) are all explained via one or more of the four pathways presented in Table 14.3. Interestingly, the conditions are always positive: no absence of a condition together with other conditions lead to a positive outcome. The first pathway is in the sense of the Multiple Streams framework and combines three indicators from each stream: first, high PERFORMANCE. This means that there is evidence that the energy performance of the city is rather high and not much effort needed anymore to comply with the gold standard requirements. This effectively seems to ease the adoption of the label, together with an own ENERGY SUPPLY company (policy stream), and the LEGISLATIVE in the hands of the left-wing parties. The second and third pathway include the presence of a LEFT department head (politics stream) together with CHANGES occurring in other subsystems (such as incentives coming from the national energy subsystem; problem stream). Additionally, the second pathway includes a policy stream factor: the GOAL

Continued

Box 14.3 (cont.)

formulation (the city makes climate change mitigation a local goal to respect); whereas the third pathway again a political indicator: the LEGISLATIVE in the hands of the left. The fourth pathway then combines PERFORMANCE; with external CHANGE; GOAL and an own ENERGY SUPPLY company present in the city.

14.4.3 Synthesis of the Two Case Studies

Swiss federal climate policy, albeit modest in ambitions and held in check as recently as June 2021, does at least allow regional and local governments to become climate pioneers and adopt both adaptation and mitigation strategies in their own right. Swiss federalism permits both innovation and experimentation. However, the climate policy field is quite different from other areas such as education or health (see Füglister 2012; Maggetti and Gilardi 2016). There, different diffusion mechanisms and also intensive inter-cantonal competition can be observed that lead to the willingness of some cantons to provide best practice examples and learn from each other. Cantons rely extensively on inter-cantonal cooperation bodies (*interkantonale Direktoren-Konferenzen*, see Vatter 2018, 73ff.), where the respective department heads meet to exchange experiences in such best practices. While there does indeed exist a Conference of Cantonal Energy Directors (EnDK), it seems that there is administrative coordination, but no real regulatory competition between or diffusion among the cantons about climate change policies (see also Sprinz and Weiss 2001). Thus, while innovation does occur, it is less likely to translate into a diffusion of best practices or even a coherent national and binding strategy.

That climate change is such a cross-cutting policy area definitely does not help either, since there are separate Conferences for Landscape (KWL), Transportation (KöV), Planning and Environment (BPUK), Agriculture (LDK), and, of course, Economy (VDK) and Finance (FDK) (Vatter 2018, 76). By consequence, it seems unlikely that higher energy or environmental protection standards get promoted jointly (see Casado-Asensio and Steurer 2016). Moreover, and with relevance mainly for climate change adaptation, climate effects also vary a lot across cantons: every canton seems best served in compiling its own portfolio of measures to fight climate change and does not compete with or rely upon experiences made in other cantons.

Finally, neither the cantons nor the cities have so far come under pressure from the central government to adopt ambitious adaptation or mitigation targets and policies. There also seems to be very little competition between the cantons and cities in this regard. At both levels, there are several cantons and cities that have introduced climate policies, but there are at least as many that have not. Our results show that climate policies are easiest to introduce when left-wing parties (Socialists or Greens) are leading the respective cantonal or municipal department or hold majorities in the parliament. Besides this political condition, existing policies also matter: as soon as the canton or city supports already formulated targets at the national level, or introduces its own targets at cantonal or municipal level, adaptation and mitigation policies are more likely to materialize. Finally, problem perception and affectedness are also part of the mix of conditions that lead to regional and local climate action. Yet vulnerability and affectedness amount to neither a sufficient nor a necessary condition: it is only in combination with being prepared (through policy goals) and having the 'optimal' leader or political support (left-wing) that regional or local climate policy materializes. But what is great news for Swiss federalism as decentralization is bad news for overall Swiss climate change governance.

14.5 Conclusions: Enhancing Intercantonal Coordination

Swiss cantons and cities only occasionally collaborate and exchange experiences when it comes to climate change adaptation and mitigation. Put differently, it is not through 'learning from others' that they start engaging in more ambitious climate change policies. This is also true for measures on energy efficiency and the promotion of renewable energies. Every canton has its own geo-topographical specificity which affects its energy portfolio as well as its vulnerability to climate change. This in turn makes some steering mechanisms more suitable or acceptable than others (Stadelmann-Steffen et al. 2020). In other words, every canton is affected differently by climate change and also possesses a different potential to promote renewables. As a result, every canton needs a different portfolio of steering mechanisms and implementation arrangements to achieve the set targets in both climate change mitigation and adaptation (Kammermann and Ingold 2019).

While this would indeed offer a favourable context for experimental policy-making, so far federalism has created a fragmented patchwork of different regional climate policies rather than a joint learning from innovative solutions and best practices. The plethora of cantonal and local solutions can also obstruct national policymaking in that not only are national directives implemented differently in the different cantons, but centralization steps are generally harder to take given the obstacle of direct democracy in combination with a still deeply ingrained federal political culture. Paradoxically, subnational pioneers can also be misused as an excuse for further delaying action at federal level: if they can do it, why should we? This feeds into a wider default-reaction against further centralization: 'The combination of federalism with direct democracy … gives the cantons high veto power and amounts to a considerable obstacle for federal innovation. One of the most common arguments against national policies is mistrust of the federal government and defence of cantonal autonomy' (Linder and Mueller 2021, 82).

Nevertheless, our research has shown that there is ample room for intensified diffusion and exchange: there do exist institutionalized platforms such as the Conference of Cantonal Energy Ministers, where experiences can be shared and evidence-based problem perception about climate change effects developed. This is even truer for the municipal level: unlike the cantons, Swiss cities face very similar challenges in this area, and thus the exchange of experiences would be even more effective at this level. In short, so that subnational policy innovations become elements of experimental learning their existence must be communicated and discussed and their effects assessed and compared.[9]

Linking the acceptance of subsidiarity as bottom-up policymaking to the role of politics identified in this chapter, leads to the conclusion that local decision-makers especially from left–green parties are able to fill the void created by both the reluctant national level and insufficient subnational coordination. At the same time, we see the price to be paid for the excessive levels of both local and regional autonomy still present in Switzerland: subnational units can afford not to learn from each other. Or rather, they can be left to believe so.

Notes

1 See www.tripartitekonferenz.ch/fr [25.10.2021].

2 See www.bafu.admin.ch/bafu/en/home/topics/climate/info-specialists/adaptation/strategy.html

3 CO_2 is stored in wood and the surrounding soil. Forests are considered CO_2 sinks when more wood grows than is used. Under the Kyoto Protocol, incremental forest growth can be counted as negative emissions.

4 The reason for giving preference to the term 'non-centralization' is that 'decentralization' implies movement away from and delegation by the centre. However, in the Swiss context the movement, if ever, is generally in the opposite direction: delegation by the cantons to the federal level (see e.g., Dardanelli and Mueller 2019, 139)

5 *Federal Act on the Reduction of CO_2 Emissions*, at www.fedlex.admin.ch/eli/cc/2012/855/en [12.7.2021].

6 An energy city is a municipality or a city that is continuously committed to the efficient use of energy, climate protection, the promotion of renewable energies, as well as environmentally compatible mobility. When fulfilling precisely defined targets in these areas, it receives the 'energy city label' from the sponsoring association. This is re-evaluated every four years. For more details, consult: www.local-energy.swiss/programme/energiestadt#. The vision and finally the concept of a '2000-Watt Society' was developed at the Swiss Federal Institute of Technology (ETH) in Zürich. It is a model for energy policy, which demonstrates how it is possible to consume only as much energy as worldwide energy reserves permit and which is justifiable in terms of the impact on the environment. It is possible when every person in every society limits their energy consumption to a maximum of 2,000 watts. So, the overall average primary energy usage should be lowered to 2,000 watts (i.e., 2 kWh per hour or 48 kWh per day) by the year 2050. Today, the primary energy consumption per capita worldwide is on average 2,500 watts – with enormous country-specific

differences. At present, each Swiss inhabitant uses about 4,700 watts. For more details, consult: www.2000watt.swiss/english.html.

7 Aargau (AG), Basel-City (BS), Basel-Landschaft (BL), Geneva (GE), Grisons (GR), Schaffhausen (SH), Solothurn (SO), Ticino (TI), Uri (UR), Vaud (VD), Valais (VS), Zurich (ZH).

8 High problem perception is defined as being above the Swiss mean. Low problem perception is accordingly the opposite.

9 Ironically, in its renewed proposal to revise the CO_2-Act rejected by the people in summer 2021, the federal government referred to three cantonal popular decisions that had taken place in the meantime to make its case for a consistently strong popular demand for stricter state-wide measures (FOEN 2021, 5).

References

BFE – Bundesamt für Energie. 2019. *Schweizerische Elektrizitätsstatistik 2019* (Art-Nr. 805.005.19). www.bfe.admin.ch/bfe/de/home/versorgung/statistik-und-geodaten/energiestatistiken/elektrizitaetsstatistik.html.

BFS – Bundesamt für Statistik - BFS. 2018. *Mehr als 60% des Schweizer Treibhausgas-Fussabdrucks entsteht im Ausland.* BFS. https://naturwissenschaften.ch/service/publications/107735-mehr-als-60-des-schweizer-treibhausgas-fussabdrucks-entsteht-im-ausland.

BFS – Bundesamt für Statistik [Federal Statistical Office]. 2019. *Bundesamt für Statistik.* www.bfs.admin.ch.

Bolleyer, N. 2009. *Intergovernmental Cooperation: Rational choices in federal systems and beyond*. Oxford: Oxford University Press.

Burck, J., Thea Uhlich, Christoph Bals, et al. 2021. *Climate Change Performance Index Results 2022*. Germanwatch, at https://ccpi.org.

Casado-Asensio, J., and R. Steurer 2016. 'Mitigating Climate Change in a Federal Country Committed to the Kyoto Protocol: How Swiss federalism further complicated an already complex challenge'. *Policy Sciences* 49(3): 257–79.

Dardanelli, P., and S. Mueller. 2019. 'Dynamic De/Centralization in Switzerland, 1848–2010'. *Publius* 49(1): 138–65. https://doi.org/10.1093/publius/pjx056.

Dusa, A. 2019. *QCA with R: A comprehensive description*. Cham: Springer.

EPA – Eidgenössisches Personalamt. 2019. *Das Reporting Personalmanagement—Berichtsjahr 2019*. Retrieved 3 January 2020 from www.epa.admin.ch/epa/de/home/themen/personalpolitik/reporting_personalmanagement.html.

Fishcr, D. R. 2013. 'Understanding the Relationship between Subnational and National Climate Change Politics in the United States: Toward a theory of boomerang federalism'. *Environment and Planning C: Government and Policy* 31(5): 769–84.

FOEN – Swiss Federal Office for the Environment. 2014. *Swiss Climate Policy at a Glance: Status and perspectives on the basis of Switzerland's 2014 report to the United Nations Climate Change Secretariat.*

FOEN – Swiss Federal Office for the Environment. 2020a. *Data: Greenhouse gas emissions in Switzerland.* Retrieved 1 June 2020 from www.bafu.admin.ch/bafu/en/home/topics/climate/state/data.html.

FOEN – Swiss Federal Office for the Environment. 2020b. *National Inventory Report of Switzerland 2020 (Switzerland's Greenhouse Gas Inventory 1990–2018)*. www.bafu.admin.ch/bafu/en/home/topics/climate/state/data/climate-reporting/latest-ghg-inventory.html.

FOEN – Swiss Federal Office for the Environment. 2020c. *Switzerland's Fourth Biennial Report under the UNFCCC*. FOEN. www.bafu.admin.ch/bafu/en/home/topics/

climate/state/data/climate-reporting/national-communications-and-biennial-reports.html.

FOEN – Swiss Federal Office for the Environment. 2021. *Revision des CO_2-Gesetzes: Erläuternder Bericht zur Vernehmlassungsvorlage*. Bern, 17 December 2021.

Füglister, K. 2012. *Policy Laboratories of the Federal State? The role of intergovernmental cooperation in health policy diffusion in Switzerland*. Baden Baden: Nomos.

Ingold, K., and G. Pflieger. 2016. 'Two Levels, Two Strategies: Explaining the gap between Swiss national and international responses toward climate change'. *European Policy Analysis* 2(1): 20–38. https://doi.org/10.18278/epa.2.1.4.

Kammerer M., K. Ingold, and J. Dupuis. 2021. 'Switzerland: International commitments and domestic drawbacks'. In R. K. W. Wurzel, M. S. Andersen and P. Tobin, eds, *Climate Governance Across the Globe: Pioneers, leaders and followers*, 235–56. London: Routledge.

Kammermann, L. 2018. 'Factors Driving the Promotion of Hydroelectricity: A qualitative comparative analysis'. *Review of Policy Research*, 35(2): 213–37. https://doi.org/10.1111/ropr.12274.

Kammermann, L., and K. Ingold. 2019. 'Die Akzeptanz energiepolitischer Instrumente in den Kantonen'. In I. Stadelmann-Steffen, K. Ingold, S. Rieder, et al. eds, *Akzeptanz erneuerbarer Energie*. Universität Bern, Interface Politikstudien Forschung Beratung, EAWAG.

Keeler, A. G. 2007. 'State Greenhouse Gas Reduction Policies: A move in the right direction?' *Policy Sciences* 40 (4): 353–65.

Ladner, A., N. Keuffer, H. Baldersheim, N. Hlepas, et al. 2019. *Patterns of Local Autonomy in Europe*. Cham: Palgrave Macmillan.

Liefferink, D., and R. K. W. Wurzel. 2017. 'Environmental Leaders and Pioneers: Agents of change?' *Journal of European Public Policy* 24(7), 951–68. https://doi.org/10.1080/13501763.2016.1161657.

Linder, W., and S. Mueller. 2021. *Swiss Democracy: Possible solutions to conflict in multicultural societies*, 4th ed. Cham: Palgrave Macmillan.

Maggetti, M., and F. Gilardi. 2016. 'Problems (and Solutions) in the Measurement of Policy Diffusion Mechanisms'. *Journal of Public Policy* 36(1): 87–107. https://doi.org/10.1017/S0143814X1400035X.

Medzihorsky, J., I. Oana, C. Schneider, and M. Quaranta. 2018. 'Package "Set-Methods": Functions for set-theoretic multi-method research and advanced QCA'. Version 2.3.1. https://cran.rproject.org/web/packages/SetMethods/SetMethods.pdf.

MeteoSwiss – Federal Office of Meteorology and Climatology. 2020. *Climate Change in Switzerland*. www.meteoswiss.admin.ch/home/climate/climate-change-in-switzerland.html.

Mueller, S. 2015. *Theorising Decentralisation: Comparative evidence from sub-national Switzerland*. Colchester: ECPR Press.

Mueller, S., and A. Fenna. 2022. 'Dual Versus Administrative Federalism: Origins and evolution of two models'. *Publius* (early view). doi: 10.1093/publius/pjac008.

Mueller, S., and M. Hechter. 2019. 'Centralization through Decentralization? The crystallization of social order in the European Union'. *Territory, Politics, Governance*, 9 (1): 1–20. https://doi.org/10.1080/21622671.2019.1676300.

NCCS – National Centre for Climate Services. 2018. *CH2018 – Climate Scenarios for Switzerland* [Technical Report]. NCCS. www.nccs.admin.ch/nccs/en/home/climate-change-and-impacts/swiss-climate-change-scenarios/technical-report.html.

Ragin, C. C. 2008. *Redesigning Social Inquiry: Fuzzy sets and beyond*. Chicago: University of Chicago Press.

Reich, Johannes. 2021. 'Federalism and Mitigating Climate Change: The merits of flexibility, experimentalism, and dissonance'. *Transnational Environmental Law* 10 (2): 263–91.

Schmid, S. 2018. *Unter welchen Bedingungen greifen Schweizer Energiestädte nach GOLD? Eine qualitative vergleichende Analyse zur Identifikation fördernder Bedingungskonstellationen für die Einführung des Energiestadt GOLD-Labels.* University of Bern.

Sprinz, D., and T. Vaahtoranta. 1994. 'The Interest-Based Explanation of International Environmental-Policy'. *International Organization* 48 (1): 77–105.

Sprinz, D. F., and M. Weiss. 2001. 'Domestic Politics and Global Climate Policy'. In U. Luterbacher and D. F. Sprinz, eds., *International Relations and Global Climate Change*, 67–94. Cambridge, MA: MIT.

Stadelmann-Steffen, I., and C. Dermont. 2018. 'The Unpopularity of Incentive-Based Instruments: What improves the cost–benefit ratio?' *Public Choice* 175 (1–2): 37–62. https://doi.org/10.1007/s11127–018-0513-9.

Stadelmann-Steffen, I., Ingold, K., Rieder, S., et al. (2018). *Akzeptanz erneuerbarer Energie*. Universität Bern, Interface Politikstudien Forschung Beratung, EAWAG.

Stadelmann-Steffen, I., S. Rieder, and C. Strotz. 2020. 'The Politics of Renewable Energy Production in a Federal Context: The deployment of small hydropower in the Swiss cantons'. *Journal of Environment & Development* 29 (1): 75–98.

Statistik Stadt Zürich. 2017. 'Kapitel 13: Verwaltung, Kommunale Einrichtungen'. In *Statistisches Jahrbuch der Stadt Zürich 2017*. www.stadt-zuerich.ch/prd/de/index/statistik/publikationen-angebote/publikationen/Jahrbuch/statistisches-jahrbuch-der-stadt-zuerich_2017/JB_2017_kapitel-13.html.

Swissvotes 2021. CO_2-Gesetz. At https://swissvotes.ch/vote/644.00.

Vatter, A. 2018. *Swiss Federalism: The transformation of a federal model*. London & New York: Routledge.

Wieser, E. 2018. *Many Impacts and Even More Responses: Climate adaptation in Swiss cantons: Explaining the diversity of political answers*. University of Bern.

15

Climate Governance and Federalism in the United States

BARRY RABE AND HANNAH SMITH

15.1 Introduction

This chapter examines the enduring political challenges of adopting and sustaining climate policy in the American federal system. It notes the substantial carbon footprint of the United States and its ongoing struggle to secure federal-level political support for durable domestic emission reduction commitments or sustained engagement in international processes, including the Paris Agreement. The American separation-of-powers system creates numerous obstacles to either federal legislation or ratification of international agreements via treaty, both of which require legislative and executive branch assent. As a consequence, presidents frequently avoid working with Congress in favour of unilateral executive action, including reinterpretation of the federal air quality legislation to address climate concerns. These policies, however, face numerous durability challenges once a presidential term ends.

State policy adoption has been highly uneven (Bromley-Trujillo and Holman 2020; Hultman et al. 2019). Numerous jurisdictions began to adopt ambitious climate mitigation policies in the 1990s and have continually expanded the boundaries of policy innovation (Karapin 2016). This includes California, the most populous state, which has established policies in multiple sectors, maintained formidable regulatory bodies that drive implementation, and routinely prods the federal government to take added steps through either litigation or unique levers it controls (Vogel 2018). At the same time, many states do not adopt climate policies and oppose most proposed federal climate policies. This can include active resistance to compliance and multistate litigation coalitions involving elected attorneys general (Nolette 2015). In some respects, state opposition coalitions have represented an ongoing check to presidential climate policy efforts, usually involving the party opposite the president, while Congress remained gripped by prolonged inertia on climate change and other environmental issues. Texas, the second most populous state, has been a leader in state efforts to block federal climate policy, a polar opposite to California in many respects.

Partisan divides have created enormous uncertainty and conflict in the American political system, leaving a very uneven set of federal and state policies and no clear policy path for meeting Paris reduction targets. The very issue of Paris participation became unclear given policy shifts between the Barack Obama and Donald Trump presidencies, although Joe Biden prioritized Paris re-engagement upon succeeding Trump and announced bold new emission reduction goals. In turn, there has been minimal sustained discussion of adaptation policy at either federal or state levels. Biden's 2020 election alongside narrow Democratic Party control of both Congressional chambers opened the possibility of both executive actions and legislation that could give the federal government a more far-reaching role in both mitigation and adaptation while encouraging states to consider bolder steps.

Federalism can play a compensatory role at times whereby state policy adoption and implementation can partially offset federal inertia. However, state policy has faced enduring limitations, both in terms of horizontal diffusion across regions and in vertical diffusion informing and driving federal policy. In turn, states often play an active role in undermining federal policy initiatives, particularly those launched from the executive branch. Federal capacity to build upon state models and best practices has been confounded by these enduring state divides as well as growing patterns of hyper-partisanship that deter cross-party collaboration. This has been most notable in prolonged periods of Congressional inability to address climate change or other pressing environmental issues.

15.2 Climate Change in the United States

15.2.1 Contributions to Climate Change and Its Impacts

As the second largest national contributor globally of greenhouse gas emissions (CIAT 2019), the United States' climate footprint is considerable. In 2019, the country's total greenhouse gas emissions were an estimated 6,558.3 million metric tons CO_2 eq. (EPA GHG Inventory Data Explorer). Evaluating the emissions trend since the early 1990s, total greenhouse gas emissions steadily rose through the 1990s and first part of the 2000s, reaching a high point in 2007 that was 15.6 per cent higher than 1990 levels. Following 2007, emission levels generally declined but with more fluctuation between years. In 2019, they were 1.8 per cent higher than 1990 levels, and indicators of decline in 2020 were largely attributable to the global pandemic (EPA 2021).

Carbon dioxide remains the most prominent greenhouse gas emitted, accounting for about 80 per cent of total emissions in the last decade. The main sectoral sources of carbon dioxide are transportation and electric power generation. The electricity sector has experienced a greater decline in emissions due to increasing renewable energy production and shift from coal-fired to natural gas systems. The transportation sector, on the other hand, has remained centred on petroleum fuel and in recent years has produced more GHG emissions annually than the electricity sector (EPA 2021: ES-7).

Methane accounted for about 10 per cent of total greenhouse gas emissions in 2019 (EPA GHG Inventory Data Explorer), of which the primary sources are associated with energy production, agriculture, and livestock. Reported methane emissions have generally declined in the past few decades, decreasing by 7 per cent since 2005 and 18.1 per cent since 1990 (EPA 2021), although mounting evidence from increasingly sophisticated analyses indicates that federal estimates of these emissions routinely fall well below actual levels (Alvarez et al. 2018).

There are numerous ways that the USA already is, or increasingly will be, affected by the effects of climate change, many of which pose threats to the country's current infrastructural, economic, and environmental systems. While the increase in temperature poses a nation-wide threat, different parts of the country are vulnerable to particular elements of climate change in different ways (US Global Change Research Program 2018). Coastal cities face the threat of sea level rise and need resilience and adaptation strategies for impacts to buildings and infrastructure. Increase in the severity and frequency of severe precipitation and weather events poses a large threat in the Midwest, while drought intensity in the Southwest is increasing (NASA 2020). The rise in intensity of severe weather events and natural disasters affects all areas of the USA, from forest fires in western states and flooding in the Midwest to hurricanes in southern states. Agricultural systems across the nation are facing changing growing conditions and regional shifts in growing seasons. Low-income groups are particularly vulnerable and disproportionately at risk from the effects of climate change. They tend to be more exposed to severe weather such as heat and cold or precipitation and drought events, and often lack resources easily accessible to respond or adapt to changing conditions or events.

15.2.2 Commitment Relating to Climate Change

The US Constitution divides responsibility for international agreements between executive and legislative branches, while denying states any formal role in these matters. It states that the president 'shall have Power, by and with the Advice and Consent of the Senate, to make Treaties, provided two thirds of the Senators present concur'. More than 1,500 treaties were approved in this manner during the first 200 years of the federation, with only twenty-one rejections, but ratification has subsequently become far more difficult. Environmental agreements such as the Law of the Sea Convention, the Montreal Aviation Protocols, and the Kyoto Protocol, among others, have been rejected or withdrawn, reflecting the challenges of securing Senate super-majorities in a body in which every state holds a pair of seats regardless of population. The United States ratified the 1988 Montreal Protocol on Ozone Depleting Substances and approved four subsequent amendments. Congress took a major step on HFC (hydrofluorocarbons) transition through adoption of 2020 legislation that placed the United States on a timetable consistent with international phase-down goals and ratified the Kigali Amendments in 2022.

This divide between governmental branches has greatly complicated full American engagement in climate-related treaties. President George H. W. Bush signed the Earth Summit agreement but refused to sign the 1992 Convention on Biological Diversity. His successor, President Bill Clinton, signed this convention in 1993, but was unable to secure Senate ratification, along with a growing body of other international agreements (Blomquist 2002). This phenomenon repeated itself after the signing of the Kyoto Protocol, when the Clinton Administration never submitted Kyoto to the Senate for ratification given formidable opposition. In 2001, President George H. W. Bush formally withdrew the USA from the Kyoto process.

Congressional inability to ratify treaties led Obama to consider an alternative path, an executive agreement, whereby a president may articulate American support for international collaboration without Senate support for binding commitment. Both the Copenhagen Accord and Paris Agreement were negotiated by the Obama Administration with this in mind, as well as North American climate agreements with neighbouring Canada and Mexico (Riccucci 2018). The executive agreement approach needed for American involvement contributed to the Paris shift toward emphasizing volunteered and non-binding emission reduction targets, or Nationally Determined Contributions (NDC). This flexibility allowed Obama to pledge under Paris that the United States would reduce its annual GHG emissions by 26-to-28 per cent from 2005 levels by 2025, without Senate ratification or supportive legislation (Milkoreit 2019).

America's Paris NDC commitment relied heavily upon Obama executive actions – primarily electricity and transportation sector performance standards under the 1990 *Clean Air Act*, and a compilation of existing state emission reduction policies. Applicable state policies included carbon pricing (twelve states), renewable electricity standards (thirty states), energy efficiency standards (twenty-one states), and renewable fuel standards (thirteen states), among others. It also took advantage of major shifts in American electricity production from coal to natural gas produced through hydraulic fracturing. By including modest methane emission estimates, the administration could claim significant electricity sector emission reductions.

All recent administrations have continued to attend major multilateral and transnational climate deliberations. Absence of treaty status gave Trump latitude to withdraw from Paris without Congressional consultation, although the agreement's withdrawal process meant that final departure could not occur until the very end of his term (Leggett 2019). This delay allowed the incoming Biden administration to restore American engagement in Paris upon taking office in 2021 and take an active role in the Glasgow COP meetings, although no effort was made to secure Senate ratification.

15.3 Climate Change and Federalism in the United States

15.3.1 General Practice of American Federalism

American climate policy is forged and implemented within a system that blends federalism with formal separation of powers between executive, legislative, and judicial branches at both federal and state levels. American presidents and state governors have some structural similarities, operating alongside bicameral legislative chambers and multitiered courts. Their interaction has routinely generated conflict in the formation and implementation of American climate policy, including numerous policy reversals following initial adoption.

15.3.1.1 Division of Responsibilities

At the federal level, executive power vests considerable authority over international affairs and domestic policy in an elected president. However, the chief executive is not selected through popular vote but rather an Electoral College in which each state receives the sum of its members in the Senate (two per state regardless of population) and House (based on population). Popular vote in each generally leads to a winner-take-all model whereby a victorious candidate receives all state electoral votes. A president may lose the total electoral vote nationally but win the election through victory in enough states to produce an Electoral College majority, as occurred in 2000 and 2016 with far-reaching climate policy consequences.

Legislation must pass in both Senate and House and be approved by the president to become law. The Constitution gives Congress a lead role in taxation and spending as well as broad powers to pre-empt state policy addressing cross-state commercial activity. These powers have increasingly been applied to the environmental policy arena over the course of American political history (Kincaid 2019) Congressional authority can be restricted either by unilateral presidential powers (such as executive orders) or federal court rulings overturning legislation that violates Constitutional principles.

State governments retain authority to establish their own constitutions and parallel governance systems. Article 10 of the federal Constitution explains that 'powers not delegated to the United States by the Constitution, nor prohibited by it to the States, are reserved to the States respectively to the people'. States maintain separate executive, legislative, and judicial branches and active processes to amend or replace their constitutions (Dinan 2018). They have been historically dominant players in areas such as education and public health. However, many areas of public policy have shifted during the past half-century from state domination to shared state and federal authority (Kincaid 2019). This includes medium-based statutes for air and water quality, which markedly expanded federal authority during active periods of Congressional output in the 1970s and 1980s (Lowry 1997).

15.3.1.2 Contentiousness

This balance of power has not dampened tensions between federal and state governments, much less local authority, which exists largely at the discretion of individual states. Both federal and state governments have experienced mounting 'hyper-partisanship' between Republican and Democratic parties during recent decades, often resulting in policy gridlock in cases where different parties formally share power between executive and legislative branches. Climate change has fallen into this increasingly partisan divide. Earlier state-level patterns of bipartisan engagement in the late 1990s and early 2000s have subsequently declined precipitously. As a result, exclusive Democratic Party control across state government levels has increasingly become a strong predictor of active climate policy engagement, reflected in such major states as California, New York, and Massachusetts (Bromley-Trujillo and Holman 2020). Partial or exclusive Republican Party control across state government levels has increasingly posed an obstacle to climate mitigation policies, reflected in such major states as Texas, Florida, and Ohio.

These partisan divides have been exacerbated by the geological distribution of fossil fuel resources that states can potentially tap. The realization that vast shale deposits in many American regions could be unlocked via hydraulic fracturing and horizontal drilling had a profound impact on American climate policy. The fracking boom that began in the early 2000s would by the late 2010s restore America's standing as the world's leading producer of oil and natural gas. More than thirty states produce such energy and many leading production states have experienced dramatic increases in employment and tax revenue linked to this expansion, often leading them to take vigorous stands opposing climate policy threatening continued production. Most state climate policy leaders, in contrast, lack such fossil-based energy development opportunities.

15.3.2 Climate Governance in the American Federal System

15.3.2.1 Constitutional Climate Authority

Both federal and state levels possess broad constitutional powers to pursue far-reaching climate policy if they choose to do so. Multiple forms of carbon pricing, energy procurement mandates, performance standards, and other policies have been adopted in numerous states over the past quarter century, facing few federal pre-emption or other constitutional threats. The modest federal role on both national and international scales reflects political and partisan divides rather than formal constitutional impediments.

15.3.2.2 Respective Policy Development and Implementation Roles

At the federal level, Congressional gridlock is reflected in the fact that no major federal environmental legislation has been adopted in the United States since the 1990 *Clean Air Act* Amendments, approved by a predominantly Democratic Congress and signed into law by Republican President George H. W. Bush. This legislation did not expressly identify greenhouse gases as an environmental threat, instead reflecting a Congressional decision to delay in addressing climate change until later (Carlson and Burtraw 2019). No federal legislative window on climate opened sufficiently for passage during the subsequent three decades, even though considerable climate legislation was introduced into every two-year legislative session. The House did pass far-reaching climate legislation in 2009 but failed to secure Senate approval, though the 2020 elections reopened the door to renewed consideration of major climate legislation. Congress eschewed carbon pricing or regulatory standards, instead focusing on proposed expenditures and tax incentives to accelerate clean energy transition.

This Congressional inertia has shifted authority to the federal executive branch and the states, generating considerable political and legal conflict and expanding the federal court role. All presidents serving since 1990 have utilized unilateral executive powers on climate policy, filling gaps left by Congress. These 'administrative presidency' initiatives have included efforts to interpret or reinterpret ways to apply existing legislation to climate mitigation (Clinton, Obama, and Biden) or avoid doing so despite pressure from some states to act (George H. W. Bush and Trump).

States have divided in their responses to these federal initiatives. Some routinely fall in line and support implementation, whereas others pursue either resistance in implementation or outright opposition via litigation in federal courts (Merriman 2019). State partisan alignment often predicts its approach in any given instance. Forty-three states elect their attorneys general on a partisan basis, and this office is frequently a platform for future career advancement to higher state or federal office. State attorneys general can individually or in partnership with colleagues from other states attempt to delay or overturn federal executive decisions via litigation. They dramatically expanded their efforts on climate change during the 2010s (Nolette 2015; Nolette and Provost 2018). State attorneys general of one party can provide core climate policy opposition to a president of the other party, particularly salient in an era where Congress remains hamstrung in adopting legislation or counter-balancing executive branch power (Thompson, Wong, and Rabe 2020).

One early manifestation of this conflict was a landmark 2007 Supreme Court decision, *Massachusetts v US Environmental Protection Agency*. This case was brought by attorneys general from Massachusetts and other Democratic-led states to challenge the George H. W. Bush administration unwillingness to apply administrative presidency powers to climate change. It confirmed that states possessed legal standing to sue the federal government in federal court for inaction, affirming that the well-being of their citizens had been 'endangered' by federal failure to take measures to mitigate climate harms. This case focused on key *Clean Air Act* provisions, whereby California and allied states prodded the Bush administration to extend vehicle emission standards to cover carbon. The five-to-four court majority concurred that the federal government and its lead environmental agency, the Environmental Protection Agency (EPA), needed to take mitigation steps unless it was able to advance 'some reasonable explanation as to why it cannot or will not exercise its discretion to determine whether they do' (Thompson, Wong, and Rabe 2020, 30). This decision formally legitimized a state role in climate change, including formal legal challenges brought against alleged federal inaction. However, it did not resolve the issue of how the federal government would respond to the court challenge, with EPA offering very different responses during the Bush, Obama, Trump, and Biden presidencies.

15.3.2.3 State Inclusion within NDCs

States lack formal authority to participate in international negotiations or introduce emission reduction commitments that would be internationally accepted, even in cases of federal disengagement. However, state climate policies can be influential in two potential ways linked to American involvement in Paris. First, the Obama administration referenced them in explaining how the United States would achieve its Paris reduction commitments, noting anticipated emission impacts of various state-sponsored programmes. The Biden administration restored this approach in the 2021 COP meetings in Glasgow. Second, the Trump administration's 2017 decision to withdraw from Paris was met with considerable outcry from state climate policy leaders, including many governors and legislators. This reaction fostered formation of the 'We're Still In' movement and launch of the US Climate Alliance, whereby governors pledged to honour their portion of previous national NDCs, either through existing commitments or additional ones they would develop. This roster of states grew initially from thirteen to twenty-four, including additions following 2018 mid-term elections and shifts toward expanded Democratic control in states such as Illinois, New Mexico, and Wisconsin.

15.4 Case Study of American Climate Governance

15.4.1 State Climate Change Action

By contrast with the federal government's prolonged inability to produce climate legislation, many states have adopted and implemented both regulatory and more market-based climate mitigation policies. Most of these policies are specific to individual states, albeit with potential to diffuse through emulation and adaptation in other jurisdictions (Grumbach 2018; Shipan and Volden 2021; Stokes 2020). Renewable electricity standards were first adopted in Iowa in 1991 and expanded to thirty states as of 2021 with seven others maintaining voluntary versions. There is also some precedent for multiple states to adopt and implement policies that can be formally linked across jurisdictional boundaries. In carbon pricing, for example, the Regional Greenhouse Gas Initiative (RGGI) features nine partnership states maintaining a regional cap-and-trade system focused on electricity sector emissions. Auction revenue is commonly allocated for state energy transition expenses (Raymond 2016). RGGI expanded in 2019–20, adding New Jersey and Virginia as formal members while North Carolina and Pennsylvania also considered joining.

States diverge widely in their willingness and capacity to develop climate policies. California leads the nation in population and ranks second in total greenhouse gases, with particularly large releases from its transportation and agricultural sectors. It has long been a national leader on environmental policy, particularly air quality issues that have proven particularly vexing, and it has long prodded other states and the federal government to take more aggressive stances (Vogel 2018). Substantial climate change engagement began in the early 1990s, and California maintains a multisector cap-and-trade programme (operated jointly with the Canadian province of Québec) that allocates auction revenues for climate mitigation and adaptation. California also implements a wide range of renewable energy, energy efficiency, and biofuels programmes, often delegating considerable administrative authority to the formidable California Air Resources Board.

In turn, Texas leads the nation in total greenhouse gas emissions and ranks second in population. It has been a laggard in many climate policy areas and energy sector emissions have surged due to massive oil and gas production expansion since 2005 linked to hydraulic-fracturing techniques, particularly in the abundant Permian Basin. State leaders have long expressed doubts about the existence of climate change and have regularly led opposition to federal executive efforts to adopt climate policy. At the same time, Texas has abundant wind resources and has experienced substantial growth in its wind energy deployment, jumping from less than 1 per cent of total state electricity generation in 2000 to nearly 20 per cent in 2019. It has also invested heavily in renewable energy transmission infrastructure through fees added to electricity bills.

State policy divergence has greatly complicated federal attempts to use executive power to develop climate policy nationally, including Obama's tenure. Alongside failure to consummate proposed cap-and-trade legislation in 2009, Obama advanced multiple administrative presidency initiatives linked to separate *Clean Air Act* provisions. All of these, including those focused on the electricity and transportation sectors, required extensive federal–state interaction and built on state model cases. They required extensive rule-making procedures set forth in federal legislation and could theoretically be adopted and implemented by EPA without Congressional input or approval (Belton and Graham 2019). Collectively, they represented cornerstones in the Obama administration's plan to achieve proposed Paris Agreement emission reductions just as they would subsequently under the Biden administration. States led by California actively inspired these efforts and actively endorsed them. States led by Texas actively opposed them, taking formal steps to reverse or undermine them.

For electricity, Obama sought to revise *Clean Air Act* provisions overseeing State Implementation Plans (SIPs) for compliance with various air emissions control requirements. This involved multiyear development of the Clean Power Plan (CPP). The power sector had already registered major emission reductions due to substantial coal-to-gas transition and numerous state policies promoting renewables. The nation was already heading towards net emission reductions in the sector greater than the overall 26-to-28 per cent levels established nationally under the Paris process. The CPP was intended to ensure that it exceeded those targets, compensating for other sectors less likely to achieve such extensive reductions. Each state was given a numeric emissions reduction target set federally. Many states that had already made or planned for major reductions were given more modest reduction targets and they generally proved supportive of the CPP. Many others generally opposed the very idea of a federally mandated CPP, particularly once they received more demanding reduction targets.

For transportation, the Obama administration embraced unilateral California legislation as a driving force behind a national strategy to reduce vehicular carbon emissions. Under federal air quality legislation first adopted in 1967, California was given unique authority to seek federal waivers in cases when it wanted to set higher vehicle emission standards than federal ones. Waivers are not formally established in the federal Constitution but have emerged as an intergovernmental tool that can allow or encourage one or more states to pursue policy innovation, including the case of mobile sources where single-state innovation has prompted regional and then national policy adoption. Waivers have not, however, been widely used in other areas of climate policy. The vehicle emissions case reflected California's acute air quality issues linked to transportation and its active policy development in this area long before initial federal legislation (Vogel 2018). On more than 120 occasions since the late 1960s, California waiver requests were approved by EPA. At that point, other states are allowed, under 1977 provisions, to adopt the California standard. This so-called federalism 'bandwagon effect' regularly created momentum whereby the federal government ultimately harmonized its national standards with those of California and allied states (Carlson and Burtraw 2019). The Obama administration worked independently of Congress to embrace the California carbon waiver and merge it with separate federal fuel economy standards that pre-empted state action, thereby setting ambitious tailpipe emission reduction targets for multiple vehicle classes.

15.4.2 Adoption versus Implementation, and Policy Evolution

Implementation of these two federal attempts to use executive power through distinct federalism strategies for climate mitigation proved highly contentious, reflected in aggressive opposition from many states. In the CPP, more than a dozen Republican state attorneys general filed suit in federal court within days of final rule issuance. They claimed that the CPP represented an unconstitutional reinterpretation of air quality legislation that did not address climate. They contended that it outlined remedies linked not only to energy production but 'outside-the-fence' considerations, ultimately designed to eliminate fossil fuel use without legislation. This state opposition coalition eventually expanded to Texas and twenty-six other states, primarily jurisdictions with Republican leadership and significant production and use of fossil fuels within their boundaries. In contrast, California and seventeen other states, primarily with Democratic leadership and less-intensive carbon profiles, embraced the CPP. These states contended that CPP compliance would be highly feasible given growing renewable energy cost-competitiveness and could contribute significantly to American climate mitigation efforts.

This executive strategy drew no formal Congressional response but faced an unusual Supreme Court stay, suspending implementation until after 2016 elections. Trump's election generated an executive order launching CPPs replacement with a far more modest Affordable Clean Energy rule that was suspended by a federal court in 2021. It was generally expected to have negligible impact on reducing emissions and may have actually increased them in some states through extended coal plant operation (Keyes et al. 2019). However, it faced its own durability challenges in the Biden administration.

Trump was not as outspoken during his campaign against the vehicle emissions programme as the CPP. But his administration took rapid steps to begin the unprecedented process of reversing the waiver that California (and bandwagon states) had received, substituting a far more modest plan that precluded any state role through federal pre-emption of the process. Trump elevated his waiver opposition to a particularly intensive level upon learning that four major vehicle manufacturers pursued quiet negotiation of higher standards with California, including vulgar presidential diatribes and threats against these firms and the state. As with CPP, Trump did not entirely eliminate Obama's regulatory effort but rather reconfigured it, producing more modest emission reduction targets and more flexible implementation. This approach would make it difficult for any successor to reverse it, while constraining state influence on policy design. Nonetheless, Trump efforts were not finalized prior to Biden administration arrival, which moved to restore the earlier waiver and build upon it in advancing its own executive climate strategy.

15.4.3 Factors Facilitating or Hindering State Actions

The federal government role on climate policy has largely remained confined to the executive branch with occasional court engagement. This reflects long-standing legislative branch inability to adopt climate or environmental legislation regardless of partisan control of Congress. The cases demonstrate limits that individual presidents and administrations face in advancing far-reaching policies that can be implemented and prove durable. The federal government has not created intergovernmental revenue transfer to encourage or support state energy transition policies or invested in boosting state implementation capacity. States nonetheless retain considerable latitude to either develop their own policies or take individual or collective action to either support or oppose federal initiatives. The case studies suggest that states may be more successful at blocking initiatives that they dislike rather than fully implementing them.

15.4.3.1 Constitutional or Devolved Authority of States

The electricity and transportation cases demonstrate the considerable latitude that states have to respond to proposed federal climate strategies that are quite ambitious in their potential scope for emissions reductions. In the electricity case, Texas played a central role in supporting opposition through litigation challenges that ultimately wounded the CPP through a Supreme Court stay when plans were advancing towards full implementation. In the transportation case, Texas opposed California's position and backed the Trump administration's efforts to weaken the federal vehicle emissions programme. California took the opposite side on these policies, actively supporting the CPP and contending that it would blend effectively with its existing entourage of climate policies. It was the impetus behind the entire vehicle emissions programme, having secured Obama administration support not only for granting its waiver request but also elevating its state-wide policy into national policy. Despite their setbacks under the Trump administration, California and allied states worked cooperatively with the incoming Biden administration on new federal policies that built upon their early efforts.

15.4.3.2 State Capacity

The divergent paths that California and Texas have taken on climate policy is further reflected in their respective development of state administrative capacity to address this issue. California revenue under cap-and-trade reached \$2.5 billion in 2019, distributed across a wide range of state and local climate mitigation and adaptation initiatives. The California Air Resource Board retains extensive staff depth and talent, rivalling the US Environmental Protection Agency in these respects. It regularly receives broad support from California's governor and has had remarkable durability in senior leadership, making it a formidable force in every arena of state climate policy development and implementation.

Texas also maintains large state agencies with jurisdiction over air quality and energy production but has pursued major staff and budget reductions for these units in recent decades. It consistently ranks far behind California on comparative measures of capacity and commitment to environmental protection. The Texas Railroad Commission is a throwback to an early era in Texas but has continued to govern most aspects of oil and gas production and has remained highly deferential to industry preference on issues such as methane mitigation. Unlike California, Texas elected officials do not encourage state agencies to identify climate change as an express concern, and many seized upon an extended period of state-wide electricity loss in early 2021 to place the blame on freezing wind turbines rather than consider broader electricity system challenges contributing to the situation.

15.4.3.3 State Paradiplomacy Engagement

States face substantial federal constitutional constraints on international policy engagement, including treaty participation. However, California and a few other states have tested those powers on climate change in recent decades. California governors have hosted global climate summits for two decades, routinely inviting national and sub-federal leaders for conferences and periodic signing of non-binding memoranda of understanding. Its efforts to build formal partners for its cap-and-trade programme secured four Canadian provincial partners in 2010, although only Québec has remained in alliance through implementation.

15.4.3.4 State Incentive for Taking Climate Action

States regularly frame climate policies as sources of economic development benefits and co-benefits such as improved air quality. California has long contended, for example, that its active engagement in the transportation sector offers climate benefits but also a substantial boost for next-generation vehicle technologies being developed by its universities and private firms as well as air quality benefits. Colorado and New Mexico have emerged as national leaders in methane mitigation linked to energy production, linking climate policy with methane capture for use as natural gas that can provide tax and royalty revenue.

15.4.4 State Compensatory or Pre-emptive Action

California exemplifies Martha Derthick's (2010) notion that states might take policy action to compensate for the absence of federal engagement, attempting to address their own emissions and also provide an example or model that can be emulated by other states or the federal government. In transportation, we see California's ongoing efforts to compensate for the slow pace of federal regulatory reforms on vehicle emissions, employing its unique waiver authority to force national consideration of carbon-sensitive standards.

Many other states have also attempted to play this type of role. In carbon pricing, RGGI formation and expansion reflected concern from participating states about federal disengagement as well as desire to create a model that could inform future federal policy. Much of the state work that contributes to the US Climate Alliance was intended to fill the gap left by the Trump administration withdrawal from Paris and reversal of Obama administration regulatory programmes (Hultman, et al. 2019).

15.4.5 State Attention to Local Expertise and Circumstances

American local governments, particularly medium-to-large cities, have made considerable efforts to address both climate mitigation and adaptation in recent decades (Hughes 2019). They face many formal limitations, however, lacking independent constitutional authority and often reliant upon state interpretation of their potential taxation and regulatory powers. Nonetheless, local governments have regularly launched climate action plans that focus on areas of influence such as building standards and public transportation. Five hundred and thirty-four American cities, counties, and tribes, including some from every state, pledged fealty to Paris reduction targets. States have not, however, consistently drawn upon local expertise or unique circumstances in developing their own climate policies, reflecting a broader pattern of state–local tension in American politics. This reflects state trends in recent decades to reduce financial transfers from revenue-sharing programmes for local government as they contended with their own fiscal challenges, including the decade following the Great Recession.

In the case of California, local government climate engagement has been considerable, and some dimensions of state policy involve local or regional entities. The state's far-reaching air quality efforts include considerable delegation of authority to local air quality districts for monitoring and compliance oversight. These bodies have played some role in climate policy implementation as well, and the state has also channelled significant cap-and-trade auction revenue to local governments, placing a growing emphasis in the past decade on disadvantaged communities that may lack resources to mount their own mitigation and adaptation strategies. This pattern is also evident in many of the RGGI cap-and-trade states and their evolving use of auction revenue in local communities. In many other states, localities with a strong climate interest may be well ahead of their state in commitment and capacity and so have to act unilaterally, albeit with formal limitations on their authority.

15.4.6 Horizontal and Vertical Convergence and Divergence

The divergent paths of American states over past decades on climate change can be demonstrated in part by distinguishing three distinct directions that clusters of states have pursued (Hultman et al. 2019, 19–20). 'First-mover' states routinely adopt new policies and have created a medley of policy responses over time. These represent 45 per cent of the total population but only 33 per cent of total greenhouse gases. California is a leading first-mover, joined by other states with predominant Democratic Party control and relatively limited oil and gas production capacity. 'Fast-follower' states lack a steady pattern of early adoption but often emulate leader states over time. These states represent 20 per cent of the total population and 20 per cent of total greenhouse gases. 'Slow-follower' states tend to delay or resist climate policy adoption. These states represent

35 per cent of the total population but 47 per cent of total greenhouse gases. Texas is the leading example, joined by a number of other states with predominantly Republican Party control as well as significant fossil fuel reserves and production.

Comparable categories can be used to divide the ways that states either formally oppose or support proposed federal policies through litigation. This suggests a significant divergence among subsets of states, one that may remain fairly consistent over time but can shift if partisan control changes or other factors emerge. Earlier expectations that significant first-mover efforts in the early 2000s would foster widespread 'horizontal diffusion' of multiple policies have largely not been realized, despite a long-standing American history of diffusion of environmental and other policy innovations over the last century (Baldwin, Carley, and Nicholson-Crotty 2019; Karch 2007; Shipan and Volden 2021). Neither have earlier projections that a critical mass of states might adopt a climate policy and then tip the federal government into adopting this policy on a national basis via 'vertical diffusion' (Posner 2010).

In turn, not all state climate policies endure, reflected in cases of 'reverse diffusion' following initial adoption. Carbon cap-and-trade was adopted in two Northeastern states in the early 2000s but spread by 2010 to twenty-three states lodged in regions of the Northeast, Midwest, and Pacific West. This was widely seen as setting the stage for additional horizontal diffusion and eventual federal vertical diffusion that would build on these experiences, but political support quickly ebbed. Thirteen states subsequently withdrew, including many in the Midwest and Mountain West, although there was a slight uptick a decade later. These reversals reflected Congressional rejection of cap-and-trade, increased oil and gas production, intensified state partisan cleavages, and shifts from Democratic to Republican control in many states (Rabe 2018).

15.4.7 Climate Governance Conflict and Cooperation

Intergovernmental patterns of conflict and cooperation have waxed and waned over the course of recent decades. In general, federal–state conflict differs significantly depending upon partisan control of the presidency and Congress at a given time (Karapin 2020; Rabe 2011). In instances where Republicans dominate the federal government, states tend to more actively pursue their own climate policy development and challenge federal pre-eminence. These can be considered periods of 'state domination' whereby they attempt to fill perceived gaps in federal engagement and often find themselves in conflict with federal authorities. During Democratic presidencies, however, the tables turn. State climate policy development can slow in anticipation of new federal policies, as they did during the Obama presidency. Many Republican states entered into formal opposition to federal policy initiatives during this period, including litigation.

There is considerable opportunity for state clusters to cooperate in the American federal system, including inter-state compacts that the Constitution allows with Congressional consent (Bowman 2004). These historically have involved numerous policy areas, including water quality and energy production. No such compacts have been approved in recent decades related to climate change and two regional cap-and-trade initiatives either disappeared or withered. However, RGGI has proven a durable regional programme that has endured major adjustments in its emissions cap, implementation of an auction-and-invest system for revenue use, and membership changes (Raymond 2016).

15.4.8 Coercion, Collaboration, Competition, and Emulation in Climate Policy

The CPP represents a leading example of an executive effort to develop a federal climate strategy to operate with considerable room for single or multistate innovation. The CPP adapted the *Clean Air Act* state implementation plan process that sets federal standards but gives states considerable latitude to develop their preferred response, working either independently or collaboratively with neighbouring states. As long as the federal government was satisfied that a state had developed a credible plan, it would delegate implementation authority to the lead state agencies, much as had been done for conventional pollutants (Carlson and Burtraw 2019).

Multistate *collaboration* offered one alternative for states to work together, likely through creation of cap-and-trade systems comparable to RGGI. This would draw on prior RGGI lessons and considerable state experience operating an emissions trading system for sulphur dioxide, offering a flexible path to minimize compliance costs for regulated parties and states. Federal officials outlined this option in early briefings around the nation with state officials. There was also an option of *competition* whereby each state would design its own strategy, using cap-and-trade or other performance-based compliance options. Under this model, each state would receive its emissions reduction target but select its preferred option and thereby try to out-compete neighbours in containing compliance costs. Finally, there was considerable expectation that either collaborative or competitive strategies would ultimately foster policy *emulation*, as states learned from their experiences over time given their divergent approaches, ultimately adapting and refining their state implementation plans.

In the end, none of these aspirations were achieved, given the aggressive state political opposition that emerged, leading to the 2016 court-mandated implementation freeze and followed by a federal executive transition that eviscerated the programme. The Obama administration launched the CPP with the implicit assumption that the president would be succeeded by a supportive Democrat committed to seeing it into full implementation. Opposition states played a pivotal role in CPP downfall, alleging that it represented federal *coercion* of federal legislation designed solely to reduce conventional air contaminants, although the federal courts and national election results dealt the death blows. The Biden administration faced the challenge of revisiting *Clean Air Act* application to electricity sector carbon emissions once it became clear in 2021 that Congress opposed a federal clean electricity standard despite its widespread state use.

15.4.9 Policy Divergence or Convergence and Dynamics of Conflict and Cooperation

Full implementation of existing and emerging state policy commitments as portions of a federal strategy may move the United States within striking distance of initial Paris reduction targets despite federal policy disengagement. However, state-led action alone would not enable the nation to approach Biden era emission reduction proposals for 2030. Innumerable effectiveness, equity, and efficiency questions continue to emerge in examining the American emissions trajectory and likely future course, barring a major shift towards emboldened federal policy. These equity challenges include the general absence of a consistent carbon pricing strategy or a sustained funding source for energy transition as well as enormous differences in the costs of electricity and energy in various states and regions. The state-driven American experience demonstrates that climate policies can serve to reduce emissions but only in an uneven matter that exacerbates broader American political and economic challenges (Kettl 2020). These factors have served to accentuate concerns in the United States over climate and energy justice, particularly for communities that have long faced substantial environmental threats and lack resources to address them.

15.5 Conclusion

This chapter demonstrates the challenge of adopting and implementing an effective and durable strategy to address climate change in a political system that combines federalism with formal separation of powers at both federal and state levels. The past quarter-century has demonstrated considerable capacity for single-state innovation, consistent with the 'laboratory of federalism' hypothesis. This is reflected in a range of state policy initiatives and varying degrees of horizontal diffusion for particular policy tools, including renewable portfolio standards and cap-and-trade. However, vertical policy diffusion has largely failed to emerge from these state efforts. In turn, state engagement has been and remains highly uneven, concentrated most heavily in states lodged along oceanic coasts that tend towards Democratic Party control of state government and have few fossil-fuel resources. State policy adoption tends to be most active during periods when the Republican Party controls the presidency. Some states remain unable to either launch or sustain initial policy commitments.

Deep divides between states and regions have profoundly complicated efforts to develop federal climate policy strategies either through legislation or executive channels. Veto points include the Senate and the ability of states from the political party opposite the president to pursue litigation and other strategies to undermine federal policy proposals. Consensus on climate mitigation strategies across regions, branches of government, and political parties remained elusive and climate adaptation was rarely addressed through policy. The dramatic expansion of oil and gas production linked to fracking technology has only emboldened pro-fossil fuel interests at the state and federal levels and deters serious examination of federal climate policy options.

Courts have played a generally supportive role in climate policy development, including the historic *Massachusetts* case that legitimized state standing to call for federal action. But their roles do not extend to policy adoption or forcing any particular policy action. In the end, decades of Congressional disengagement from most areas of environmental policy have posed a fundamental challenge that must be surmounted before any far-reaching federal climate policy strategy becomes politically feasible. In 2020, the growing possibilities of a national election favourable to climate policy interests, particularly within the Democratic Party, unleashed a wide range of proposals that began to outline a possible shift in the future Congressional role. Early steps by the incoming Biden administration in 2021 emphasized unilateral executive action while also moving towards possible legislative steps that might prove viable while his Democratic Party retained narrow control of both chambers of Congress.

American federal experience to date underscores its profound challenges in adopting and sustaining climate policy on a national scale and building on the considerable body of innovation that has been pursued within some individual state laboratories. States have partially offset federal inertia by attempting to reduce emissions within their boundaries through innovative policies. But many remain hostile or indifferent to climate policy, particularly those with substantial fossil fuel reserves, and they often lead opposition to new federal policy proposals considered by Congress as possible legislation or advanced by unilateral executive actions. In turn, the climate change policy era has coincided with mounting hyper-partisanship at both levels of government, further discouraging broader state policy development or federal policy adoption through legislation. The American federal system thus features some capacity for states to play a compensatory role when the federal government cannot engage, but this is inconsistent across states and has proven insufficient to foster politically feasible and durable policy on a national basis in recent decades.

References

Alvarez, Rámon, Daniel Zavala-Araiza, David R. Lyon, et al. 2018. 'Assessment of Methane Emissions from the US Oil and Gas Supply Chain', *Science* 361 (3698): 186–88.

Baldwin, Elizabeth, Sanya Carley, and Sean Nicholson-Crotty. August 2019. 'Why Do Countries Emulate Each Others' Policies? A global study of energy policy diffusion', *World Development* 120: 29–45.

Bednar, Jenna. 2009. *The Robust Federation: Principles of design.* Cambridge: Cambridge University Press.

Belton, Keith, and John Graham. 2019. *Trump's Deregulatory Agenda.* Bloomington: American Council for Capital Formation.

Blomquist, Robert. 2002. 'Ratification Resisted: Understanding America's response to the Convention on Biological Diversity, 1989–2002'. *Golden Gate University Law Review* 32(4): 493–586

Bowman, Ann. 2004. 'Horizontal Federalism: Exploring interstate interactions'. *Journal of Public Administration Research and Theory* 14(4): 535–46.

Bromley-Trujillo, Rebecca, and Mirya Holman. 2020. 'Climate Change Policymaking in the States: A view at 2020'. *Publius* 50(3): 446–72.

CAIT Climate Data Explorer. 2019. *Country Greenhouse Gas Emissions.* Washington, DC: World Resources Institute. Retrieved from http://cait.wri.org.

Carlson, Ann, and Dallas Burtraw, eds. 2019. *Lessons from the Clean Air Act: Building durability and adaptability into US climate and energy policy.* New York: Cambridge University Press.

Derthick, Martha. 2010. 'Compensatory Federalism'. In *Greenhouse Governance: Addressing climate change in America*, ed. Barry Rabe, 58–72. Washington, DC: Brookings Institution Press,

Dinan, John. 2018. *State Constitutional Politics.* Chicago: University of Chicago Press.

Environmental Protection Agency. 2021.*Greenhouse Gas Inventory Data Explorer.* Retrieved from https://cfpub.epa.gov/ghgdata/inventoryexplorer.

Environmental Protection Agency. 2021. *Inventory of US Greenhouse Gas Emissions and Sinks, 1990–2019.* epa.gov/ghgemissions/inventory-us-greenhouse-gas-emissions-and-sinks-1990-2019.

Grumbach, Jacob. 2018. 'From Backwaters to Major Policymakers: Policy polarization in the states, 1970–2014'. *Perspectives in Politics* 16(2): 416–35.

Hughes, Sara. 2019. *Repowering Cities.* Ithaca, NY: Cornell University Press.

Hultman, Nathan, et al. 2019. *Accelerating America's Pledge.* New York: Bloomberg Philanthropies.

Karapin, Roger. 2016. *Political Opportunities for Climate Policy.* New York: Cambridge University Press.

Karapin, Roger. 2020. 'Federalism as a Double-Edged Sword: The slow energy transition in the United States'. *Journal of Environment & Development* 29(1): 26–50.

Karch, Andrew. 2007. *Democratic Laboratories: Policy diffusion among the American states.* Ann Arbor: University of Michigan Press.

Kettl, Donald. 2020. *The Divided States of America: Why federalism doesn't work.* Princeton: Princeton University Press.

Keyes, Amelia, Kathleen Lambert, Dallas Burtraw, et al. 2019. 'The Affordable Clean Energy Rule and the Impact of Emissions Rebound on Carbon Dioxide and Criteria Air Pollutant Emissions', *Environmental Research Letters* 14 (4): 1–10.

Kincaid, John. 2019. 'Dynamic De/Centralization in the United States, 1790–2010', *Publius* 49 (1): 166–93.

Leggett, Jane. 2019. *Political Implications of US Withdrawal from the Paris Agreement on Climate Change.* Washington, DC: Congressional Research Service.

Lowry, William. 1997. *Dimensions of Federalism: State governments and pollution control policies.* Durham, NC: Duke University Press.

Merriman, Ben. 2019. *Conservative Innovators: How states are challenging federal power.* Chicago: University of Chicago Press.

Milkoreit, Manjana. 2019. 'The Paris Agreement on Climate Change: Made in USA?' *Perspectives in Politics* 17 (4): 1019–37.

NASA. *The Effects of Climate Change.* Last updated 27 July 2020. Retrieved from https://climate.nasa.gov/effects/#:~:text=Increased%20heat%2C%20drought%20and%20insect,coastal%20areas%20are%20additional%20concerns.

Nolette, Paul. 2015. *Federalism on Trial: State attorneys general and national policy-making in contemporary America.* Lawrence: University Press of Kansas.

Nolette, Paul, and Colin Provost. 2018. 'Change and Continuity in the Role of State Attorneys General in Obama and Trump Administration'. *Publius* 48 (3): 469–94.

Portney, Kent. 2013. *Taking Sustainable Cities Seriously.* Cambridge, MA: MIT.

Posner, Paul. 2010. 'The Politics of Vertical Diffusion: The States and climate change'. In *Greenhouse Governance: Addressing climate change in America,* ed. Barry Rabe, 73–100. Washington, DC: Brookings Institution Press.

Rabe, Barry. 2011. 'Contested Federalism and American Climate Policy'. *Publius* 41 (3): 494–521.

Rabe, Barry. 2018. *Can We Price Carbon?* Cambridge, MA: MIT.

Raymond, Leigh. 2016. *Reclaiming the Atmospheric Commons: The regional greenhouse gas initiative and a new model of emissions trading.* New York: Penguin Random House.

Riccucci, Norma. 2018. *Policy Drift: Shared powers and the making of US law and policy.* New York: New York University Press.

Shipan, Charles, and Craig Volden. 2021. *Why Bad Policies Spread (and Good Ones Don't).* Cambridge: Cambridge University Press.

Stokes, Leah. 2020. *Short Circuiting Policy: Interest groups and the battle over clean energy and climate policy in the American states.* Oxford: Oxford University Press.

Thompson, Frank, Kenneth Wong, and Barry Rabe. 2020. *Trump, the Administrative Presidency, and Federalism.* Washington, DC: Brookings Institution Press.
US Global Change Research Programme. 2018. *Fourth National Climate Assessment.* globalchange.gov/nca4.
Vogel, David. 2018. *California Greenin': How the golden state became an environmental leader.* Princeton: Princeton University Press.

16

Reflections on Climate Governance and Federalism

ALAN FENNA

This book has brought together accounts of the relationship between climate change governance and federal or federal-type arrangements across a wide range of cases. Despite the considerable variation among them, it is possible to make some observations about that relationship. Specifically, we wanted to get an insight into the way a system structured into two (or occasionally three) orders of government with powers divided and shared between them might either facilitate or obstruct action addressing climate change. In Chapter 1, we noted that a considerable literature highlights the 'double-edged' character of federalism in this regard.

The diversity of cases is manifest in many ways, with our sample including countries at very different levels of economic development; federations based on a dual or an administrative division of powers; ethno-federations and those with a single national culture; parliamentary and presidential systems; centralised and quite decentralised federations; and regimes that range from established liberal democracies to autocracies. In addition, three of them are federations within a supra-national federation, the EU, itself one of our cases. Finally, we also included two unitary states with some form of devolved governance.

The key characteristic of federalism here is the way it creates 'states within states' – an arrangement that provides (a) a second set governmental actors, and (b) greater scope for territorial diversity in policymaking. In turn, this focuses attention on the way powers, responsibilities and resources are divided or shared between those constituent units and the central government, and the way in which the different governments relate to one another. How those will play out is in turn dependent on the political, economic and social characteristics of each federal system.

Countries around the world have been under pressure for many years now to mitigate climate change by reducing their greenhouse gas (GHG) emissions and to introduce adaptive measures. For most of our cases, the mitigation challenge is to transition away from hydrocarbon fuel sources – coal, petroleum and natural gas – to renewables, notably wind and solar. A number of the federations represented here are rich in fossil fuels and all of them have built industrial and consumption structures around those energy sources. Another important characteristic of those resource endowments is that they tend to be regionally concentrated within each federation. Germany has its coal states, Canada its petroleum provinces, the United States its coal and petroleum states. For those less-industrialised federations, the issue is not fossil fuels, but land use. The GHG contributions of Ethiopia, Indonesia and Brazil are much more tied to deforestation and agricultural practices. In Brazil that is also regionally concentrated, while in Ethiopia and Indonesia emissions associated with agriculture and deforestation are more broadly prevalent.

16.1 Federalism as Facilitator

In Chapter 1, we canvassed three ways in which federalism might facilitate policymaking: allowing locally tailored responses, providing opportunities for compensatory action, and opening the door to policy experimentation and learning. Climate change governance would seem particularly advantaged by such potentialities since, as is often noted, it encompasses a range of possible measures that can be implemented and operate at very different governance scales.

16.1.1 Locally-Tailored Policymaking

A virtue of federalism is often seen as being the way it invites policymaking that is customised to local circumstances, needs and preferences – the '*decentralization theorem*', as Oates (1972, 35; italics in original) called it. Federal systems provide scope for such customisation to a degree unlikely in a unitary state, where 'one-size-fits-all' approaches may prevail.

This requires, of course, that the constituent units enjoy sufficient jurisdictional authority – and the studies in this book suggest that in climate change policy they generally do. Such a degree of autonomy unsurprisingly exists in the European Union's (EU) supra-national proto-federal system, as Alberton emphasises. Even in Germany, though, with its administrative division of powers where the constituent units have more of an implementational role and less policy autonomy than in dual federations, Eckersley and colleagues make it clear that the *Länder* have had considerable scope to choose their own climate policy direction.

However, this is not the case everywhere. Pillai and Dubash emphasise the degree to which the highly centralised nature of India's federal system reduces the scope for independent action by the constituent units. Although India's states control several important areas of climate policy, financial and capacity constraints limit prospects for bottom-up action. Meanwhile, de Visser and du Plessis show that what they characterise as South Africa's 'quasi-federal' system leaves the provinces with very little policy role at all, and the cities without the resources to pursue the climate action they contemplate. As Di Gregorio and Moeliono explain, the Indonesian case is similar: lacking the constitutional guarantees provided by federalism, the local role in forestry management there has been assumed by central government authorities, with adverse implications for outcomes.

The cases covered in this book provide numerous instances of regionally tailored climate change policy. Constituent units in Australia, Brazil, Canada, Germany, Mexico, Spain, Switzerland, the United States and other countries, have been able to adopt approaches consistent with their specific circumstances and preferences. Most important is how dependent the different constituent units are upon emissions-intensive economic activity – and therein lies the rub. As discussed below, while such flexibility has long been seen as a virtue of federalism, it has more contentious consequences for climate change governance.

16.1.2 Compensatory Federalism

The redundancy of having more than one order of government also creates potential for the constituent units to step up when the central government fails to act; they can play a 'compensatory' role (Derthick 2010). This is something that sets federal states clearly apart from unitary ones. Given, as noted above, that constituent units typically hold a number of mitigation cards, the potential for compensatory federalism is undoubtedly there. We must be mindful, though, that the question is not whether compensatory action by constituent units is as effective or efficient as an optimal set of policies implemented across the country by the central government. It is simply whether constituent unit initiatives go some way to filling a void created by central government inaction.

The chapters in this book provide evidence of such compensating action occurring across a number of federations. Partisanship has been at the heart of the matter – fiercely so in the United States, but quite evidently so in other federations as well. All policy has its ideological inflection, and climate change is no exception. When central governments have been in the hands of those opposed to climate change action, constituent units have often taken the initiative – typically, those of a different ideological complexion. This has been the case in the United States, as Rabe shows; in Australia; in Canada as Harrison shows; in Switzerland as Kammerer and colleagues show; and Brazil as Barbi and Rei show.

For good reason, California is the best-known case of constituent unit environmental activism, but it is anomalous. There are certainly 'leader' jurisdictions in other federations – such as the Mexican state of Yucatan, as

Corral and Heredia show – but these do not exercise the kind of systemic leverage that California has enjoyed. Prominent among the measures adopted at constituent unit level have been policies to encourage investment in renewable energy and to penalise emissions through cap-and-trade systems. Occasionally constituent units can go further and endeavour to force the central government to act, as a group of US States did via a legal challenge in 2007. However, there is little evidence of this in our other cases.

Does this compensatory action have a perverse effect by letting central governments off the hook? There is some suggestion this might be the case in Switzerland, but in general it would seem not. In the case of the EU, Alberton describes the leadership role played by those more ambitious member states, setting the agenda for the Union as a whole. To what extent has compensatory federalism filled the void created by inaction at the centre? That is a question addressed below.

Institutional realities are also a factor in some of our cases. Rabe reminds us how the presidential separation-of-powers system in the United States has helped ensure that 'no major federal environmental legislation has been adopted in the United States since the 1990 *Clean Air Act* Amendments'. Kammerer and colleagues point to Switzerland's unique system of direct democracy as playing an inhibiting role as far as the federal government is concerned. However, as the Australia case shows, it is quite possible to have inaction prevail at the federal level in a straight parliamentary federation if conservative parties hold office there. In India, the central government has taken the position that the country should not have to compromise on economic development for the sake of climate change mitigation, and has refused to implement such policies unless there are clear co-benefits. Given that India is the third largest emitter, this is clearly an issue. Pillai and Dubash show that while the states have compensated somewhat, their ability to do so is limited by the highly centralised nature of India's federal system.

16.1.3 Laboratory Federalism

Federalism's greatest promise for policymaking lies in the potential provided by the existence of multiple governments for policy innovation and accompanying interjurisdictional learning. Learning can take place either horizontally among the constituent units, or vertically from one or more constituent units up to the central government. In climate change governance, though, there is little reason to expect a great deal of genuine policy innovation: most of the mitigation measures have been well known for some time and the issue is not coming up with new techniques, but simply making the decision to adopt and implement them (Engel 2015). In that regard, federalism still provides conditions for second-order policy learning – or 'political learning' (May 1992) – in that pioneering jurisdictions can demonstrate the political and economic feasibility of what might be otherwise eschewed as excessively adventurous initiatives.

The accounts provided here are largely consistent with this more modest interpretation of the laboratory federalism idea. There is little mention of genuine policy innovation, but numerous examples of pioneering adoptions that provide a model for other jurisdictions. Even in as centralised a federation as India, there has been some vertical diffusion of state innovations, note Pillai and Dubash. However, there are certainly examples of innovations that failed to inspire emulation. As Harrison notes, British Columbia's message that a provincial carbon tax can work fell on deaf ears across the rest of Canada. Québec joined with California in an emissions-trading scheme, but other provinces were not queuing up. In Switzerland, the great variation in circumstances between the cantons discourages interjurisdictional learning as far as climate change policy is concerned. In China, Yi and Cao conclude that there have been many instances of innovation and leadership but not necessarily an accompanying diffusion. Leaders without followers does not make for laboratory federalism. In general, we can say that the experience here reflects the chronic challenges that generally constrain interjurisdictional policy learning: the length of time before the 'success' of a policy becomes evident; the absence of an objective measure of success; the role of local conditions, circumstances and interests; the impact of ideology and partisanship.

16.2 Federalism as Hindrance

On the other side of the coin, Chapter 1 suggested three ways in which federalism may hinder policymaking. First, a system of divided jurisdiction creates various potential veto points whereby initiatives can be blocked. Second, if left to the constituent units, the result may be a motley collection of policies, collectively suboptimal, or even a situation where some jurisdictions negate the efforts of other. Third, there may be competitive pressure holding all constituent units back from taking action. In the latter two scenarios, 'vertical coordination' is required, one scholar has argued (Gordon 2015, 122–3). It is easy to see how climate change governance may be complicated by any or all of these, particularly given the upfront costs mitigation measures may involve and the degree of ideological conflict surrounding them.

16.2.1 Veto Points

Federalism has been seen as obstructing policymaking because either the relevant government has inadequate jurisdictional authority or institutions of constituent-unit representation in the central government have a veto power. These cases have highlighted some instances of such obstruction. In Spain, Rodrigo and colleagues note the high degree of jurisdictional conflict that has characterised environmental issues. In the EU, Alberton notes the retarding effect of reluctant member states. In Canada, Harrison points to the ability of status quo provinces to block pan-Canadian action because of a convention requiring intergovernmental unanimity. And when the federal government did eventually take unilateral action, passing the *Greenhouse Gas Pollution Pricing Act* in 2018, that might have succumbed to constitutional challenge (though it was upheld by the Supreme Court of Canada). Underpinning these dynamics is the degree to which Canada stands out as an unusually decentralised federation – a consequence in the largest part of the degree to which it is bicommunal, with Québec having long played a powerfully centrifugal role (Gagnon and Simeon 2010; Lecours 2019). Other than this and some jurisdictional clashes between the states and the federal government in the USA, though, the federations surveyed in this book reveal little by way of such obstacles to effective climate governance.

Indonesia, though, is a different story. There, Di Gregorio and Moeliono describe a situation where two decades of decentralisation have resulted in a misalignment of roles, responsibilities and resources that hamstrings effective action. As we emphasised in Chapter 1, Indonesia is not a federation and indeed has long evinced an 'aversion' to federalism. Its experimentation with decentralised governance is recent and the resulting system must be seen as a work in progress.

While Canadian provinces were able to block central government action because of their power in the federation, they have no formal veto authority through representation in the Canadian parliament. The situation is very different in Germany, where the second chamber of the federal parliament, the *Bundesrat* (Federal Council) is made up of delegated representatives of the *Länder* governments. There, the *Länder* do enjoy a formal veto power and a coalition of coal states could stand in the way of national mitigation strategies. Eckersley and colleagues find little evidence to this point, though, of the much-discussed *politikverflechtungsfalle* or 'joint-decision trap' being an obstacle to federal climate change policymaking in Germany – though they suggest it could become so. The *Bundesrat* is, however, a very unusually 'federal' second chamber, and thus we would not expect this kind of veto point to be a factor in other federal systems. The only other case here with German-style Council governance is the EU, where similar dynamics can be found.

16.2.2 Policy Patchwork

While constituent units may well step up and compensate for central government inaction, quite possibly only some of them will – and to varying degrees. This pattern is evident from the case studies, particularly the USA and Canada, where several jurisdictions have taken notable climate change initiatives, but, equally, a number remained steadfastly attached to the status quo. Eckersley and colleagues describe a similar situation in Germany, Alberton does likewise for the EU, Rodrigo and colleagues do for Spain, and Yi and Cao do for China – reflecting what the latter describe as 'the conflict between local economic interests and national goals for climate governance'.

In many cases the problem goes beyond the mere existence of diverse responses. Adela and colleagues describe a situation in Ethiopia where spillover problems between ethnically defined constituent units jealous of their powers and interests undermine mitigation and adaptation efforts. Oates (1972) noted that the decentralisation theorem is predicated on costs and benefits of a given policy being internalised, and there being no significant external effects, or spillovers. Decentralisation becomes more problematic the more serious those externalities.

In Canada, the issue is the degree to which those jurisdictions that have not taken action are the ones where action is most needed. As Harrison emphasises, no matter how concertedly Québec or British Columbia might work to reduce their emissions, those efforts cannot compensate for the large and increasing contribution from Alberta. 'With only 12 percent of the population, Alberta contributes over 40 percent of Canada's emissions' (Harrison, Chapter 4 this volume), and its economic welfare is tied to its high-emission industries. As a consequence, Canada's emissions have continued to rise. Similarly, as Barbi and Rei show, the efforts of the southern states in Brazil cannot compensate for the large and rising emissions of their Amazonian counterparts.

In Chapter 1, we noted the view that such policy discrepancies will be fatal. 'Attempts to reduce greenhouse gas ... emissions by one jurisdiction are meaningless if others allow emissions to increase by an equal (or greater) amount. An effective response ... requires vertical coordination' (Gordon 2015, 122–23). This would, however, seem to exaggerate. If emissions increase in some jurisdictions on a business-as-usual basis while other members of the federation are implementing policies to reduce theirs, those reductions are nonetheless reductions in the net federation-wide output, just as they would be if there were no shirkers. While, as Harrison emphasises, mitigation efforts by BC, Québec and Ontario are overshadowed by Alberta's large and growing emissions and thus the leaders cannot compensate for the laggards, Canada's net emissions are still lower than they would be if those three provinces had also continued on a business-as-usual path. None of the achieved reductions is 'meaningless'. But that's cold comfort, Harrison argues; in a federation such as Canada's, the central government's heavy hand is required if emissions are to be reduced. This would seem to be the case for any federation where there are jurisdictions wedded to emissions-intensive industries of such scale.

A very different scenario is represented by the Australian case, where states have varied in their commitment but not such that some jurisdictions effectively undo the efforts of the others. Because of their coal and LNG exports, Western Australia and Queensland play a role analogous to Alberta's, but not to the same degree. The difference lies not in the respective federal systems, but in the type and distribution of resources within those systems. Both the fossil fuel resources and the renewable alternatives are more evenly distributed in Australia and none are as dirty to produce as the oil from Alberta's tar sands.

16.2.3 The Collective Action Problem

Given that no constituent unit 'acting alone, is even capable of adopting emission controls that would make a dent in global emissions', and assuming that mitigation efforts are costly, there is good reason to think that all jurisdictions would be tempted to 'free ride' on the efforts of others (Adler 2008, 448). The studies in this book provide little of evidence, though, of such an effect prevailing. That seems consistent with findings that jurisdictional action on climate change is determined more by internal dynamics than by the external collective action dilemma (e.g., Aklin and Mildenberger 2020; Matisoff 2008).

16.3 Conclusion

Federalism has undoubtedly enhanced climate change governance in several of the countries examined here. It has done so primarily through its inbuilt redundancy – providing the opportunity for constituent units to step up and 'fill the void' should the central government fail to act. This 'compensatory' potential has been most evident in the Australian case, but also across other federations canvassed here: Canada, the United States and Brazil. 'Fill the void', however, almost inevitably overstates things. There is no example of the net effect of constituent unit action equalling what a whole-of-federation programme equivalent to that undertaken in the more ambitious constituent units would have achieved. The void gets at best only partially filled. The extent to which this occurs is powerfully influenced by the geopolitical economy of the federation in question. Constituent units whose economic self-interest is heavily tied to high-emission patterns of economic activity are very unlikely to follow the lead of those with a different resource base and economic profile.

Importantly, dysfunctionality has been much less a problem of federalism than of *insufficient* federalism. It has been in the highly centralised systems or those in the developmental stage of decentralisation that we have seen the most persistent challenges to climate change governance. In those cases, insufficient autonomy, resources or capabilities at the constituent unit level or a misalignment of responsibilities and incentives have presented significant obstacles.

The accounts provided in this book confirm that federalism and federal-type arrangements work in cross-cutting ways, facilitating climate governance in some respects, hindering it in others. Divided jurisdiction's effects vary according to a range of institutional, political, social, economic and geographic factors. Some patterns have emerged, though, and not all of federalism's mooted advantages or disadvantages have been important factors.

References

Adler, Jonathan H. 2008. 'Hothouse Flowers: The vices and virtues of climate federalism'. *Temple Political & Civil Rights Law Review* 17 (2): 443–64.

Aklin, Michaël, and Matto Mildenberger. 2020. 'Prisoners of the Wrong Dilemma: Why distributive conflict, not collective action, characterizes the politics of climate change'. *Global Environmental Politics* 20 (4): 4–27.

Derthick, Martha. 2010. 'Compensatory Federalism'. In *Greenhouse Governance: Addressing climate change in America*, ed. Barry G. Rabe, 58–72. Washington, DC: Brookings Institution.

Engel, Kirsten H. 2015. 'The Enigma of State Climate Change Policy Innovation'. In *The Law and Policy of Environmental Federalism: A comparative analysis*, ed. Kalyani Robbins, 169–87. Cheltenham: Edward Elgar.

Gagnon, Alain-G., and Richard Simeon. 2010. 'Canada'. In *Diversity and Unity in Federal Countries*, ed. Luis Moreno and César Colino. Montreal & Kingston: McGill–Queen's University Press.

Gordon, David J. 2015. 'An Uneasy Equilibrium: The coordination of climate governance in federated systems'. *Global Environmental Politics* 15 (2): 121–41.

Lecours, André. 2019. 'Dynamic De/Centralization in Canada, 1867–2010'. *Publius* 49 (1): 57–83.

Matisoff, Daniel C. 2008. 'The Adoption of State Climate Change Policies and Renewable Portfolio Standards: Regional diffusion or internal determinants?' *Review of Policy Research* 25 (6): 527–46.

May, Peter J. 1992. 'Policy Learning and Failure'. *Journal of Public Policy* 12 (4): 331–54.

Oates, Wallace E. 1972. *Fiscal Federalism*. New York: Harcourt Brace Jovanovich.

Index

Note: Page numbers followed by *f* indicate a figure on the corresponding page. Page numbers followed by *t* indicate a table on the corresponding page.

adaptation policy
  Australia, 20, 26
  Brazil, 50, 53*t*
  China, 89, 92, 95–6, 99
  Ethiopia, 109–10, 114, 117, 119–20, 123
  European Union, 138–41
  Germany, 162, 165–6, 168–9
  India, 178–9, 182, 184, 186, 189
  Indonesia, 203
  Mexico, 218, 221–2, 225–31, 234
  South Africa, 243, 246–55, 257
  Spain, 264, 267, 273–4
  Switzerland, 293–301, 295–6*t*, 297*t*, 299*t*
  United States, 307–8, 314–16, 324
administrative federalism, 7–8, 130, 178, 181
Affordable Clean Energy rule (US), 317
African Development Bank, 113
African National Congress (ANC), 241, 244
agropastoral communities, 112
air pollution, 189, 221
Alberta Climate Leadership Plan, 74, 76
Alberta New Democratic Party (NDP), 74
aluminium manufacturing, 15–16
Amazon forest protection, 44–5
Australian Capital Territory Labor government, 24
Australian Energy Market Operator (AEMO), 19
Australian Labor Party (ALP), 21–2
Australia's climate governance and federalism
  adaptation policy, 20, 26
  analysis and assessment, 28–30
  bicameralism, 23
  carbon emissions, 15–18
  carbon tax in, 24–6
  climate change governance, 18–29, 335
  coal energy, 15–16, 21
  Coalition parties, 23
  collective state action, 24
  cooperative federalism, 19–20
  COVID-19 pandemic, 19–20
  division of powers, 18–19
  energy production, 15–17
  experimentation and learning, 29
  federalism in, 18–20, 28–30
  global warming, 17
  greenhouse gas (GHG) emissions, 14–18
  ideology and partisanship, 21–2
  international commitments, 17–18
  introduction to, 14–15
  Kyoto Protocol, 17
  overview, 18–20, 28–30
  Paris Agreement, 17, 28
  regional variation, 22–3
  state-level partisan divide, 23–4, 26–7
  summary of, 30–1
Austria's climate governance and federalism, 6, 9–10
Autonomous Communities (ACs) in Spain, 263–4, 268–79, 275–6*t*, 279*t*

bicameralism, 8–9, 23, 29
Biden, Joe, 307, 313–14, 318, 323
Bio Carbon Fund (Indonesia), 208
Bolsonaro, Jair, 45, 57
Brazil Center on Climate, 56
Brazilian Alliance for Climate Action (ACA), 56
Brazilian Association of Municipalities (ABM), 57
Brazilian Association of State Environmental Entities (ABEMA), 56
Brazil's climate governance and federalism
  adaptation policy, 50, 53*t*
  carbon emissions, 43–4
  climate change in, 43–50, 47–9*t*, 51–2*t*, 53*t*
  commitment to Amazon protection, 44–5
  COVID-19 pandemic, 43
  decentralized policy-making, 42, 50–5

Brazil's climate governance and federalism (cont.)
experimental policy-making, 56–8
federalism in, 42–3
greenhouse gas (GHG) emissions, 41, 43–4, 46
introduction to, 41–2
Kyoto Protocol, 44
mitigation policies, 50, 53*t*
municipal-level climate change, 46–50, 51–2*t*
Paris Agreement, 44–5
practice of federalism, 42–3
state-level climate change, 45–6, 47–9*t*
at subnational level of government, 50–8
summary of, 58–9
brown coal (lignite), 15, 26
Bush, George H.W., 309, 312
Bush, George W., 70
business-as-usual (BAU), 110, 200

California Air Resources Board, 314, 318
Canada's climate governance and federalism
carbon emissions, 64, 72–80
carbon pricing and, 72–80
carbon tax, 73
climate change in, 65–72
coal energy, 67, 73–4
as decentralised, 8
federalism in, 9–10, 75–80, 332–4
fossil fuels, 64–5, 71
global warming, 66–7
greenhouse gas (GHG) emissions, 65–6
introduction to, 64–5
joint decision trap, 64, 69–71
Kyoto Protocol, 67, 70, 72
oil production, 69, 73, 76–7
Pan-Canadian Framework on Clean Growth and Climate Change, 75–80
provincial governments and, 71–2
summary of, 80–1
Canadian Council of Ministers of the Environment, 69
cap-and-trade system, 2–3, 71–2, 128–9, 314
carbon emissions
Australia, 15–18
Brazil, 43–4
Canada, 64, 72–80
China, 15, 87–8
decarbonization, 179
Ethiopia, 114–15
European Union, 130
impact of, 2
India, 179
Indonesia, 199
South Africa, 242, 247
United States, 15, 306–8, 316
Carbon Pollution Reduction Scheme (CPRS) (Australia), 25
carbon sequestration, 3, 58, 200
carbon tax
Australia, 24–6
Canada, 73
impact of, 3
Mexico, 222–3, 230–1
South Africa, 242, 247, 257
Switzerland, 292
carryover credits, 18
Central Inspection System for Ecological and Environmental Protection (CISEEP) (China), 94, 102–3
centralised division of powers, 8
Central Party Committee, 92
Centre for Research on Epidemiology of Diseases (CRED) (Ethiopia), 111
China Minmetals Group, 102
China's climate governance and quasi-federalism
carbon emissions, 15, 87–8
Central Inspection System for Ecological and Environmental Protection, 94, 102–3
climate change in, 86–90
climate mitigation efforts, 95–103, 96–7*t*, 101*t*
COVID-19 pandemic, 90
division of powers, 90–2
divisions of power, 90–2, 91*t*
fiscal incentives, 94
greenhouse gas (GHG) emissions, 88–9, 97–9
introduction to, 86
local environmental agencies, 93–4
Paris Agreement, 89
public participation in climate governance, 95
quasi-federalism in, 7, 86, 90–5, 332
summary of, 103–4
target-responsibility system, 93–4, 104
China's climate mitigation case study
agenda setting, 97
analytical framework, 95–6, 96*t*
formulation and adoption, 97–9
implementation of, 99–102, 101*t*
low-carbon pilot initiative, 96–103, 97*t*
supervision and evaluation, 102–3
Chinese Communist Party (CCP), 90
Chretien, Jean, 70
Clean Air Act Amendments (1990) (US), 312–13, 315, 323, 331
Clean Power Plan (CPP) (US), 315–17, 322–3
Climate and Energy Package (EU) (2020), 128–9
Climate Change Act (Australia), 26
Climate Change Act (Germany), 156
Climate Change Act (UK), 134
Climate Change and Energy Transition Law (Spain), 267, 277–8
climate change mitigation. *See also* China's climate mitigation case study
Brazil, 50, 53*t*
China, 95–103, 96–7*t*, 101*t*
defined, 1, 6
European Union, 133–8

forest-based mitigation, 205–13
global efforts, 328–9
Indonesia, 199–201, 205–13
Mexico, 225–32
politics of, 14
Switzerland, 6, 286–8, 292–3, 296–300, 299*t*
United States, 314–23
Climate Change Mitigation System (South Africa), 247
Climate Change Strategy (South Africa), 248
Climate Protection Act (Germany), 156–64
Climate Resilient Green Economy (CRGE) Strategy (Ethiopia), 109–10, 112–24
Clinton, Bill, 309
CO_2-Act (2013) (Switzerland), 285
coal energy
Australia, 15–16, 21
brown coal, 15, 26
Canada, 67, 73–4
Germany, 159, 161–2
coastal erosion, 151–4
compensatory federalism, 5, 28–9, 330–1
Conference of Cantonal Energy Directors (EnDK) (Switzerland), 300
Conservation and Development of Urban Trees Law (Mexico), 232
constituent unit action, 5–6
Convention on Biological Diversity (Spain), 266
cooperative federalism, 19–20, 157–8, 181, 244
Copenhagen Accord, 67, 309
Copenhagen Conference of Parties (COP), 179
Council for the Australian Federation, 24
Council of Australian Governments (COAG), 20, 24–6
Covenant of Mayors for Climate & Energy (EU), 139
Covenant Territorial Coordinators (CTCs) (EU), 137
COVID-19 pandemic
Australia, 19–20
Brazil, 43
China, 90
Germany, 158
India, 189
customary forests (*hutan adat*), 206

decarbonization, 179
decentralization
division of powers, 8
federalism and, 329–30
Indonesia, 201–13
of policies in Brazil, 42, 50–5
Spain, 273
deforestation, 41, 44–5, 198–201
Delivery Unit for Development Monitoring (UKP4), 209
Department of Environment, Forestry and Fisheries (DEFF), 243
desertification risk in Spain, 264–5
Disaster Management Act (South Africa), 248
diversity in Swiss federalism, 290–2
division of powers
Australia, 18–19
centralised/decentralised, 8
China, 90–2, 91*t*
climate change governance, 19–20
in dualist federations, 7–8
Ethiopia, 116–17
European Union, 131–3
Germany, 155
India, 181
Indonesia, 201–2
South Africa, 249–51
Spain, 269, 280, 286–7*t*, 286
Switzerland, 291
United States, 310–11
Draft Climate Change Bill (South Africa), 256
droughts
Australia, 17
China, 88
Ethiopia, 111, 118
Indonesia, 198–9
Mexico, 220
South Africa, 242
Spain, 265
Switzerland, 288
United States, 308
dryland agroecology, 112
Durban Climate Change Strategy (2014), 248

Ecological Planning Program (Mexico), 233
Ecosystem Restoration Concessionaires (Indonesia), 210
Education Law (Mexico), 232
Effort Sharing (ES) instrument, 128–9
Ekurhuleni Climate Change and Energy Strategy (South Africa), 248
Electricity Act (India) (2003), 180, 187
electricity generation, 15–16, 182, 256, 289–90
Electricity Regulations on New Generation Capacity of the Electricity Regulation Act (South Africa), 249
electric vehicles (EVs), 187
El-Niño events, 199
Emissions Trading Scheme (ETS), 29–30, 72, 128–9
Energiewende strategy (Germany), 150, 152, 159
Energy Efficiency Directive (EU), 136
energy imports, 136, 164–5
environmental and social impacts (ESIA) report, 121–2
environmental consciousness, 263
environmental disaster risk management, 46
environmental federalism, 86, 90, 91*t*, 103, 177
Environmental Fund Management Agency (BPLDH), 209
environmental impact assessments (EIAs), 250
environmental performance in promotions, 94
environmental protection, 43
Environmental Protection Law (Mexico), 232
Environmental Protection Tax Law (China), 94

Environmental Secretaries of Brazilian Capital Cities, 56
Environment Fund Management Agency (BPDLH) (Indonesia), 203
Environment Protection Act (India) (1986), 181
equal status rule of federalism, 90
Ethiopian People's Revolutionary Democratic Front (EPRDF), 115
Ethiopia's climate governance and federalism
  adaptation policy, 109–10, 114, 117, 119–20, 123
  carbon emissions, 114–15
  climate change and, 110–17
  Climate Resilient Green Economy (CRGE) Strategy, 109–10, 112–24
  division of power, 116–17
  ethnic federalism, 116–24
  federalism, 9–10, 115–24
  greenhouse gas (GHG) emissions, 114–15
  impact of, 111–12
  introduction to, 109–10
  Kyoto Protocol, 114
  policy and institutional framework, 112–13
  responses to, 113–15
  summary of, 124–6
  trends and patterns, 110–11
ethnic federalism, 116–24, 244
ethno-federalism, 10, 115–16
EU Covenant of Mayors, 137, 139
European Climate Adaptation Platform (Climate-ADAPT), 138–40
European Climate Law (EC), 154
European Commission, 131, 135, 137–8, 154
European Union's climate governance and federalism
  adaptation action, 138–41
  carbon emissions, 130
  climate change and, 128–33
  climate mitigation and, 133–8
  competitive cooperation, 135–7
  cross-cutting policy, 140–1
  division of powers, 131–3
  federalism in, 7–9, 329
  greenhouse gas (GHG) emissions, 128–30
  initiatives and financial assistance, 138–40
  institutional architecture of EU, 131
  integration process and, 130–3
  introduction to, 128
  multilevel reinforcing mechanisms, 137–8
  opportunity structure of, 133–5
  Paris Agreement, 134
  quasi-federalism, 130–1
  summary of, 141–2
executive federalism, 9, 69

Federal Commission for Protection against Health Risks (*Comisión Federal para la Protección contra Riesgos Sanitarios* or COFEPRIS), 220
Federal Constitution (CESOP 2006) (Mexico), 223
Federal Environmental Agency (Germany), 154
federalism
  administrative federalism, 7–8, 130, 178, 181
  Australia, 18–20, 28–30
  bicameralism, 8–9
  Brazil, 42–3
  Canada, 9–10, 75–80, 332–4
  centralised/decentralised division of powers, 8
  climate change governance and, 3–7
  collective action and, 335
  compensatory federalism, 5, 28–9, 330–1
  conflict and cooperation, 9
  cooperative federalism, 19–20, 181, 244
  decentralization and, 329–30
  defined, 1, 5
  dividing powers, 7–8
  environmental federalism, 86, 90, 91*t*, 103, 177
  Ethiopia, 9–10, 115–24
  ethnic federalism, 116–24, 244
  ethno-federalism, 10, 115–16
  executive federalism, 9, 69
  as facilitator of governance, 5–6, 329–32
  Germany, 9–10, 154–66, 333
  as hinderance to governance, 6–7, 332–5
  India, 180–91, 329–31
  integrated federalism, 8–9
  laboratory federalism, 5, 103, 331–2
  local-level policy making, 329–30
  Mexico, 222–31
  presidential or parliamentary, 9
  reflections on, 328–9, 335–6
  separation-of-powers system and, 9, 306, 331
  societal and economic factors in, 9–10
  South Africa, 243–54
  Spain, 7, 267–79
  Switzerland, 286, 290–301, 331
  United States, 307, 310–14, 330–1
Federal Ministry of the Environment (SEMARNAT) (Mexico), 222, 224
Federal Office for the Environment (FOEN) (Switzerland), 294
feed-in-tariffs (FiTs), 152
fiscal incentives, 86, 94–5, 100
floods
  Ethiopia, 111, 118
  European Union, 138, 140
  Germany, 151–4, 166
  India, 180, 188
  Indonesia, 198–9
  Mexico, 220
  South Africa, 242
  Spain, 265
  Switzerland, 294
  United States, 308
food security, 111, 203, 221, 248
forest-based mitigation, 205–13
Forest Management Units (FMUs), 205
Forum of Environmental Secretaries of Brazilian Capital Cities, 56

fossil fuels, 16, 21, 64–5, 71, 150, 152
fragmentation, defined, 14
Fund for Natural Disasters (FONDEN) (Mexico), 220–1
Fund for the Prevention of Natural Disasters (FOPREDEN) (Mexico), 220–1

Gandhi, Indira, 181
general allocation grants, 201
General Law on Climate Change (GLCC) (Mexico), 218, 222–6
geothermal power, 293
Germany's climate governance and federalism
  administrative federalism in, 7–8
  Bund-Länder associations, 158
  city states and, 165–6
  climate change, 151–66
  coal energy, 159, 161–2
  cooperative federalism in, 157–8
  COVID-19 pandemic, 158
  discussion on, 166–8
  division of powers, 155
  Energiewende strategy, 150, 152, 159
  energy imports, 164–5
  federalism, 9–10, 154–66, 333
  greenhouse gas (GHG) emissions, 151–4, 153–4*t*, 163, 165
  integrated federalism in, 8–9, 157–8
  introduction to, 150–1
  Kommunalrichtlinie initiative, 157
  Länder policy, 150–8, 156*t*, 158–61, 159–60*t*, 161–2*t*, 162*t*, 166–8
  nuclear power, 152, 163
  resources, 10
  solar energy, 163
  summary of, 168–9
  wind energy, 163–4
Gillard, Julia, 25
glacial melt, 151–4
Global Climate Risk Index (Mexico), 219
Global Covenant of Mayors for Climate and Energy (GCMCE) (Brazil), 57
Governors Climate and Forest Task Force (GCFTF) (Indonesia), 207–8, 210–13
Great Recession, 320
Green Climate Cities Program (Brazil), 55
Green Deal (EU), 154
greenhouse gas (GHG) emissions
  attempts to reduce, 6, 328–9
  Australia, 14–18
  Brazil, 41, 43–4, 46
  Canada, 65–6
  China, 88–9, 97–9
  Ethiopia, 114–15
  European Union, 128–30
  Germany, 151–4, 153–4*t*, 163, 165
  governance and, 1–2
  India, 179
  Indonesia, 198–9, 205
  Mexico, 218–19, 224, 228–9
  South Africa, 242
  Spain, 265–6, 266*t*
  United States, 307–8, 315
Greenhouse Gas Pollution Pricing Act (2018) (Canada), 77, 333
Green Party (EU), 152, 159, 161

Harper, Stephen, 70
hydrocarbons, 10, 21–2
hydrometeorological monitoring, 221
hyper-partisanship, 311

independent power producers (IPPs), 249
Indian Administrative Service (IAS), 188
India's climate governance and federalism
  carbon emissions, 179
  central frameworks, 8, 185–8
  climate change and, 178–80
  co-benefits experiments, 188–90
  COVID-19 pandemic, 189
  division of powers, 181
  Electricity Act, 180, 187
  environmental federalism, 177
  federalism and, 180–91, 329–31
  Finance Commissions, 184
  finance flows and, 184–6
  financial experiments, 185–6
  introduction to, 177–8
  Paris Agreement, 179–80
  policy innovations, 190–1
  political translation, 188–91
  solar energy, 190
  state climate policies, 186–7
  state-level bureaucratic capacity, 187–8
  summary of, 191–2
  wind energy, 190
Indonesia's climate governance and federalism
  carbon emissions, 199
  climate change and, 198–201
  climate change mitigation and, 199–201, 205–13
  decentralization processes, 201–13
  district level, 204
  division of powers, 201–2
  federalism in, 7, 333
  Governors Climate and Forest Task Force, 207–8, 210–13
  greenhouse gas (GHG) emissions, 198–9, 205
  introduction to, 198
  jurisdictional approaches, 207–13
  land use decisions, 198–201, 205–6
  Low Emission Development, 208, 210, 213
  national level, 209–10, 212–13
  provincial level, 204, 211–13
  recentralization processes, 201–2
  REDD+ Investment Program, 200, 204, 207–13
  summary of, 213–14

Integrated Coastal Management Act (South Africa), 247
Integrated Development Plans (IDPs) (South Africa), 251–2
Integrated Development Regions (RIDEs) (Brazil), 54
Integrated Energy Plan (South Africa), 247
integrated federalism, 8–9, 157–8
Integrated Resources Plan (IRP), 247
Integrated Urban Development Framework (IUDF), 250–1
INTERACT-Bio Project, 54
Inter-Administrative Cooperation on Climate Change and Energy (Spain), 271
Intercultural Biocultural Board of Inter-Municipal Decentralised Public Organisations of Puuc (JIBIOPUUC) (Mexico), 234–5
Intergovernmental Committee on Climate Change (IGCCC), 253
Intergovernmental Panel on Climate Change (IPCC), 2, 180
Intergovernmental Relations Framework Act (South Africa), 252–3
intergovernmental relations (IGR), 9
Interministerial Commission on Climate Change (Mexico), 226–7, 232–3
Interministerial Commission on Global Climate Change (Brazil), 50
Interministerial Committee on Climate Change (Brazil), 50
Interministerial Committee on Climate Change (IMCCC) (South Africa), 253
International Energy Agency (IEA), 87
Irish Red Cross Society, 111

joint decision trap, 64, 69–71

Kigali Amendments (US), 309
Kommunalrichtlinie initiative (Germany), 157
KwaZulu-Natal (Inkatha Freedom Party), 244
Kyoto Protocol
  Australia, 17
  Brazil, 44
  Canada, 67, 70, 72
  Ethiopia, 114
  United States, 75, 309

laboratory federalism, 5, 103, 331–2
land for other uses, 206
Land Use, Land Use Change and Forestry (LULUCF) Regulation (EU), 128–9
land use decisions in Indonesia, 198–201, 205–6
Law of the Sea Convention (US), 309
Leaders' Climate Summit (Brazil), 56
Leyen, Ursula von der, 128
local environmental agencies, 93–4
Long Term Forestry Management Plan (Indonesia), 205
Low Emission Development (LED) (Indonesia), 208, 210, 213

Management by Objectives (China), 94
mandatory renewable energy target (MRET), 23, 29
Martin, Paul, 70, 72
Massachusetts v. U.S. Environmental Protection Agency (2007), 313, 324
Mayors Adapt (EC), 139
Mexico City Climate Commission, 228
Mexico's climate governance and federalism
  adaptation policy, 218, 221–2, 225–31, 234
  background and context of, 219–22
  carbon tax, 222–3, 230–1
  climate change and, 219–22
  climate mitigation and, 225–32
  federalism and, 222–31
  General Law on Climate Change, 218, 222–6, 235–6
  greenhouse gas (GHG) emissions, 218–19, 224, 228–9
  institutional arrangements, 226–7
  introduction to, 218–19
  legal framework, 225–6
  National Strategy for Climate Change, 219, 226
  Paris Agreement, 221
  planning instruments, 226
  state climate policy, 227–31
  summary of, 235–6
  Yucatan case, 232–5
Ministry of Ecology and Environment (MEE) (China), 92
Ministry of Environment and Forestry (MoEF) (Indonesia), 202–3, 205–6, 209
Ministry of National Development Planning (BAPPENAS) (Indonesia), 203–4
Ministry of the Environment of Czech Republic, 140
Montreal Aviation Protocols, 309
Montreal Protocol on Ozone Depleting Substances, 309
Morrison, Scott, 21
multi-level governance structure, 41, 93, 130, 136, 192, 198, 212

National Action Plan Addressing Climate Change (RAN-PI) (Indonesia), 199
National Action Plan for the Reduction of GHG Emissions (RAN-GRK) (Indonesia), 200
National Action Plan on Climate Change (NAPCC) (India), 179, 186
National Act on Climate Change Policies (PNMC) (Brazil), 44
National Adaptation Plan (Brazil), 45
National Adaptation Plan of Action (NAPA) Ethiopia (NAP-ETH), 114
National Adaptation Plan (PNA) (Brazil), 53

National Association of Municipal Environmental Entities (ANAMMA) (Brazil), 56
National Centre for Disaster Prevention (CENAPRED) (Mexico), 220
National Civil Defence and Protection Policy (Brazil), 55
National Civil Defence and Protection System (Brazil), 55
National Climate Change Adaptation Strategy 2020 (NCCAS) (South Africa), 243
National Climate Change Fund (Brazil), 50
National Climate Change System (SINACC) (Mexico), 222
National Climate Process (Canada), 70
National Confederation of Municipalities (CNM) (Brazil), 57
National Council on Climate Change (DNPI) (Indonesia), 209
National Determined Contribution (NDC) (Indonesia), 200
National Determined Contribution (NDC) (Mexico), 221–2
National Development and Reform Commission (NDRC) (China), 92, 97–8
National Development Plan (NDP) (South Africa), 242
national electricity market (NEM), 19
National Emissions Trading Scheme (Australia), 24
National Emissions Trading Taskforce (Australia), 24
National Energy Guarantee (Australia), 26
National Front of Mayors (FNP) (Brazil), 57
National Greenhouse Gases and Compounds Emissions Inventory (INEGyCEI) (Mexico), 219
National Green Transport Strategy (South Africa), 247
National Institute for Ecology and Climate Change (INECC) (Mexico), 224, 226–8, 231
National Integrated Energy and Climate Plan (ENCP) (Spain), 267
Nationally Determined Contribution (NDC), 2, 124, 309
National Plan for Adaptation (NPACC) (Spain), 264, 267, 273–4
National Policy on Climate Change (Brazil), 54
national spatial development framework (NSDF), 249
National Steering Committee on Climate Change (NSCCC) (India), 182, 185–7, 189
National Strategy for Climate Change (Mexico), 219, 226
National System for Climate Change (SINACC) (Mexico), 226–7, 231
Nature Conservancy (Indonesia), 208
ND-GAIN Index (Brazil), 41
net-zero emissions, 16, 21
New South Wales Labor (NSW), 24
Notley, Rachel, 74
nuclear power, 152, 163, 293

Obama, Barack, 307, 309–10, 313–16, 318, 323
oil production, 69, 73, 76–7

Pan-Canadian Framework on Clean Growth and Climate Change (PCF), 75–80
Paris Agreement (2015)
  Australia and, 17, 28
  Brazil and, 44–5
  China and, 89
  European Union and, 134
  India and, 179–80
  introduction to, 2–4
  Mexico, 221
  South Africa and, 242
  Spain and, 266
  United States and, 307, 309–10, 313–15
parliamentary federalism, 9
pastoral communities, 112
Plan for Deforestation Prevention and Control in Amazonia (Brazil), 45
Plan for the Prevention and Control of Deforestation and Burning Practices in the Cerrado Region (Brazil), 45
presidential federalism, 9
private climate finance, 200
ProAdapta Project (Brazil), 55
Protocol to the Barcelona Convention on Integrated Coastal Zone Management (Spain), 266
Provincial Council on Climate Change (Indonesia), 208
Provincial Low Carbon Development Plans (Indonesia), 200
provincial spatial development frameworks (PSDFs), 249
public participation in climate governance, 95
public transportation policy, 54

rain-fed mixed crop production, 109
rainforests, 199
recentralization processes, 201–2
Reconecta Program at Campinas Metropolitan Region (Brazil), 54
REDD+(Reducing Emissions from Deforestation & forest Degradation) Investment Program
  Brazil, 113
  Indonesia, 200, 204, 207–13
  Mexico, 234
Regional Adaptation Strategy for the Yucatan Peninsula (Mexico), 234
Regional Climate Change Commission (Mexico), 233–4
Regional Greenhouse Gas Initiative (RGGI) (US), 314, 322
Renewable Energy Directive (EU), 136
Renewable Energy Heat Act (EU), 156
renewable energy target (RET), 25
renewable energy zones, 16
Rudd, Kevin, 21, 25

Sarkaria Commission Report (India), 180
sea-level rise, 17
self-rule in Ethiopia, 123–4
Sendai Framework for Disaster Risk Reduction (EU), 266
separation-of-powers system, 9, 306, 331
social-ecological resilience, 125
societal factors in federalism, 9–10
solar energy
  Australia, 16
  Germany, 163
  India, 190
  Switzerland, 293
South African Local Government Association (SALGA), 252–3
South Africa's climate governance and federalism
  adaptation policy, 243, 246–55, 257
  administrative federalism in, 8
  carbon emissions, 242, 247
  carbon tax, 242, 247, 257
  as centralised, 8
  climate change and, 242–3, 246–57
  distribution of power, 244–6
  division of power, 249–51
  federalism and, 243–54
  greenhouse gas (GHG) emissions, 242
  historical context, 243–4
  institutional actors, 251–2
  integrated development planning, 252–4
  intergovernmental coordination and planning, 252–4
  introduction to, 241
  law and policy framework, 246–52
  local level, 255–7
  municipal level, 248–9
  national level, 247–8
  Paris Agreement, 242
  provincial level, 248–9, 254–5
  quasi-federalism and, 246
  summary of, 257–8
Spain's climate governance and federalism
  adaptation policy, 264, 267, 273–4
  Autonomous Communities in, 263–4, 268–79, 275–6*t*, 279*t*
  central government and, 273
  challenges to, 264–5
  climate change and, 264–79
  Climate Change and Energy Transition Law, 267, 277–8
  decentralization, 273
  division of powers, 269, 280, 286–7*t*, 286
  federalism and, 7, 267–79
  greenhouse gas (GHG) emissions, 265–6, 266*t*
  institutional framework, 269–72
  international commitment, 266–7
  introduction to, 263–4
  law and policy objectives, 267
  local level, 278–9, 279*t*
  National Plan for Adaptation, 264, 267, 273–4
  Paris Agreement, 266
  summary of, 280
  tax revenues, 272
Spanish Constitution (1978), 268–9
spatial development frameworks (SDFs), 252–4
Spatial Planning and Land Use Management Act (SPLUMA) (South Africa), 249
Special Action Program on Climate Change (Mexico), 232
Special Climate Change Program (Mexico), 220
Specified Gas Emitters Regulation (SGER) (Canada), 73–4
State Action Plans on Climate Change (SAPCCs) (India), 179, 182, 186–8, 191
State Implementation Plans (SIPs) (US), 315
subsidiarity principle, 4, 213, 251, 290–2
Supreme Court of Canada, 78
Sustainable Forest Development Law (Mexico), 232
Sustainable Rural Development Law (Mexico), 232
Switzerland's climate governance and federalism
  adaption measures, 293–301, 295–6*t*, 297*t*, 299*t*
  carbon tax, 292
  climate change and, 286, 286–7*t*, 288–90, 289*t*
  climate mitigation and, 6, 286–8, 292–3, 296–300, 299*t*
  core principles, 290–2
  as decentralised, 8
  direct democracy system, 8–9
  division of powers, 291
  federalism and, 286, 290–301, 331
  intercantonal coordination, 301–2
  introduction to, 285–8, 286–7*t*
  municipal level, 295–301, 299*t*
  qualitative comparative assessment, 295
  resources, 10

target-responsibility system, 93–4, 104
tar sands, 67, 71, 335
tiao-kuai governance structure, 90–2
top-down policy process, 92
Trans Mountain Expansion Project (Canada), 76–7
transnational municipal networks (TMNs), 55–6
Treaty of Lisbon, 131
Treaty on European Union (TEU), 131
Treaty on the Functioning of the European Union (TFEU), 131–3
Trump, Donald, 307, 313–14, 318

UN Agenda for Sustainable Development, 266
UN Framework Convention on Climate Change Conferences of the Parties (UNFCCC COP), 285, 292
UN Framework Convention on Climate Change (UNFCCC)
  Brazil, 44
  Canada, 67
  Ethiopia, 113–14, 124

India, 185
Mexico, 221
South Africa, 251
Spain, 266
unilateralism, 65, 77–8, 81
United Conservative Party (Canada), 77
United Nations Conference on Environment and Development Conference, 113
United Nations Convention to Combat Desertification, 266
United States climate governance and federalism
adaptation policy, 307–8, 314–16, 324
carbon emissions, 15, 306–8, 316
Clean Air Act Amendments, 312–13, 315, 323, 331
Clean Power Plan, 315–17, 322–3
climate change and, 307–14
climate change mitigation, 314–23
compensatory or pre-emptive action, 319
conflict and cooperation, 321–3
constitutional climate authority, 312, 317–18
contentiousness, 311–12
convergence and divergence, 320–1
division of responsibilities, 310–11
federalism and, 307, 310–14, 330–1
greenhouse gas (GHG) emissions, 307–8, 315
introduction to, 306–7
Kyoto Protocol, 75, 309
local level, 319–20
Paris Agreement, 307, 309–10, 313–15
policy development and implementation, 312–13, 316–17
Regional Greenhouse Gas Initiative, 314, 322
separation-of-powers system, 9, 306
state administrative capacity, 318
state level, 313–23
state paradiplomacy engagement, 318–19
summary of, 323–5
U.S. Constitution, 308, 310–11
US Environmental Protection Agency (EPA), 313, 315, 318

venue shopping, 5
vertical coordination, 6, 14, 332, 334
Victoria's Renewable Energy Act (2006), 24

water management policy, 54, 190
Western Climate Initiative (WCI) (Canada), 73
wind energy
Australia, 16
Germany, 163–4
India, 190
Switzerland, 293
World Bank, 111
World Banks' Forest Carbon Partnership Facility, 208

Yucatan Peninsula Climate Change Fund (Mexico), 234

zero-emission electricity, 16

www.ingramcontent.com/pod-product-compliance
Lightning Source LLC
LaVergne TN
LVHW062206200926
848484LV00006B/1331
9781009249652